D1289197

CHEMICAL and STRUCTURAL APPROACHES to RATIONAL DRUG DESIGN

CHEMICAL and STRUCTURAL APPROACHES to RATIONAL DRUG DESIGN

Edited by

David B. Weiner
William B. Williams

University of Pennsylvania
Philadelphia, Pennsylvania

CRC Press
Boca Raton   Ann Arbor   London   Tokyo

Library of Congress Cataloging-in-Publication Data

Chemical and structural approaches to rational drug design / edited by David B. Weiner, William V. Williams.
    p.   cm. — (Pharmacology and toxicology)
   Includes bibliographical references and index.
   ISBN 0-8493-7818-4
   1. Drugs—Design. I. Weiner, David B. II. Williams, William V.
III. Series: Pharmacology & toxicology (Boca Raton, Fla.)
   [DNLM: 1. Drug Design. QV 744 C5155 1994]
RS420.C48 1994
615.19—dc20
DNLM/DLC
for Library of Congress   94-2866
                                                      CIP

183209

INTRODUCTION

Rational drug design? Many would question the thesis of this book, stating that this is a goal, not a reality with today's technology. Yet, this is a question of definitions. Let's take the terms literally (from *Webster's Unabridged 20th Century Dictionary*). *Rational:* of, based on, or derived from reasoning. *Drug:* any substance used as a medicine, or in the preparation of medicines, or chemical mixtures. *Design:* to contrive; to project with an end in view; to form an idea, as a scheme. Taken together, rational drug design is simply defined as a reasoned approach to developing medicines. This does not suggest that the design process is always immediately successful, but instead that it is based on reason. By this broad definition, the age of rational drug design is clearly upon us.

A more stringent definition might also be applied: a rational approach to drug development for a disease based on understanding the molecular pathophysiology of the disease process. This could imply intervening at a critical metabolic step, perturbing a specific receptor system, or administration of a specific stimulatory/inhibitory molecule. In its purest form, rational drug design would allow the development of a specific, non-toxic drug based on the molecular and structural information of a target molecule. This is the goal many are currently striving for, with varying levels of success. However, the structural approach is not the sole approach to drug development, and is not the only rational method for drug development. Many biological and chemical approaches to therapeutic design are currently available, and the application of several complementary technologies to the drug design process is perhaps the most "rational" approach. Based on these considerations, this book is divided (inaccurately) into two sections: Chemical Approaches to Rational Drug Design and Structural Approaches to Rational Drug Design, and is a companion volume to *Biological Approaches to Rational Drug Design*. While many of the chapters utilize overlapping technologies, this broad classification highlights the predominant technology exploited by the authors. The content of the book is summarized here.

Chemical Approaches to Rational Drug Design deals with predominantly synthetic chemical approaches to drug design. Medicinal chemistry is the most widely used approach to the successful development of drugs today. Standard techniques are more and more being supplemented with biological and structural approaches. This is apparent in most of the chapters of this section. While it is beyond the scope of this book to review medicinal chemistry as a field, its application to rational drug design is highlighted in this section.

"Design, Synthesis and Antitumor Activity of an Inhibitor of Ribonucleotide Reductase" by James R. McCarthy and Prasad S. Sunkara describes the use of structure-activity relationships in the design of novel inhibitors of ribonucleotide reductase. This enzyme is essential for cell growth, and inhibitors also inhibit tumor growth. The authors employed novel synthetic strategies to arrive at compounds that are analogs of ribonucleotide reductase, with potent inhibitory activity. One of these is described in detail, and inhibits growth *in vitro* and *in vivo* of several tumors. Along with rational design, the rational clinical application of these inhibitors is also described. As ribonucleotide reductase inhibitors result in accumulation of cells in DNA synthesis or S phase of the cell cycle, the inhibitor was used in conjunction with agents that inhibit cells in S phase or later in the cell cycle during mitosis (M phase). This resulted in lower toxicity and greater efficacy for the combined regimens.

"The Development of Human Immunodeficiency Virus Type 1 Reverse Transcriptase Inhibitors" by Andrew Stern provides an interesting contrast between the "rationally designed" nucleoside analogs and the "rationally discovered" non-nucleoside reverse transcriptase (RT) inhibitors. The nucleoside analogs, such as AZT, ddI and ddC are structural homologs of the natural substrates for reverse transcriptase. As such, their development was quite "rational" and has led to the first effective therapeutics against HIV-1 infection. Several of the more recently described non-nucleoside RT inhibitors were developed from a rational screening approach where nontoxic compounds were screened for *in vitro* inhibitory activity to derive lead compounds. Thus, these contrasting approaches highlight the utility of both the rational methods and the more traditional screening procedures.

"The Development of Potent Angiotensin II Receptor Antagonists Using a Novel Peptide Pharmacophore Model" by Richard M. Keenan, Joseph Weinstock, Judith C. Hempel, James M. Samanen, David T. Hill, Nambi Aiyar, Eliot H. Ohlstein, and Richard M. Edwards describes the use of an overlay hypothesis strategy in the design of a potent series of angiotensin II receptor antagonists. The structural approach utilizes a weak non-peptide antagonist and the angiotensin II peptide agonist to arrive at geometric constraints on candidate antagonists. This chapter exemplifies the utility of comparing divergent ligands for a common binding site in determining critical contacts with the receptor. By correlating structural information with biological activity, several overlay hypotheses are presented and optimized in developing a high affinity antagonist.

"Endothelin Structure and Development of Receptor Antagonists" by Annette M. Doherty details the many exciting developments over the past few years in identifying this novel vasoconstrictor and its receptor. Structure function studies of the endothelin, a small peptide, have combined with molecular structure information in determining the active site of the ligand, and several antagonists have been developed. The discovery of non-peptide antagonists by traditional blind screening is exploited by developing overlay hypotheses that in turn lead to additional insights regarding contact residues. This in turn has led to additional more rationally designed antagonists.

"Molecular Targeting Chemistry in Rational Drug Design" by Alan Fritzberg, Linda Gustavson, Mark Hylarides, and John Reno describes many of the chemical approaches available for delivering radiopharmaceuticals, imaging agents, and toxins to specific target cells. These utilize a targeting molecule, such as an antibody, peptide or drug, which binds to specific sites on cells. The coupling chemistry approaches described allow derivitization of these targeting molecules with radionuclides or toxins for a variety of diagnostic and therapeutic applications. These chemistries have become increasingly more specific and can be combined with molecular biological approaches to increase the specificity of targeting.

Structural Approaches to Rational Drug Design describes the application of the tools of structural biology to pharmaceutic development. Structural information is avidly being applied to the drug design process by many groups. This includes primary structure information, secondary and three-dimensional structure, global protein structure and specific active site analysis. Technologies range from crystallographic to predictive models.

"The Role of X-ray Crystallography in Structure-Based Rational Drug Design," by Alexander McPherson, provides both an overview of structure-based rational drug design, and an in-depth look at X-ray crystallographic approaches. The use of crystallography in protein structure determination is a widely used technology, and is being pursued avidly for drug design. This chapter describes the most commonly used techniques, including differential Fourier transfer and co-crystalliza-tion/diffusion techniques for structural evaluation of receptor-ligand/enzyme-substrate complexes.

"Biocomputational Approaches to Protein-Based Drug Design" by Katherine Prammer, Matthew Wiener, and Thomas Kieber-Emmons is a comprehensive overview of computational chemistry and biocomputational methods applied to drug design. These approaches stem from determination and analysis of protein three-dimensional structures, active site analysis, with compound design based on structural information. Biocomputational approaches discussed are already actively employed by many pharmaceutical companies. This chapter also highlights the use of nuclear magnetic resonance spectroscopy and antibody-based drug design techniques in drug design.

"Architecture and Design of Zinc Protein-Ligand Complexes," by David W. Christianson and Anastasia M. Khoury Christianson provides an in-depth examination of zinc-protein structural biology. The many available zinc-protein structures are exploited in predicting structural features of similar complexes where the structure is unknown. The insights provided allow evaluation of the structural determinants of enzyme activity and hormone action. Concepts including induced fit and metalloprotein-drug interactions are examined with regard to rational design strategies for pharmacophore development.

"Design of Immunomodulatory Peptides Based on Active Site Structures" by Anil B. Mukherjee and Lucio Miele provides both an overview of active-site based peptide design and an in-depth look at the successful use of this technology in developing novel therapeutics. The general principles used in designing oligopeptides from protein active site structures are described. Then, the application of these techniques to the development of the antiflammins is detailed. These immunomodulatory peptides were designed based on active site analysis of large protein enzyme inhibitors. Their development and optimization has led to current clinical trials in ocular disease. This provides an excellent example of bench-to-bedside application of molecular structure analysis and peptide design.

Together, these chapters provide both a broad overview of the current methodologies being applied to both chemical and structural approaches to rational drug design, and in-depth analyses of progress in these specific fields. The reader with specific therapeutic targets in mind should be able to explore this strategy for drug development. Those with interest in specific technologies should be able to view both this volume and *Biological Approaches to Rational Drug Design* in the broader context of the currently available technologies. The extensive reference lists provided by the authors will allow those with specific targets and technologies to explore their interests in greater depth.

David B. Weiner, Ph.D.
William V. Williams, M.D.

Pharmacology and Toxicology: Basic and Clinical Aspects

Mannfred A. Hollinger, Series Editor
University of California, Davis

Published Titles

Inflammatory Cells and Mediators in Bronchial Asthma, 1990,
    Devendra K. Agrawal and Robert G. Townley
Pharmacology of the Skin, 1991, Hasan Mukhtar
In Vitro *Methods of Toxicology*, 1992, Ronald R. Watson
Basis of Toxicity Testing, 1992, Donald J. Ecobichon
Human Drug Metabolism from Molecular Biology to Man, 1992, Elizabeth Jeffreys
Platelet Activating Factor Receptor: Signal Mechanisms and Molecular Biology,
    1992, Shivendra D. Shukla
Biopharmaceutics of Ocular Drug Delivery, 1992, Peter Edman
Beneficial and Toxic Effects of Aspirin, 1993, Susan E. Feinman
Preclinical and Clinical Modulation of Anticancer Drugs, 1993, Kenneth D. Tew,
    Peter Houghton, and Janet Houghton
Peroxisome Proliferators: Unique Inducers of Drug-Metabolizing Enzymes, 1994,
    David E. Moody
*Angiotensin II Receptors, Volume 1: Molecular Biology, Biochemistry,
    Pharmacology, and Clinical Perspectives*, 1994, Robert R. Ruffolo, Jr.
Chemical and Structural Approaches to Rational Drug Design, 1994,
    David B. Weiner and William V. Williams
Direct Allosteric Control of Glutamate Receptors, 1994, M. Palfreyman, I. Reynolds,
    and P. Skolnick

Forthcoming Titles

*Alcohol Consumption, Cancer and Birth Defects: Mechanisms Involved in
    Increased Risk Associated with Drinking*, Anthony J. Garro
Animal Models of Mucosal Inflammation, Timothy S. Gaginella
Brain Mechanisms and Psychotropic Drugs, A. Baskys and G. Remington
*Development of Neurotransmitter Regulation and Function: A Pharmacological
    Approach*, Christopher A. Shaw
Drug Delivery Systems, V. V. Ranade
Endothelin Receptors: From Gene to Human, Robert R. Ruffolo, Jr.
Genomic and Non-Genomic Effects of Aldosterone, Martin Wehling
Human Growth Hormone Pharmacology: Basic and Clinical Aspects,
    Kathleen T. Shiverick and Arlan Rosenbloom
Immunopharmaceuticals, Edward S. Kimball
Neural Control of Airways, Peter J. Barnes
Pharmacological Effects of Ethanol on the Nervous System, Richard A. Deitrich
Pharmacology in Exercise and Sport, Satu M. Somani
Pharmacology of Intestinal Secretion, Timothy S. Gaginella
*Phospholipase A2 in Clinical Inflammation: Endogenous Regulation and
    Pathophysiological Actions*, Keith B. Glaser and Peter Vadas
Placental Pharmacology, B. B. Rama Sastry
Placental Toxicology, B. B. Rama Sastry
Receptor Characterization and Regulation, Devendra K. Agrawal
Ryanodine Receptors, Vincenzo Sorrentino
Serotonin and Gastrointestinal Function, Timothy S. Gaginella and
    James J. Galligan
Stealth Liposomes, D. D. Lasic and F. J. Martin
Targeted Delivery of Imaging Agents, Vladimir P. Torchilin
TAXOL: Science and Applications, Matthew Suffness
Therapeutic Modulation of Cytokines, M.W. Bodner and Brian Henderson
CNS Injuries: Cellular Responses and Pharmacological Strategies, Martin Berry

THE EDITORS

David B. Weiner, Ph.D., is an Assistant Professor of Pathology and Laboratory Medicine and Adjunct Assistant Professor of Medicine, at the University of Pennsylvania School of Medicine; and is Director of Biotechnology in the Institute for Biotechnology and Advanced Molecular Medicine at the University of Pennsylvania.

Dr. Weiner earned his B.S. in Biology from the State University of New York at Stony Brook in 1978; and obtained both M.S. and Ph.D. degrees in Developmental Biology from the University of Cincinnati in 1985. He completed postgraduate training in the Division of Immunology, Department of Pathology, at the University of Pennsylvania, in 1988.

Following his postdoctoral fellowship, Dr. Weiner was appointed in the Pathology Department at the University of Pennsylvania, then in the Wistar Institute. In 1989, Dr. Weiner was appointed as Director of Biotechnology at the Wistar Institute. He has presented more than 40 invited talks at international and national meetings, and has chaired sessions at the American Society for Microbiology, Clinical Immunology Society, BioEast, GHI-Vaccines and Gene Therapy Meetings. Dr. Weiner has published over 110 peer reviewed manuscripts and 10 review chapters. He currently serves on the editorial board of *Viral Immunology,* and as Senior Editor of *DNA* and *Cell Biology.*

In 1992, Dr. Weiner was appointed the Director of Biotechnology of the newly formed Institute for Biotechnology and Advanced Molecular Medicine (IBAMM). This Institute has been developed in collaboration with the Technology Council of Greater Philadelphia and the University of Pennsylvania to foster the growth of Biotechnology in the Delaware Valley by providing expertise on critical issues to a wide range of academic and industrial interests. In 1993, he became appointed in the Department of Pathology at the University of Pennsylvania as an Assistant Professor, and as an Adjunct Assistant Professor in the Department of Medicine. He has gained international recognition for his work in the area of genetic inoculation, molecular characterization of retroviral pathogenesis, and in the characterization of human retroviral envelope and regulatory genes. His studies in genetic inoculation have led to the development of novel prophylactic and immunotherapeutic vaccines against HIV infection. His current research interests include genetic pharmaceutics, regulatory gene structure and function, as well as mechanisms of retroviral pathogenesis.

William V. Williams, M.D., is an Assistant Professor of Medicine, in the Rheumatology Division, University of Pennsylvania School of Medicine; an Assistant Professor of Medicine in Pediatrics, Children's Hospital of Philadelphia; and Head of Molecular Medicine, Biotechnology Department, Institute for Biotechnology and Advanced Molecular Medicine.

Dr. Williams received his undergraduate training from Massachusetts Institute of Technology, from 1972—1976, and was granted Sc.B. degrees in Biology and Chemistry, graduating Phi Beta Kappa. He attended Tufts University School of Medicine from 1976—1980, receiving an M.D. degree. Following his clinical training, Dr. Williams distinguished himself as an Arthritis Foundation Postdoctoral Fellow, studying mechanisms of autoimmune disease pathogenesis under the auspices of Helen Mullen, Ph.D. He then spent 2 years as a National Institutes of Health Postdoctoral Fellow, studying Molecular Immunology, at the University of Pennsylvania with Mark Greene, M.D., Ph.D. In 1988, Dr. Williams became an Assistant Professor in the Rheumatology Division, Department of Medicine, at the University of Pennsylvania. In 1989, he was appointed Assistant Professor of Medicine in Pediatrics, Children's Hospital of Philadelphia. In 1992, he was asked to head the Molecular Medicine Division of the Biotechnology Department at the Institute for Biotechnology and Advanced Molecular Medicine at the University of Pennsylvania.

Dr. Williams is a member of the American Association for the Advancement of Science, the American College of Physicians, the American College of Rheumatology, and the American Federation for Clinical Research. He has authored over 100 papers, and has been the recipient of several research grants. He serves as a Senior Editor for the journal *Antibody.*

Dr. Williams has gained international recognition for his work in the development of biologically active peptides based on antibody structure. His research interests include molecular analysis of ligand-receptor interactions (particularly antibody-receptor interactions), molecular mechanisms of immunoregulation, molecular diagnostics, and rational drug design.

CONTRIBUTORS

Nambi Aiyar
Research and Development
 Division
SmithKline Beecham Pharmaceuticals
King of Prussia, Pennsylvania

Anastasia M. Khoury Christianson
Zeneca Pharmaceuticals Group
Wilmington, Delaware

David W. Christianson
Department of Chemistry
University of Pennsylvania
Philadelphia, Pennsylvania

Annette M. Doherty
Director
Cardiovascular Medicine
 and Chemistry
Parke-Davis Pharmaceutical
 Research
Ann Arbor, Michigan

Richard M. Edwards
Department of Pharmacological
 Sciences
SmithKline Beecham Pharmaceuticals
King of Prussia, Pennsylvania

Alan R. Fritzberg
Vice President
NeoRx Corporation
Seattle, Washington

Linda M. Gustavson
NeoRx Corporation
Seattle, Washington

Judith C. Hempel
Department of Medicinal
 Chemistry
SmithKline Beecham Pharmaceuticals
King of Prussia, Pennsylvania

David T. Hill
Research and Development Division
SmithKline Beecham Pharmaceuticals
King of Prussia, Pennsylvania

Mark D. Hylarides
NeoRx Corporation
Seattle, Washington

Richard M. Keenan
Department of Medicinal
 Chemistry
SmithKline Beecham Pharmaceuticals
King of Prussia, Pennsylvania

Thomas Kiber-Emmons
Assistant Professor
The Wistar Institute
Philadelphia, Pennsylvania

John A. Lipani
Centocor, Inc.
Malvern, Pennsylvania

James R. McCarthy
Director—Medicinal Chemistry
Neurocrine Biosciences, Inc.
San Diego, California

Alexander McPherson
Department of Biochemistry
University of California
Riverside, California

Lucio Miele
Human Genetics Branch
National Institutes of Health
Bethesda, Maryland

Anil B. Mukherjee
Human Genetics Branch
National Institutes of Health
Bethesda, Maryland

Eliot H. Ohlstein
Research and Development Division
SmithKline Beecham Pharmaceuticals
King of Prussia, Pennsylvania

Katherine Prammer
Department of Structural Biology
The Wistar Institute
Philadelphia, Pennsylvania

John M. Reno
NeoRx Corporation
Seattle, Washington

Stephen Roth
Neose Pharmaceuticals, Inc.
Norsham, Pennsylvania

James M. Samanen
Department of Medicinal Chemistry
King of Prussia, Pennsylvania

Andrew M. Stern
Department of Biological Chemistry
Merck Research Laboratories
Merck & Co., Inc.
West Point, Pennsylvania

Prasad S. Sunkara
Senior Director
Cancer Biology
 and Cancer Product Development
Cell Therapeutics, Inc.
Seattle, Washington

David B. Weiner
Assistant Professor
Department of Pathology
 and Laboratory Medicine
University of Pennsylvania
 School of Medicine
Philadelphia, Pennsylvania

Matthew Wiener
The Wistar Institute
Philadelphia, Pennsylvania

Joseph Weinstock
Department of Medicinal Chemistry
SmithKline Beecham
Pharmaceuticals
King of Prussia, Pennsylvania

William V. Williams
Assistant Professor
Rheumatology Division
University of Pennsylvania
School of Medicine
Philadelphia, Pennsylvania

ACKNOWLEDGMENTS

The editors would like to acknowledge support received from the Institute of Biotechnology and Advanced Molecular Medicine, the National Institutes of Health for DBW and WVW during the editing of this book, and the contributions of all the authors. We also would like to acknowledge the following people for their inspiration and support over the years: Elof Carlson, Carolyn Trunca, Henry Wortis, and Helen Mullen; our wives, Abby I. Phillipson and Lorraine M. Williams, and our children, Rebecca, Lindsey, Elizabeth, Ron, Chris, and Jon, for their forebearance and understanding.

TABLE OF CONTENTS

Section I:
Chemical Approaches to Rational Drug Design

Chapter 1

DESIGN, SYNTHESIS, AND ANTITUMOR ACTIVITY OF AN INHIBITOR OF RIBONUCLEOTIDE REDUCTASE

James R. McCarthy and Prasad S. Sunkara

CONTENTS

I. INTRODUCTION

Ribonucleotide reductases catalyze the rate-limiting step in the *de novo* synthesis of 2′-deoxyribonucleoside triphosphates (dNTPs) which are essential precursors for DNA synthesis. The normal substrates for this enzyme are ribonucleoside diphosphates (**1**). However, recently it was shown that ribonucleoside triphosphates were found to be substrates in certain bacteria.[1,2] Deoxyribonucleoside diphosphates (**2**), the products of this enzymatic reaction (Scheme 1) are

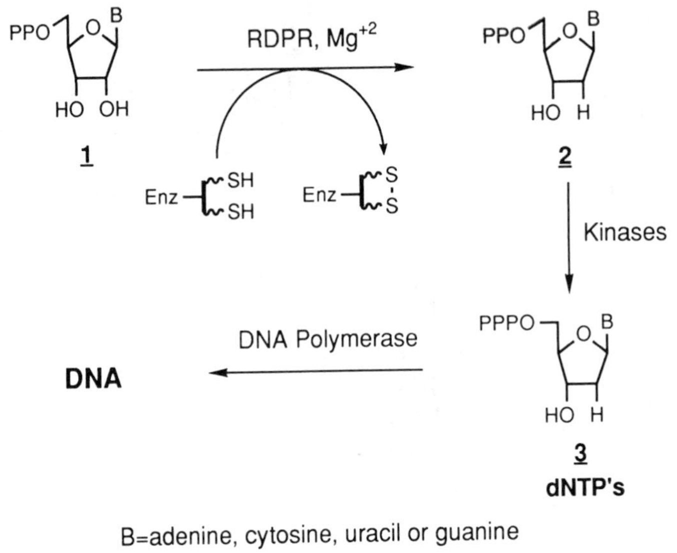

B=adenine, cytosine, uracil or guanine

SCHEME 1.   Reactions catalyzed by ribonucleoside diphosphate reductase (RDPR) and DNA polymerase.

further phosphorylated to dNTPs (**3**) and incorporated into DNA by the catalytic action of DNA polymerase (Scheme 1). Inhibition of ribonucleotide reductase interrupts DNA synthesis within seconds due to the limited intracellular levels of dNTPs.[3] The discovery of this critical role for reductases was made in the 1950s and led to intense research activity to determine the structure and mechanism of action, and obtain potent inhibitors of the enzyme. Although considerable progress has been made in elucidating the structure and mechanism of action of the enzyme, little progress has been made in the design and synthesis of potent inhibitors of the enzyme, until recently.

Hydroxyurea, a weak inhibitor of the mammalian enzyme, has been used with some success for the treatment of human leukemias and malignant melanomas.[4] A program has been initiated to discover potent inhibitors of the mammalian enzyme as antitumor agents. Hence, the focus of this chapter is the design, synthesis, and antitumor activity of a mechanism-based inhibitor of ribonucleotide diphosphate reductase (RDPR). A brief review of ribonucleotide reductases and inhibitors of the *Escherichia coli* enzyme is also included in this chapter. However, for more detailed information, the reader is referred to several recent reviews.[5-7]

A. RIBONUCLEOTIDE REDUCTASE CHARACTERISTICS

Ribonucleotide reductases are found in all living cells and catalyze the reduction of ribonucleoside diphosphates to deoxyribonucleoside diphosphates. There are at least four different forms of the reductase and specific metal centers and organic cofactors are required for each. The *E. coli* reductase has served as the prototype for the mammalian enzyme and has a binuclear iron center and a tyrosyl radical.[8] Other eukaryotic, prokaryotic, and viral ribonucleotide reductases with similar biochemical characteristics include herpes simplex virus and phage T4. A second form of reductase from lactobacilli requires 5′-deoxyadenosylcobalamin (coenzyme B_{12}) as a cofactor and ribonucleoside triphosphates as substrates.[1] A third reductase contains a binuclear Mn(III) complex [which is isomorphous with the binuclear Fe(III) complex] and is found in *Brevibacterium ammoniagenes*.[9] Recently, a fourth type of enzyme that requires *S*-adenosylmethionine and an as yet unidentified metal cofactor for activity was identified in anaerobically grown *E. coli*.[2] It is important to note that even with the major differences in the cofactor requirements for the four classes of the enzyme, the mechanism of ribonucleotide reduction is apparently similar and involves the formation of a 3′ radical on the substrate nucleotide.[10]

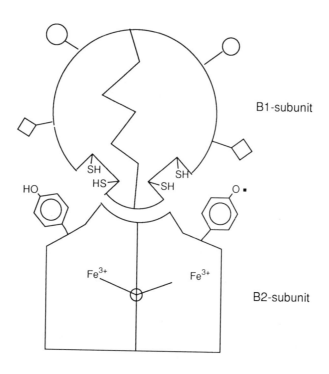

FIGURE 1. Model of *E.coli* RDPR. (Adapted from Reference 8.)

The active form of the *E. coli* reductase is believed to be a one-to-one complex of the B1 and the B2 subunits (Figure 1). The B1 subunit is a homodimer and consists of two 761-amino acid polypeptide chains ($\alpha\alpha'$), as is the B2 protein, which is composed of two identical polypeptides of 375 amino acid residues ($\beta\beta$). The α chains of the B1 subunit each contain two thiol groups that undergo a redox reaction during the conversion of the ribonucleotide to the deoxyribonucleotide. In addition, the subunit contains allosteric effector sites and substrate binding sites. Both the B1 and B2 subunits contribute to the active site of the enzyme. The active site of the enzyme recognizes cytidine-, guanosine-, adenosine-, and uridine diphosphates. The B2 protein contains the binuclear iron center that generates a tyrosyl radical via an oxygen-dependent reaction.[10] The tyrosyl radical is essential for the formation of the 3′ radical on the substrate. Site-directed mutagenesis has shown that the tyrosyl radical is derived from the tyrosine at position 122 on either of the homodimeric parts of the B2 subunit, but not both simultaneously.[11] Recently, two of the three intermediates proposed in the formation of the tyrosyl radical were observed.[12] The three-dimensional structure of the B2 subunit has been published, clearly showing the position of the iron centers and the tyrosyl 122 radical.[13] The tyrosyl radical is located 10 Å within the B2 subunit, while the binding site for the nucleotide substrate is on the B1 subunit. Therefore, in order to abstract the 3′-hydrogen of the substrate, the tyrosyl radical must generate a protein radical, probably via long-range electron transfer.

Stubbe has carried out a number of elegant experiments to determine the biochemical mechanism for the conversion of ribonucleotide diphosphates to the corresponding deoxynucleotides.[5] Originally, Stubbe proposed the mechanism outlined in Scheme 2. This mechanism stimulated our interest in the design of inhibitors of RDPR. A key part of this mechanism was the formation of a radical at the 3′ position of the nucleotide with subsequent loss of water at the 2′ position to form a radical cation. Hydride attack on the radical cation was proposed for the formation of the deoxynucleotide. Recently, the transfer of hydrogen to the 2′ position was proposed to involve a

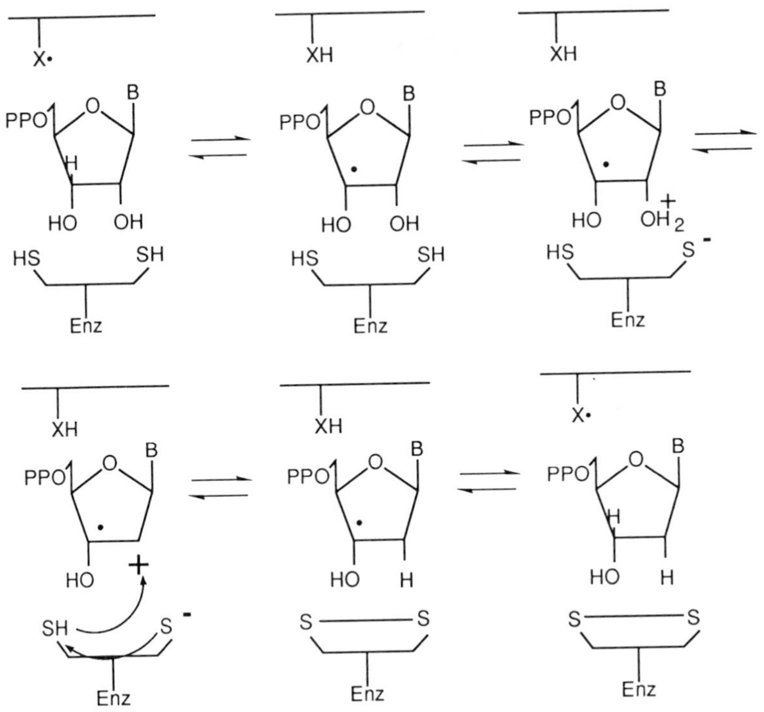

SCHEME 2. Original mechanism proposed by Stubbe (Reference 5) for the conversion of ribonucleoside diphosphates to deoxyribonucleoside diphosphates.

series of single electron transfer reactions instead of hydride transfer.[5] However, the salient features of the mechanism (i.e., 3′ radical and radical cation intermediates) remained the same. The structure and mechanism of action of the mammalian enzyme are similar to the *E. coli* enzyme described above.[7]

B. RDPR AS A TARGET FOR INHIBITION

All prokaryotic and eukaryotic cells require a continuous supply of dNTPs for DNA synthesis. dNTPs are obtained mainly from the *de novo* pathway by the action of ribonucleotide reductase and to a small extent from the salvage pathway.[3] The pool size of dNTPs in cells is limited and rapid synthesis of these intermediates is required for cell replication. Synthesis and accumulation of sufficient quantities of dNTPs is only possible via *de novo* synthesis, as was first demonstrated by Rose and Schweigert in 1953.[14] Therefore, ribonucleotide reductase plays an essential role in DNA biosynthesis. The activity of this enzyme varies with the cell cycle and is greatest in late G_1 to early S-phase, corresponding to the initiation of DNA replication, and least in early G_1 and M-phase.[15,16] Further, there is a positive correlation between the growth rate of tumors and the activity of ribonucleotide reductase.[15,17] Hence, RDPR was selected as a target for the design of novel antitumor agents.

C. INHIBITORS OF RDPR

Several review articles describe inhibitors of RDPR as antineoplastic agents.[8,18–21] Hydroxyurea (**4**), a weak inhibitor of the enzyme, is the only compound that shows clinical utility in the treatment of leukemia and malignant melanoma.[21] Three approaches for the design of inhibitors of RDPR have been taken. The first approach has utilized molecules such as hydroxyurea (**4**) or picolinaldehyde thiosemicarbazone (**5**) (Figure 2), which are single electron reductants of the tyr-122 radical. A second and more recent approach directed at herpes simplex virus RDPR is to prevent binding of

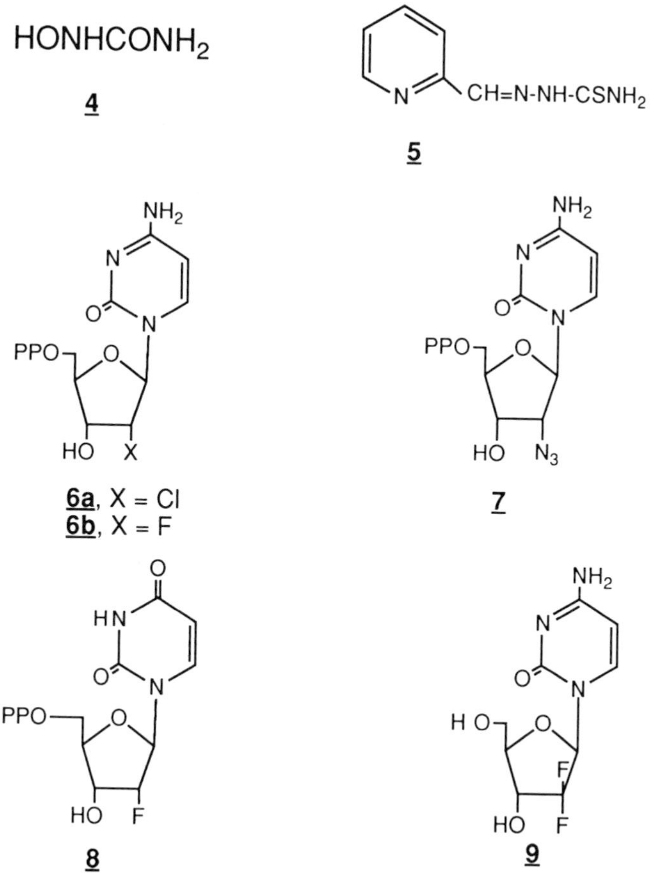

HONHCONH$_2$

4

5

6a, X = Cl
6b, X = F

7

8

9

FIGURE 2.   Inhibitors of RDPR.

the two subunits (H1 and H2).[22] Two groups independently reported that a nonapeptide derived from the C-terminal end of the H2 subunit was a potent inhibitor of viral reductases.[23,24]

The third and most promising approach utilizes mechanism-based inhibitors that are nucleoside analogs of the natural substrates.[25] Thelander et al.[26] initially reported that 2'-chloro-2'-deoxycytidine 5'-diphosphate (**6a**) and the corresponding uridine analog are inhibitors of RDPR. In the same article, it was reported that 2'-azido-2'-deoxycytidine 5'-diphosphate (**7**) inactivates the enzyme and destroys the tyrosyl radical. Stubbe studied these compounds in detail and determined that they were mechanism-based inhibitors of the enzyme and proposed mechanisms of inactivation.[5,27] 2'-Azido-2'-deoxycytidine 5'-diphosphate rapidly inactivates the enzyme (in a nonlinear time-dependent manner) with concomitant destruction of the tyr-122 radical. Incubation of RDPR with radiolabeled 3'-[^{3}H]-2'-azide derivative demonstrated cleavage of the 3'-carbon-hydrogen bond; in addition, one equivalent of inorganic pyrophosphate, nitrogen (N$_2$), and uracil were formed.[28] Inactivation of RDPR with 5'-[^{3}H]-2'-azido-2'-deoxycytidine 5'-diphosphate resulted in covalent modification of the B1 subunit. A significant observation was the production of a new stable radical from 2'-azido-2'-deoxycytidine 5'-diphosphate (**7**) that has eluded structural assignment.[5] However, the new inhibitors discussed below (i.e., **15** and **16**) should form stable radicals amenable to structural elucidations.

The inactivation of RDPR by 2'-chloro-2'-deoxycytidine 5'-diphosphate (**6a**) was proposed to involve the formation of a 3' radical, with subsequent loss of chloride anion and the formation of a radical cation intermediate (**10a** and **10b**) (Scheme 3). Reprotonation of the radical results in the formation of a 3'-ketonucleotide (**11**) and subsequent elimination of cytosine. The resulting proposed

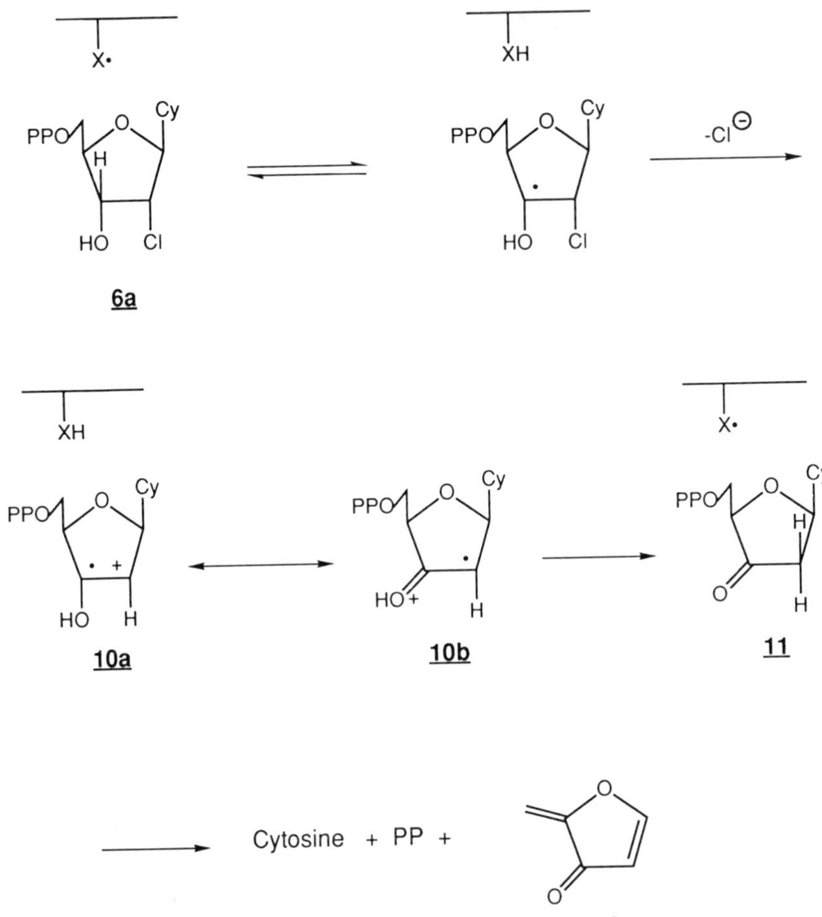

SCHEME 3. Proposed mechanism for the inactivation of RDPR by **6a** (Reference 5).

electrophile (**12**) possesses a chromophore which was separately shown to absorb near 320 nm as was observed in the ultraviolet spectrum of the mixture of *E. coli* enzyme and **6a**. Subsequently, 2′-fluoro-2′-deoxycytidine 5′-diphosphate (**6b**) and 2′-fluoro-2′-deoxyuridine 5′-diphosphate (**8**) were shown to be mechanism-based inhibitors of RDPR.[5] However, no antitumor activity has been reported for these compounds. Recently, 2′,2′-difluoro-2′-deoxycytidine (gemcitabine, **9**) (Figure 2) has been shown to be the bioprecursor to a mechanism-based inhibitor of RDPR.[29] This compound is currently in Phase II clinical trials for the treatment of various tumors.[30]

II. RATIONALE FOR DESIGN OF INHIBITOR

Based on the mechanism of action for RDPR, it was proposed that *(E)*- and *(Z)*-2′-fluoromethylene-2′-deoxycytidine (**13** and **14**) (Figure 3) would serve as bioprecursors for the inactivation of RDPR. Following conversion to the diphosphates (**15** and **16**, respectively), inactivation of the enzyme could occur by a mechanism outlined in Scheme 4.[31] Significantly, the proposed mechanism of inactivation of RDPR by **15** and **16** predicts the formation of the 3′-fluoroallyl radical (**17a**) which may be identifiable by EPR spectroscopy. A major issue that was addressed before the decision to proceed with the synthesis of the target molecules was the spatial constraints for a 2′-fluoro olefin functional group in the active site of the enzyme. Molecular models indicated that the

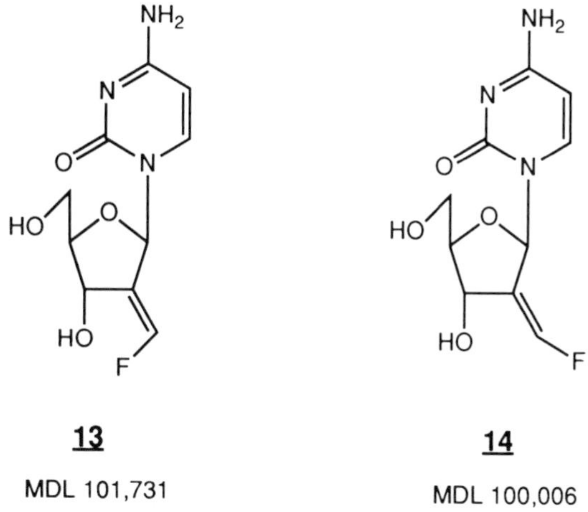

FIGURE 3. Structures of proposed mechanism-based inhibitors of RDPR.

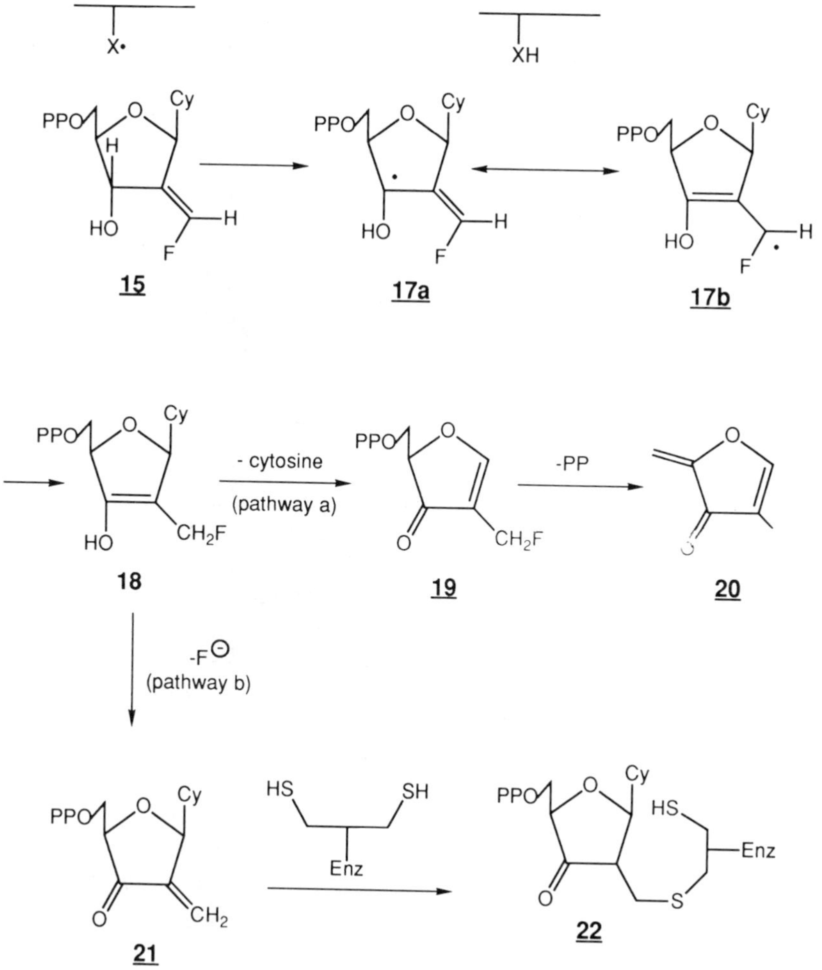

SCHEME 4. Proposed mechanism for inactivation of RDPR by **15**.

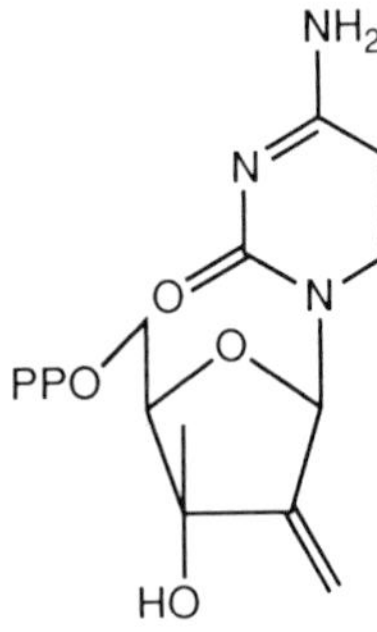

23

FIGURE 4.   Structure of **23**.

2′-fluoromethylene group would be accepted by RDPR since 2′-chloro-2′-deoxycytidine 5′-diphosphate (**6a**) is a mechanism-based inhibitor of the enzyme.[5] Furthermore, the recent disclosure that 2′-methylene-2′-deoxycytidine 5′-diphosphate (**23**) (Figure 4) is a mechanism-based inhibitor of RDPR is consistent with the original hypothesis, that an olefin at the 2′ position of a cytidine analog leads to the formation of a stabilized allylic radical intermediate while not interfering with substrate recognition.[29]

Deoxycytidine kinase exhibits broad substrate specificity for cytidine analogs.[32] The target nucleoside requires conversion to the diphosphate in the cell for recognition by RDPR; therefore, it was felt that **13** had a better chance for 5′-phosphorylation than either an adenosine or uridine-based inhibitor. The initial screen was HeLa cell proliferation. Activity in this assay depends on phosphorylation of the nucleoside. The 5′-diphosphate of the most active compounds were prepared for enzyme inhibition studies. A second enzyme that was considered to be important and is being investigated for the mechanism of action of **13** and **14** is DNA polymerase. Incorporation of **13** and **14** into DNA as the corresponding triphosphate could inhibit DNA synthesis by chain termination and provide another mechanism for antitumor activity.

III. SYNTHESIS

Scheme 5 outlines two retrosynthetic approaches that were considered for target molecules **13** and **14**. The key reaction that was envisioned for the introduction of the fluoro olefin functional group into the target molecules was the fluoro-Pummerer (or DAST-Pummerer) reaction that was initially reported in 1985.[33] This reaction introduces fluorine on an aliphatic carbon adjacent to sulfur by treatment of a sulfoxide (**26**) with diethylaminosulfur trifluoride (DAST) (Scheme 6); with aryl alkyl sulfoxides, an electron-donating group on the aryl ring facilitates the reaction. A Lewis acid, particularly antimony trichloride, catalyzes the reaction.[34] Reoxidation of the α-fluorosulfide (**27**) to the sulfoxide (**28**) and subsequent pyrolysis provides fluoro olefins (**29**) as a mixture of geometric isomers. The synthesis of the precursor nucleoside 2′-sulfoxide **24** (Scheme 5) would be required for the introduction of the 2′-fluoro olefin, via the fluoro-Pummerer reaction, in a nonconvergent synthesis.

Alternately, 2′-ketonucleoside **25** and fluoromethyl phenyl sulfone (**32**, Scheme 7) would be pivotal intermediates in a convergent synthesis of the target molecules. This route was considered more attractive and was pursued. A synthesis of **32** had been developed, using the fluoro-Pummerer reaction (Scheme 7). Treatment of methyl phenyl sulfoxide (**30**) with DAST and oxidation of the intermediate fluoromethyl phenyl sulfide (**31**) with either *meta*-chloroperbenzoic (MCPBA) or Oxone® provided fluoromethyl phenyl sulfone (**32**) in an overall yield of greater than 90%.[35,36] Over 100 g of **32** can be obtained in 2 d from methyl phenyl sulfoxide in excellent yield.[36] Sulfone **32** is

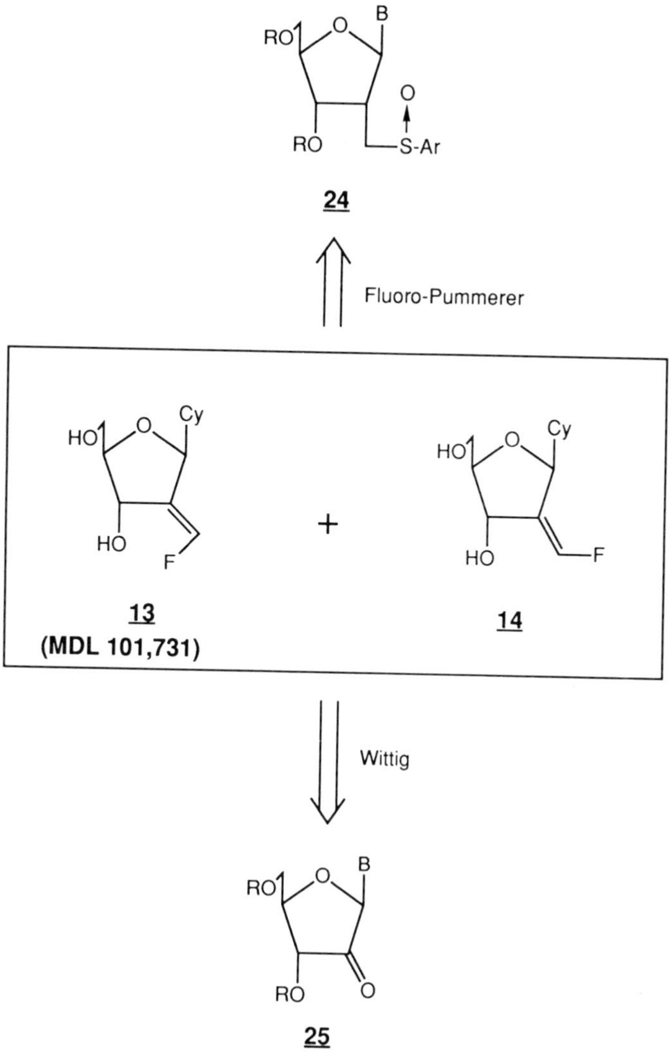

SCHEME 5. Two retrosynthetic approaches 13 and 14.

crystalline and can be easily purified. An alternate route to **32** that is cost effective for large-scale process work was developed. Fluoromethyl phenyl sulfone (**32**) was used for the preparation of β-fluorostyrenes from benzaldehydes, as outlined in Scheme 8.[37] Treatment of 4-chlorobenzaldehyde with **32** and butyllithium provided good yields of the condensed alcohol (**33**). Mesylation of **33** in the presence of triethylamine provided the α-fluorovinyl sulfone **34**. Removal of the sulfone group was accomplished with amalgamated aluminum in 90% yield to provide the β-fluorostyrene **35** as a 1:1 mixture of geometric isomers. However, when acetophenone was utilized as the starting material, the allylic fluorosulfone **36** was isolated in 82% yield, making it apparent that this method would not be applicable to the synthesis of **13** and **14**. The greater thermodynamic stability of the β,γ-saturated isomers of similar sulfones has been reported.[38,39] It was surmised that a Wittig reaction using the reagent derived from the carbanion of diethyl 1-fluoro-1-(phenylsulfonyl)methane-phosphonate (**37**) (obtained from **32**), would provide the fluorovinyl sulfone directly without isomerization; this was accomplished in 95% yield from acetophenone providing a 4 to 3 ratio of the *(E)* and *(Z)* geometric isomers (**38**, Scheme 9).[35]

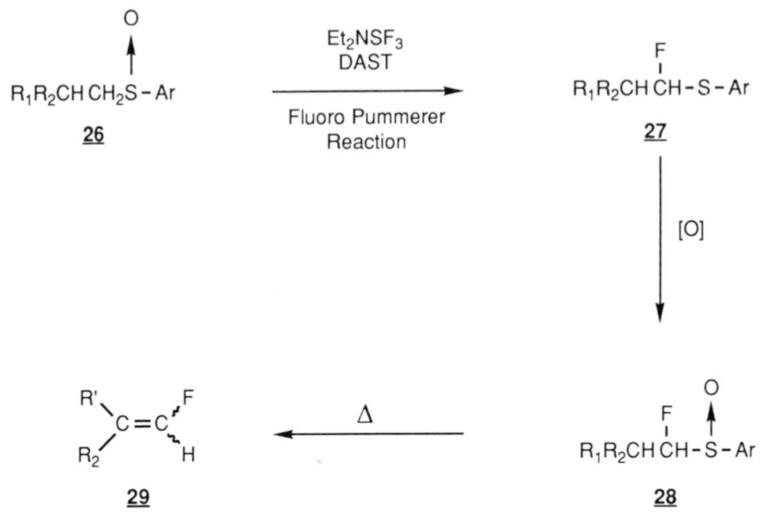

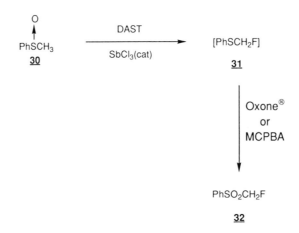

SCHEME 6. Synthesis of fluoro olefins (**29**) utilizing the fluoro-Pummerer reaction.

SCHEME 7. Synthesis of fluoromethyl phenyl sulfone (**32**)

  This method to fluoro olefins was applied to the target molecules **13** and **14**. Treatment of 2′-ketonucleoside **39**[40] with the carbanion **37**, generated from 1.5 equivalents of **32**, 1.5 equivalents of diethyl chlorophosphate and 2.7 equivalents of lithium hexamethyldisilazane provided about a 10 to 1 ratio of the *(Z)* and *(E)* isomers of the fluorovinyl sulfones **40** and **41**, respectively (Scheme 10). Ketone **39** was sensitive to excess base and this nearly led to the demise of the synthesis of target molecules **13** and **14**. When 2′-keto nucleoside **39** was initially treated with the same molar ratios of **32** and diethyl chlorophosphate, but with a slight excess (>3 equivalents) of lithium diisopropylamide (LDA), the α-xylonucleoside **42** was isolated in which both the 1′ and 3′ positions were inverted (Scheme 11). This material was assumed to be either the *(E)* or *(Z)* isomer of the desired fluorovinyl sulfone (i.e., **40** or **41**) and was converted to a white crystalline compound with elemental analysis, nuclear magnetic resonance (^{1}H NMR and ^{19}F NMR) and mass spectrum consistent for the desired product **13**. However, careful analysis of the nuclear Overhauser effects (NOEs) were inconsistent with **13**. Most importantly, this compound did not inhibit HeLa cell growth at a concentration of 10,000 ng/ml (Table 1 for comparison with **13**). The lack of cytotoxicity in the HeLa cell assay was

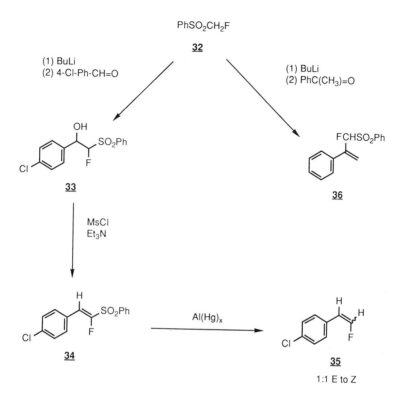

PhSO2CH2F
32
(1) BuLi
(2) 4-Cl-Ph-CH=O
(1) BuLi
(2) PhC(CH3)=O
OH
SO2Ph
F
Cl
33
FCHSO2Ph
36
MsCl
Et3N
H
SO2Ph
F
Cl
34
Al(Hg)x
H
H
F
Cl
35
1:1 E to Z

SCHEME 8. Synthesis of β-fluorostyrenes utilizing **32.**

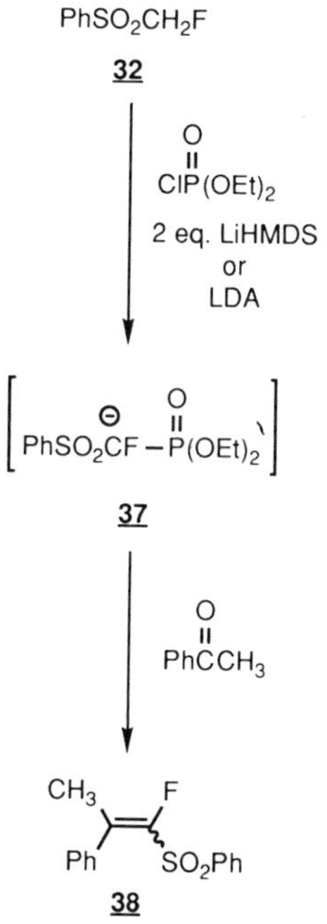

PhSO2CH2F
32
O
ClP(OEt)2
2 eq. LiHMDS
or
LDA
O
PhSO2CF − P(OEt)2
37
O
PhCCH3
CH3
F
Ph
SO2Ph
38

SCHEME 9. Synthesis of **38.**

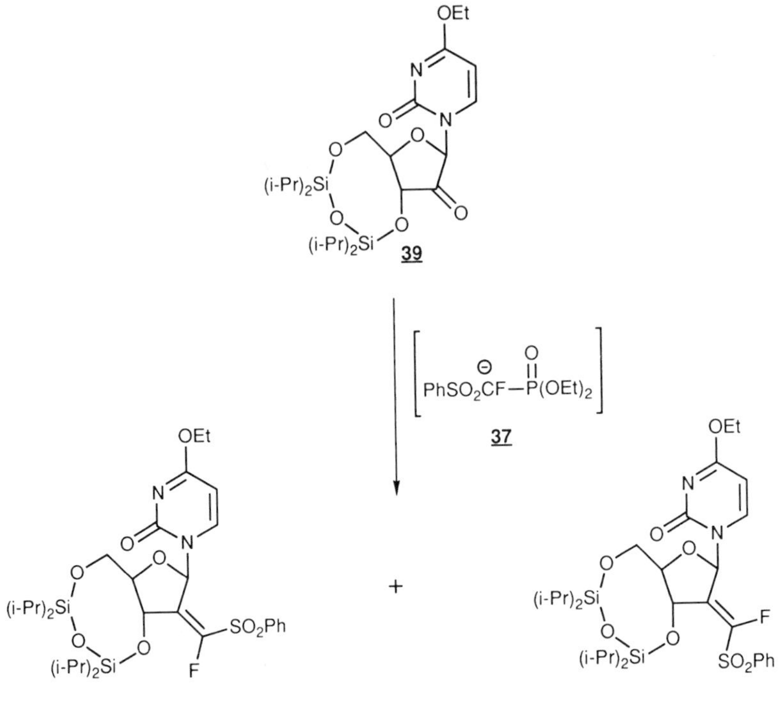

SCHEME 10. Synthesis of **40** and **41**.

discouraging, but in retrospect would be the expected result with the structure obtained. The compound was submitted for single-crystal X-ray structure determination, and to our surprise, both the 1′ and 3′ positions had inverted to provide the α-xylo nucleoside **43**. Analysis of each step in the reaction sequence indicated that the inversion occurred with the formation of **42** and was due to the excess LDA present in the reaction mixture. Before this problem was recognized, LDA and LiHMDS were used interchangeably for the synthesis of fluorovinyl sulfones but the milder base LiHMDS is preferred for sensitive molecules.

A key step in the proposed reaction sequence to the target fluoro olefins was the removal of the vinyl sulfone group with amalgamated aluminum (e.g., from **40** and **41**).[35,37] The fluorovinyl sulfone intermediates for both the cytidine and adenosine analogs were prepared and treatment of either fluorovinyl sulfone with amalgamated aluminum gave multicomponent mixtures as determined by thin layer chromatography (Scheme 12). This led to the discovery of a new stereospecific method for the synthesis of fluoro olefins. At the time this problem was encountered, there was only one general stereospecific synthesis of terminal fluoro olefins.[41] This synthesis relied on acetylenes as starting materials and provided fluoro olefins contaminated by 5 to 15% nonfluorinated olefins.

We found that fluorovinyl sulfone **40** was transformed to (fluorovinyl)stannanes **45** by treatment with two equivalents of tributyltin hydride (Scheme 13). Similarly, sulfone **41** gave exclusively stannane **46.** Watanabe and coworkers first observed the conversion of vinyl sulfones to mixtures of (E)- and (Z)-vinylstannanes on treatment with two equivalents of tributyltin hydride.[42] An electron-transfer mechanism consistent with the formation of mixtures of (E)- and (Z)-vinylstannanes was proposed. However, they did not study 2,2-disubstituted vinyl sulfones. It was observed that 2-monosubstituted fluorovinyl sulfones gave mixtures of (E)- and (Z)-(fluorovinyl)stannanes. On the basis of these observations, a radical addition-elimination mechanism was proposed in which elimination is faster than free rotation resulting in exclusive *cis*-elimination when rotation is

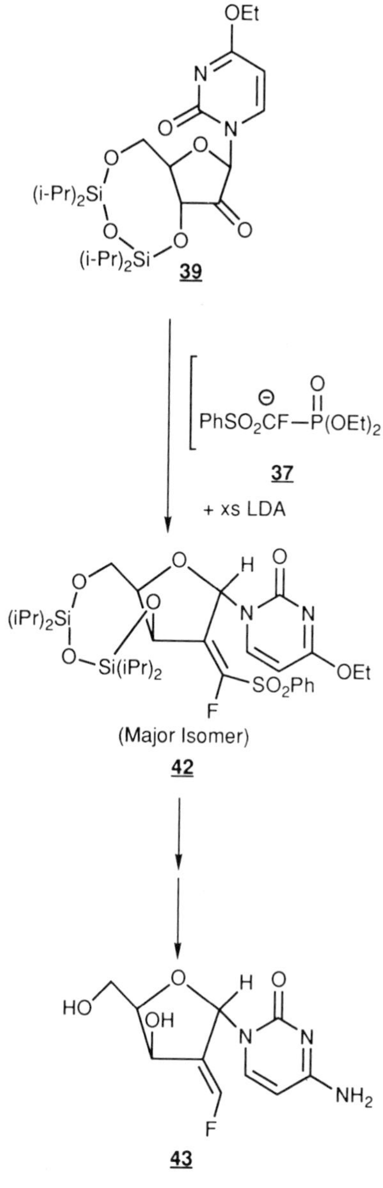

SCHEME 11. Synthesis of the α-lyxo fluoro olefin **43**.

restricted. The mechanism is consistent with observations of essentially complete retention of configuration of (fluorovinyl)stannanes derived from less constrained 2,2-disubstituted acyclic *(E)*- and *(Z)*-fluorovinyl sulfones.[31] This was the first report of a stereospecific radical reaction of this kind involving tributyltin hydride. It should be noted that (fluorovinyl)stannanes can function as fluorovinyl carbanion synthons. Thus, in addition to the stereospecific replacement of the tributyltin group with a proton, other electrophiles can be introduced stereospecifically.

 (E)- and *(Z)*-2′-deoxy-2′-(fluoromethylene)cytidines (**13** and **14**) were obtained as pure geometric isomers by separately treating **45** and **46**, respectively, with cesium fluoride and methanolic ammonia. Alternately, **45** was treated with methanolic ammonia to provide **47**, which, upon treatment with cesium fluoride, yielded **13**. The stereospecific destannylation reaction was also catalyzed by refluxing methanolic sodium methoxide or potassium fluoride. The latter routes are

TABLE 1
Antiproliferative Activity of Nucleoside Analogs against HeLa Cells in Culture

COMPOUND	GROWTH INHIBITION (IC_{50}, ng/mL)
(13) (MDL 101,731)	10 - 15
(14)	1000 - 5000
(55)	100
(56)	>10000
(57)	>10000
(58)	>10000
(59)	>10000
(60)	>10000

TABLE 1 (continued)
Antiproliferative Activity of Nucleoside Analogs against HeLa Cells in Culture

(**61**)   >10000

(**62**)   1000 - 10000

(**63**)   >10000

(**53**)   >10000

(**64**)   >10000

(**9**)   5 - 10

HYDROXYUREA (**4**)   10000

usually the methods of choice; however, in the case of the target molecules, the ammonia also displaces the 4-ethoxy group to produce the desired cytidine analog. It was proposed that catalytic amounts of methoxide present in the methanolic ammonia cleave the tin-carbon bond with concomitant protonation of the incipient vinyl carbanion by methanol regenerating methoxide. The geminal fluorine increases the susceptibility of the tin-carbon bond to cleavage, as evidenced by the fact that the tributylstannyl group was not removed from the nonfluorinated vinylstannane obtained from benzophenone under conditions (reflux MeOH/NaOMe, 16 h) where the fluorinated vinylstannane provided 1,1-diphenyl-2-fluoroethylene in 91% yield.[31]

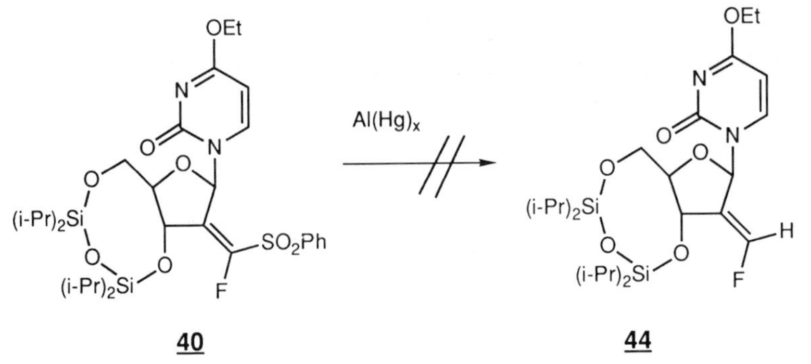

SCHEME 12. Amalgamated aluminum is ineffective for the conversion of **40** to **44**.

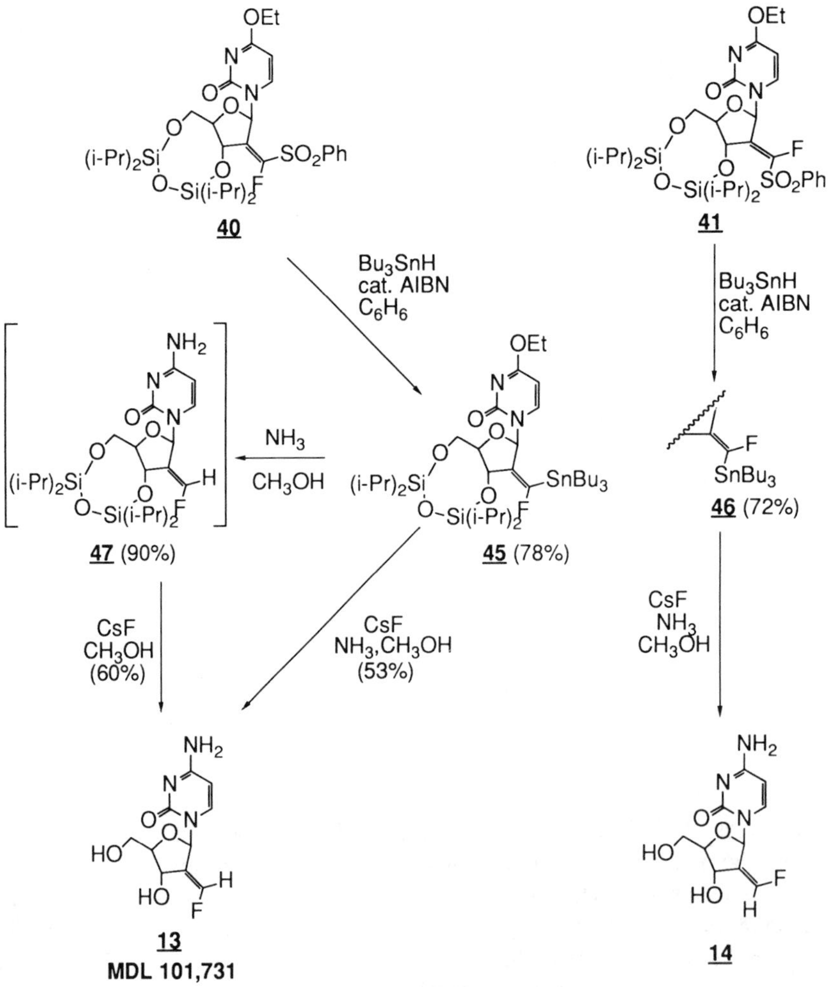

SCHEME 13. Original synthesis of **13** and **14**.

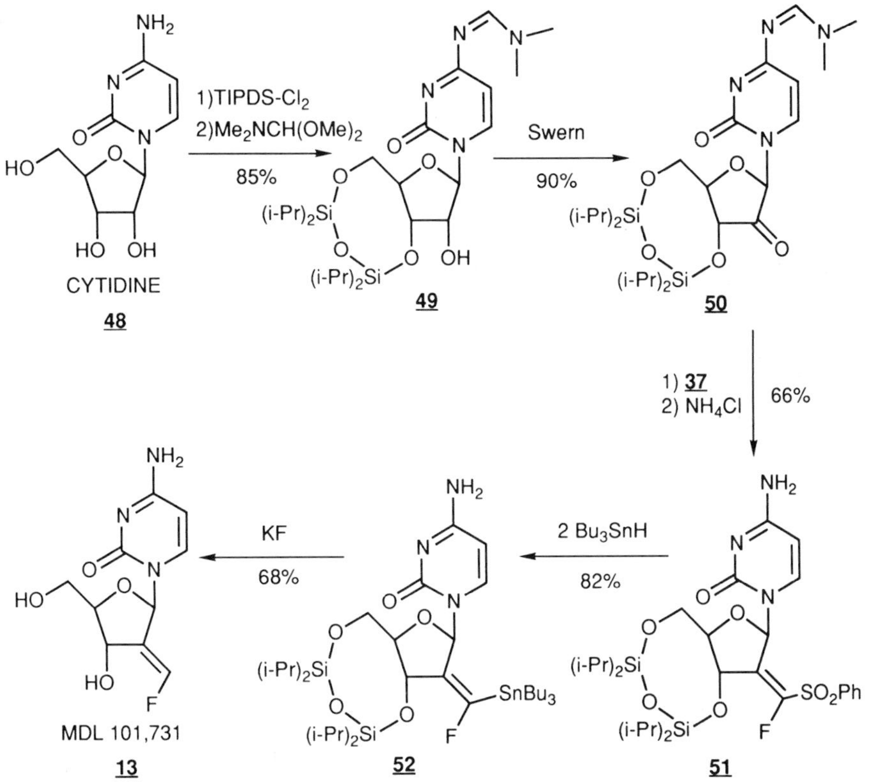

SCHEME 14. Improved synthesis of **13** and **14**.

A substantially improved synthesis of the more active *(E)* isomer **13** was developed starting with cytidine, as outlined in Scheme 14. The salient features of the improved synthesis are as follows: after protection of the 3′- and 5′-hydroxyl groups of cytidine (**48**) with the 1,1,3,3-tetraisopropyl disiloxane group (TIPDS), the N-4 amino group was protected as the *N,N*-dimethylamidine by the addition of dimethylformamide dimethyl acetal to the same reaction mixture. The protected nucleoside **49** was isolated in 85% yield after recrystallization. The ketone **50,** obtained from the Swern oxidation, was isolated by nonaqueous workup by direct crystallization in 90% yield. This ketone was treated as in the first method for the preparation of fluoro olefin **13,** except potassium fluoride was used to remove both the vinylstannane and the TIPDS protecting group. Compound **13** was obtained in an overall yield of 28% in five steps from cytidine.

IV. ENZYME INHIBITION

The enzymes involved in the metabolism of compound **13** are described in Scheme 15. It is important to note that **13** was designed as a bioprecursor that requires phosphorylation by kinases to the diphosphate **15** for inactivation of the target enzyme RDPR. Possible conversion to the triphosphate **54** by kinases may result in its incorporation into DNA with concomitant DNA chain termination. Another enzyme that was studied was cytidine deaminase because of its potential role in the inactivation of **13** by forming the uridine analog **53,** which was shown to be inactive in the HeLa cell proliferation assay (Table 1).

A. INHIBITION OF PURIFIED *E. COLI* RDPR

Some very exciting preliminary results with **15** show inactivation of the *E. coli* enzyme and the formation of a new radical concomitant with the loss of the tyrosyl radical as well as the formation

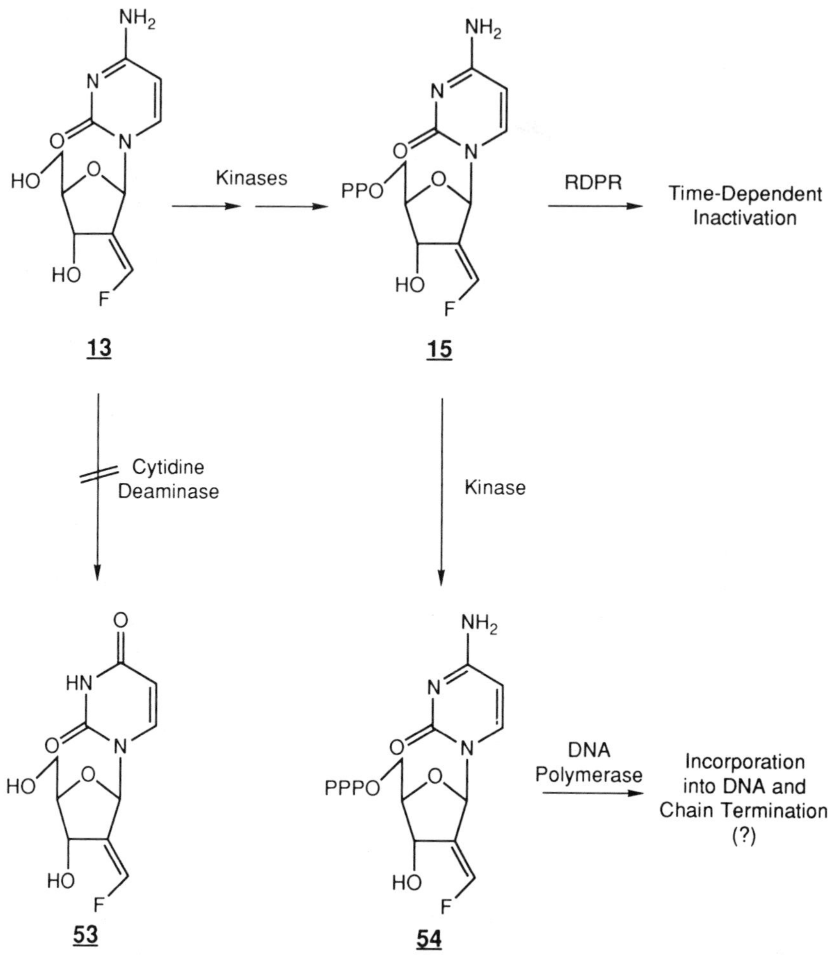

SCHEME 15. Enzymatic transformations of **13**.

of cytosine.[49] The 5′-diphosphate **15** was incubated with a 1 to 1 mixture of purified B1 and B2 subunits of *E. coli* RDPR which results in very rapid inactivation of the enzyme and loss of the tyrosyl radical with the formation of a new radical. The rate of inactivation of RDPR appears to be dose-dependent. Furthermore, it was observed that the kinetics for the loss of tyrosyl radical is biphasic. More than one equivalent of cytosine was formed per equivalent of enzyme inactivated by **15,** indicating that loss of the tyrosyl radical is only one mode of enzyme inactivation (see Scheme 4, pathway a).

B. *EX VIVO* STUDIES WITH RDPR

RDPR activity was determined *ex vivo* after a single intraperitoneal (i.p.) dose of **13** (1 mg/kg) to L1210 leukemia-bearing mice. The animals were sacrificed at 1, 2, and 3 h after drug administration. Tumor cells were collected, washed, and cell-free extracts were prepared. The enzyme was precipitated with 40% ammonium sulfate and the activity was measured after dialysis. The activity of the enzyme was 21, 12, and 3% of saline control animals at 1, 2, and 3 h postinjection, respectively. Exhaustive dialysis experiments did not lead to a return of enzyme activity. These results indicate *in vivo* inactivation of RDPR by **13** are consistent with those obtained with the purified *E. coli* enzyme.

C. CYTIDINE DEAMINASE

Both **13** and **14** were tested as substrates for partially purified cytidine deaminase from mouse kidney.[44] The deamination of cytidine results in a change in UV absorbance due to the formation of uridine. However, no significant changes in absorbance were observed when either **13** or **14** were incubated with the deaminase preparation, suggesting that neither compound is a substrate for the enzyme.

V. BIOLOGY ACTIVITY

The antiproliferative and antitumor activity of **13** (MDL 101,731) against cell cultures and experimental tumors, and synergistic activity of the compound with S-phase-specific antitumor drugs are presented in the rest of the chapter.[45–47]

A. ANTIPROLIFERATIVE ACTIVITY

1. Structure-Activity Relationship

A number of analogs of MDL 101,731 were synthesized and evaluated for antiproliferative activity against HeLa cells (Table 1). MDL 101,731 showed potent activity (IC_{50} = 10 to 15 ng/ml) whereas the geometric isomer (**14**) was inactive, suggesting the importance of the geometry of fluorine in the 2′ position of the sugar. Substitution of uracil for cytosine (**53**) resulted in loss of activity. MDL 101,731 was at least 1000-fold more active on a molar basis than hydroxyurea (**4**), the only useful inhibitor of ribonucleotide reductase in clinical use.

2. Antiproliferative Activity against Tumor Cells

The potent antiproliferative activity of MDL 101,731 against a number of mouse and human tumor cell lines is presented in Table 2. The IC_{50} values against different cell lines ranged from 2.7 to 70 ng/ml. The compound was equipotent against multidrug resistant and sensitive human KB epidermoid carcinoma (KBV1; IC_{50}: 15 to 17 ng/ml) and mouse P388[R] leukemia (P388[R]; IC_{50}: 2.7 to 3.1 ng/ml) cells in culture.

Antiproliferative and cytotoxic activity of MDL 101,731 on the growth of human cervical carcinoma (HeLa) cells is shown in Figures 5A and B. A dose-dependent inhibition of HeLa cell growth was observed (Figure 5A). In addition, MDL 101,731 was found to be a potent cytotoxic agent based on the survival of treated cells in a colony-forming assay (Figure 5B); a 100% cell kill was observed at 48 h when cells were treated with 50 ng/ml. Both DNA and RNA synthesis were inhibited by MDL 101,731 in HeLa cells, whereas protein synthesis was less affected (Figure 6).

3. Effects on dNTP Pools and DNA Synthesis in HeLa Cells

A comparison of MDL 101,731, gemcitabine (**9**), and hydroxyurea (**4**) on the antiproliferative activity, inhibition of DNA synthesis, and depletion of dNTP pools in HeLa cells is presented in Table 3. MDL 101,731 was equipotent with **9** while hydroxyurea was over 10,000-fold less active.

B. ANTITUMOR ACTIVITY IN EXPERIMENTAL TUMORS

1. Leukemias

Mouse L1210 leukemia — Antitumor activity of MDL 101,731 against mouse L1210 leukemia as determined by percent increase in survival time (% T/C) is presented in Tables 4 to 6. A dose-dependent increase in survival of leukemia-bearing animals was observed following administration of drug by both i.p. and oral routes. The compound showed a maximum % T/C value of approximately 300 at 10 mg/kg, i.p., once daily from day 1 to 9 (Table 4). At this dose, 1/5 of the animals were found to be tumor-free in each of two experiments. MDL 101,731 was also orally active (Table 5). The drug was well tolerated up to 25 mg/kg, once daily from day 1 to 9, and showed % T/C value of 277. A more effective dose and schedule of administration of MDL 101,731 is presented in Table

TABLE 2
Antiproliferative Activity of MDL 101,731 against
Several Tumor Cells in Culture

Tumor Cell Line	IC_{50} (ng/ml)
Mouse	
L1210 leukemia	6.8
P388 leukemia	2.7
P388[R] leukemia (multidrug resistant)	3.1
B16 melanoma (F10 line)	70.0
Human	
Cervical carcinoma (HeLa)	15.0
Epidermoid carcinoma (KB)	15.0
Epidermoid carcinoma (multidrug resistant, KBV_1)	17.2

Note: 1×10^5 cells/35 mm dish were plated and allowed to grow over-
night. Different concentrations of MDL 101,731 were prepared in
the respective media and added to the cells. The cells were further
incubated for 72 h. At the end of incubation, the cell number was
determined by a Coulter counter and the concentrations required
for 50% inhibition of growth compared to control (IC_{50}) were
determined.

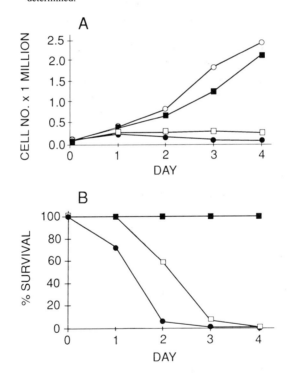

FIGURE 5.   Effect of MDL 101,731 on HeLa cell proliferation. HeLa cells (1×10^5/100 mm dish) were plated and treated
with various concentrations of MDL 101,731. Cell counts were enumerated on different days to determine the growth
inhibition, and clonogenic assays were conducted to determine cell survival. ○, control; •, 50 ng/ml; □, 25 ng/ml; and ■, 12.5
ng/ml.

6. The data suggest that twice daily, from days 1 through 14, administration of the compound at 2
to 3 mg/kg, i.p., twice daily, resulted in 60 and 80% of the animals cured of the disease (surviving
for more than 60 d after tumor inoculation), and the rest of the treated animals had a % T/C of more
than 400.

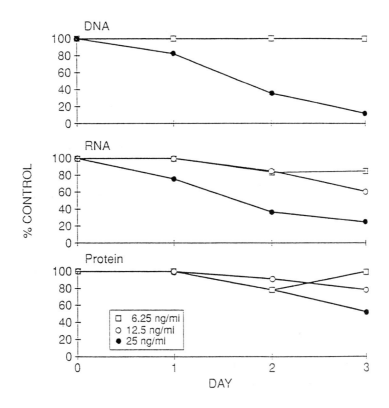

FIGURE 6. Effect of MDL 101,731 on macromolecular synthesis in HeLa cells. HeLa cells were treated with various concentrations of MDL 101,731, and, on the indicated days, the macromolecular synthetic abilities of the treated cells were determined by incorporation of radioactive precursors into DNA, RNA, and protein. •, 25 ng/ml; ○, 12.5 ng/ml; and □, 6.25 ng/ml.

Multidrug-resistant P388R leukemia — MDL 101,731 also showed significant activity against multidrug-resistant P388R leukemia. This tumor is resistant to adriamycin and vinblastine. MDL 101,731 was active by i.p., oral, and i.v. routes of administration (Table 7). Maximum % T/C values of 207, 215, and 194 were observed when the compound was administered once daily at 10 mg/kg by i.p., oral, and i.v. routes, respectively, with 20% of the animals tumor-free following i.p. administration. However, the most effective schedule and dose was found to be 3 mg/kg i.p., twice daily from day 1 to 14 (Table 8). This schedule was 38% curative in the tumor-bearing animals with an increase in % T/C of 282 in the rest of the treated animals. Hence, the curative activity and increase in survival induced by MDL 101,731 is quite significant because P388R is resistant to known antitumor agents such as adriamycin and vinblastine.

2. Solid Tumors

Murine B16 melanoma — The antitumor activity of MDL 101,731 against B16 melanoma is shown in Table 9. A dose-dependent inhibition of tumor growth was observed with 84 and 97% inhibition at 10 mg/kg, i.p. once daily from day 1 to 15, in two different studies. The compound showed marginal activity when given orally at a dose of 30 mg/kg.

Murine Lewis lung carcinoma — Lewis lung carcinoma (3LL) is an aggressive tumor that invades and metastasizes to the lungs. MDL 101,731 showed a dramatic activity against 3LL carcinoma. Treatment with MDL 101,731 at 15 mg/kg, i.p., once daily, resulted in 90% of the animals free of tumor and a 99% inhibition of tumor growth in the rest of the animals (Table 10). The compound also showed curative (4/5 animals) activity by the i.v. route. MDL 101,731 at

TABLE 3
Effect of MDL 101,731 (13) and other Inhibitors of RDPR on Cell Growth, DNA Synthesis, and dNTP Pools in HeLa Cells

	IC_{50} (ng/mL)		
COMPOUND	GROWTH INHIBITION	DNA SYNTHESIS	dNTP POOLS
(**13**)	10 - 15	0.7	25
(**9**)	5 - 10	25	25
HYDROXYUREA (**4**)	10000	100	40000

30 mg/kg by the oral route resulted in an 88% inhibition of tumor growth. At all the doses tested, complete inhibition of metastasis was observed.

Multidrug-resistant human epidermoid carcinoma (KBV1) — KBV1 is a multidrug-resistant carcinoma refractory to the treatment of vinblastine, adriamycin, and vincristine.[48] Treatment with MDL 101,731 resulted in 77 and 88% inhibition of KBV1 tumor growth in athymic nude mice at 10 mg/kg, i.p., once daily, from day 1 to 12 in two different studies (Table 11). The compound also caused regression of the established tumor based on tumor volume measurements (Figure 7). These results are quite encouraging because KBV1 is refractory to most conventional chemotherapeutic agents.

Human ovarian carcinoma (HTB-161) — HTB-161 is a drug-resistant ovarian carcinoma. The tumor was allowed to grow for 2 weeks in athymic nude mice before treatment with MDL 101,731 was initiated. Treatment of established HTB-161 with MDL 101,731 at 5 and 10 mg/kg i.p. once daily from day 1 to 23 resulted in complete inhibition of growth at the lower dose and a regression of established tumor at the higher dose (Figure 8).

TABLE 4
Antitumor Activity of MDL 101,731 by
Intraperitoneal Administration against L1210
Leukemia in Mice

Treatment	Dose (mg/kg)	Survival Time (Days) (Mean ± S.E.; n=5)	% T/C
Expt. I			
Untreated control	—		
MDL 101,731	0.5	10.2 ± 0.5	146
	1.0	13.8 ± 0.4	197
	2.0	15.8 ± 0.2[a]	226
Expt. II			
Untreated control	—		
MDL 101,731	2.5	16.8 ± 0.5	221
	5.0	20.6 ± 0.8	271
	10.0	21.3 ± 0.8[a]	280
Expt. III			
Untreated control	—	7.0 ± 0.3	
MDL 101,731	10.0	22.0 ± 1.0[a]	314
	15.0	20.3 ± 1.5[a]	290

Note: 1×10^5 L1210 cells/mouse were inoculated i.p. on day 0. The compound was dissolved in sterile water and administered i.p. once a day from day 1–9. The survival of the animals was recorded and % T/C was calculated. Animals surviving for more than 60 d were considered cured.

[a] 1/5 animals was tumor-free.

$$\% \ T/C + 100 \times \frac{\text{average days of survival in treated}}{\text{average days of survival in untreated control}}$$

C. SYNERGY STUDIES

1. Cell Culture Studies

We have studied the effect of MDL 101,731 on cell cycle traverse of HeLa cells. The data presented in Figure 9 show that at a concentration of 15 ng/ml, MDL 101,731 inhibited the growth of HeLa cells by 75% while at 20 ng/ml a complete cessation of growth was observed at the end of 4 d. Further, a 75 to 80% accumulation of cells in S-phase was observed within 24 h of treatment with MDL 101,731. At lower concentrations, the cells started progressing through the cell cycle by 72 h as evidenced by increases in cell numbers and reduction of cells in S-phase relative to the control cell population. However, at higher concentrations (20 ng/ml), a 70% accumulation of cells in S-phase was observed even after 96 h without recovery of cell growth.

Based on the above observation that MDL 101,731 treatment leads to accumulation of cells in S-phase, combination chemotherapy studies with S- and mitotic-phase-specific drugs were investigated. Treatment of HeLa cells with MDL 101,731 (15 ng/ml) for 24 h followed by treatment with cytosine arabinoside (Ara-C) or 5-fluorouracil (5-FU), adriamycin (ADR) or vinblastine (VLB) resulted in synergistic antiproliferative activity as determined by the tetrazolium (MTT) assay. The IC_{50} (µg/ml) values for Ara-C, 5-FU, VLB, and ADR when treated alone or in combination with MDL 101,731 (15 ng/ml for 24 h) are 14, 41, 87, and 0.06 compared to 3, 19, 14, and 0.04, respectively (Table 12). Treatment of HeLa cells with MDL 101,731 alone for 24 h did not have any effect on cell growth.

TABLE 5
Antitumor Activity of MDL 101,731 by Oral Administration (P.O.) against L1210 Leukemia in Mice

Treatment	Dose (mg/kg)	Survival Time (Days) (Mean ± S.E.; n=5)	% T/C
EXPT. I			
Untreated control	—	7.4 ± 0.2	—
MDL 101,731	1.0	10.0 ± 0.5	135
	2.5	14.2 ± 0.5	192
	2.0	15.8 ± 0.2[a]	226
EXPT. II			
Untreated control	—	7.6 ± 0.2	—
MDL 101,731	5.0	16.0 ± 0.8	211
	10.0	17.2 ± 0.5	226
EXPT. III			
Untreated control	—	7.0 ± 0.3	
MDL 101,731	10.0	17.6 ± 0.4	251
	15.0	19.0 ± 0.7	271
	25.0	19.4 ± 0.2	277

Note: 1×10^5 L1210 cells/mouse were inoculated i.p. on day 0. The compound was dissolved in sterile distilled water and administered orally once a day from day 1–9. The survival of the animals was recorded and % T/C was calculated. The control values in Exp. II and III in Tables 4 and 5 are from the same experiments.

[a]   1/5 animals was tumor-free.

TABLE 6
Antitumor Activity of MDL 101,731 by Multiple Dose Administration against L1210 Leukemia in Mice

Treatment (mg/kg)	Route and Schedule	Survival (Days) (Mean ± S.E.)[a]	Cures	% T/C
Untreated control		7.9 ± 0.4	0/15	
MDL 101,731				
1.0	i.p., b.i.d.	18.5 ± 1.1	1/5	264
2.0	i.p., b.i.d.	31.0	4/5	443
3.0	i.p., b.i.d.	32.3 ± 3.9	8/14	408

Note: 1×10^5 L1210 leukemia cells/animal were injected i.p. on day 0. The compound was dissolved in saline and administered as indicated from day 1–14. The survival of the animals was recorded and a % T/C was determined. Animals surviving for more than 60 d were considered cured.

[a]   Values are for animals not cured.

TABLE 7
Antitumor Activity of MDL 101,731 against Multidrug-Resistant P388^R Leukemia in Mice

Treatment	Dose (mg/kg)	Route	Survival Time (Days) (Mean ± S.E.; n=5)	% T/C
Expt. I				
Untreated control			9.8 ± 0.2	
Adriamycin	2.0	i.p.	10.4 ± 0.5	106
	2.0	p.o.	10.6 ± 0.6	108
MDL 101,731	2.5	i.p.	18.8 ± 1.7	192
	5.0	i.p.	18.6 ± 0.6	190
	10.0	i.p.	21.7 ± 0.7[a]	207
	2.5	p.o.	13.2 ± 0.2	135
	5.0	p.o	17.8 ± 1.2	182
	10.0	p.o.	19.6 ± 0.2	200
	15.0	p.o.	21.5 ± 0.5	219
Expt. II				
Untreated control	—	—	10.4 ± 0.3	
MDL 101,731	10.0	i.v.	20.2 ± 1.2	194

Note: P388^R leukemic cells (1×10^5/mouse) were inoculated i.p. into BDF1 mice on day 0. Adriamycin and MDL 101,731 were administered once daily, i.p. or p.o. from day 1–9. The survival of the animals was recorded and % T/C was calculated. Animals surviving for more than 60 d were considered cured.

[a] 2/10 animals were tumor-free.

TABLE 8
Antitumor Activity of Multiple Administrations of MDL 101,731 against Multidrug-Resistant P388^R Leukemia in Mice

Treatment (mg/kg)	Route and Schedule	Survival (Days)[a] (Mean ± S.E.)	Cures	% T/C
Untreated control		10.4 ± 0.34	0/10	
MDL 101,731				
1.0	Day 1–9 b.i.d.	22.6 ± 1.2	0/5	226
2.0	Day 1–9 b.i.d.	26.7 ± 0.5	1/5	267
3.0	Day 1–9 b.i.d.	28.7 ± 1.3	1/5	287
3.0	Day 1–14 b.i.d.	29.3 ± 4.3	6/16	282
Adriamycin				
1.0	Once daily	10.2 ± 0.2	0/10	102
Vinblastine				
0.2	Once daily	10.4 ± 0.2	0/10	100

Note: 1×10^6 P388 leukemic cells/mouse were inoculated i.p. on day 0. Compounds were dissolved in distilled water and were administered i.p. as indicated. The survival of the animals was noted and the % T/C was determined. Animals surviving for more than 60 d were considered cured.

[a] Values are for animals not cured of the tumor.

TABLE 9
Antitumor Activity of MDL 101,731 against B16-F10
Melanoma in Mice

Treatment (mg/kg)	Route	Tumor Weight (g) (Mean ± S.E.; n = 5)	% Inhibition
Expt. I			
Untreated control		4.0 ± 0.44	
MDL 101,731			
2.5	i.p.	1.17 ± 0.2	71
5.0	i.p.	0.36 ± 0.08	91
10.0	i.p.	0.11 ± 0.06	97
Expt. II			
Untreated control		7.2 ± 0.22	
MDL 101,731			
30.0	oral	3.9 ± 1.2	45
10.0	i.p.	1.15 ± 0.35	84

*Note:*B16 melanoma cells (1×10^5) were inoculated s.c. in the interscapular region of C57/BL mice on day 0. MDL 101,731 was administered once daily i.p. from day 1–15. The animals were sacrificed on day 18; tumors were dissected and weighed.

TABLE 10
Antitumor Activity of MDL 101,731 against 3LL Carcinoma in Mice

Treatment (mg/kg)	Route	Tumor Weight (g) (Mean ± S.E.)	Tumor-Free Animals	% Inhibition	No. of Foci (Mean ± S.E.)	% Inhibition
Expt. I						
Untreated control		5.3 ± 0.8	0/5		30.7 ± 8.0	
MDL 101,731						
10	i.p.	0.8 ± 0.3	0/5	84	3.2 ± 0.9	90
15	i.p.	0.02 ± 0	4/5	99.9	0	100
Expt. II						
Untreated control		4.6 ± 0.98	0/5		19.8 ± 6.0	
MDL 101,731						
10	i.p.	0	5/5	100	0	100
15	i.p.	0	5/5	100	0	100
10	i.v.	0.24	4/5	95	0	100
30	oral	0.58 ± 0.16	0/5	88	0	100

Note: 3LL cells (1×10^5) were inoculated s.c. in the interscapular region of C57/BL mice on day 0. MDL 101,731 was administered once daily, i.p. from day 1–14. The animals were sacrificed on day 15; tumors dissected and weighed.

2. Solid tumors

We have evaluated the combination of MDL 101,731 and Ara-C in two murine solid tumor models. Treatment of B16 melanoma-bearing mice at 5 mg/kg i.p. once daily from day 1 to 7 with MDL 101,731 resulted in only 10% inhibition of tumor growth at the end of the study on day 15. Ara-C, administered at 10 mg/kg i.p. once daily from day 7 to 14 did not have any effect on tumor growth. However, a combination of these treatments resulted in a 60% synergistic inhibition of tumor growth (Table 13). Similar treatment of Lewis lung carcinoma (3LL)-bearing mice resulted

TABLE 11
Antitumor Activity of MDL 101,731 against
Multidrug-Resistant Human Epidermoid
Carcinoma (KBV₁) in Athymic Nude Mice

Treatment (mg/kg)	Tumor Weight (g) (Mean ± S.E., n = 5)	% Inhibition
Expt. I		
Untreated control	3.47 ± 0.74	
MDL 101,731		
5	1.34 ± 0.18	62
10	0.8 ± 0.25	77
Expt. II		
Untreated control	1.91 ± 0.41	
MDL 101,731		
5	0.49 ± 0.06	75
10	0.26 ± 0.04	88

Note: Approximately 0.3 g of KBV₁ tumor was transplanted s.c in the flanks of athymic nude mice on day 0. The tumor was allowed to grow for 2 weeks and then drug treatment was initiated. MDL 101,731 was given i.p. once daily for 2 weeks. At the end of treatment, animals were sacrificed; tumors dissected and weighed.

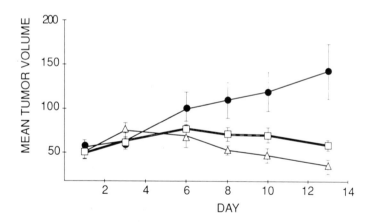

FIGURE 7.   Effect of MDL 101,731 on the growth of multidrug-resistant epidermoid carcinoma (KBV₁). Experimental protocols were similar to Table 11. The tumor volumes expressed in ×10³ cm. •, control; □, 5 mg/kg; ◁, 10 mg/kg.

in 20% inhibition by MDL 101,731, while Ara-C showed no growth inhibition. A combination of MDL 101,731 and Ara-C resulted in 70% inhibition of tumor growth with complete elimination of spontaneous metastasis (Table 14).

D. TOXICOLOGY

Preliminary toxicological studies showed that MDL 101,731 has an acute LD_{50} of more than 200 mg/kg by i.p. or oral administration in mice. In a preliminary study, i.v. administration of 100 and 300 mg/kg had no apparent toxicity in mice. Mice also tolerated 15 mg/kg i.p., 10 mg/kg i.v., and 30 mg/kg oral once daily for 15 d.

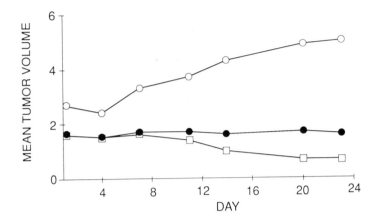

FIGURE 8. Antitumor activity of MDL 101,731 against human ovarian carcinoma (HTB-161) in athymic nude mice. Approximately 0.3 g of HTB-161 tumor was transplanted s.c. in the flanks of athymic nude mice on day 0. The tumor was allowed to grow for 2 weeks and then drug treatment was initiated. MDL 101,731 was given i.p. once daily for 2 weeks. At the end of treatment, animals were sacrificed and tumors dissected and weighed. ○, control; •, 5 mg/kg; □, 10 mg/kg.

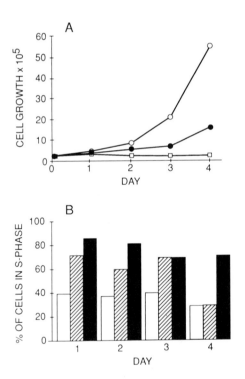

FIGURE 9. Effect of MDL 101,731 on HeLa cell proliferation and cell cycle transverse. HeLa cells (2.5×10^5/100 mm dish) were plated and treated with the various concentrations of MDL 101,731. Cell counts and labeling index were determined daily for 4 d to determine the effect of the compound on growth inhibition and accumulation of cells in S-phase of the cell cycle. (A) ○, control; •, 15 ng/ml, and □, 20 ng/ml. (B) □, control; ■, 15 ng/ml; and ▨, 20 ng/ml.

VI. SUMMARY

   MDL 101,731 (**13**) was designed as a bioprecursor of a mechanism-based inhibitor of RDPR based on the formation of a stabilized free radical intermediate during the enzymatic reaction. A new stereospecific method to fluoro olefins was developed for the synthesis of **13**. The method included

TABLE 12
Synergistic Antiproliferative Activity of MDL 101,731 in Combination with S- and M-Phase-Specific Antitumor Agents against HeLa Cells

Drug	IC_{50} (mg/ml)	
	Drug Alone	Drug + 101,731
Cytosine arabinoside (Ara-C)	13.7	3.2
5-Fluorouracil (5FU)	41.1	18.9
Vinblastine (VLB)	86.7	13.6
Adriamycin (ADR)	0.061	0.045

Note: HeLa cells were plated (2×10^3 cells/well) and allowed to grow for 18 h. MDL 101,731 (15 ng/ml) treatment was for 24 h. The compound was washed and the cells were exposed to the indicated drugs for another 72 h. The cell viability was determined by MTT assay and the IC_{50} was determined.

TABLE 13
Synergistic Antitumor Activity of a Combination of MDL 101,731 and Ara-C against B16 Melanoma in Mice

Treatment	Tumor Weight (g) (Mean ± S.E.)	% Inhibition
Control	2.6 ± 0.5	
MDL 101,731 (5 mg/kg, day 1–7)	2.4 ± 0.9	10
Ara-C (10 mg/kg, day 7–14)	3.6 ± 0.8	0
MDL 101,731 + Ara-C	1.0 ± 0.3	59

Note: B16 melanoma cells (1×10^5 cells/mouse) were injected subcutaneously on day 0. Compounds were administered i.p. once daily as indicated. The animals were sacrificed on day 15; tumors dissected and weighed.

a free radical reaction step that proceeded with complete retention of configuration at the terminal vinylfluoride carbon.

The data presented show that **13** is an inhibitor of ribonucleotide reductase with potent antitumor activity. MDL 101,731 showed significant cytotoxic activity against a number of tumor cells in culture. Furthermore, MDL 101,731 altered dNTP pools in a fashion consistent with a mechanism of action involving the inhibition of RDPR and DNA synthesis. The compound also inactivated *E. coli* RDPR with rapid loss of the tyrosyl radical. MDL 101,731, when administered to L1210 leukemia-bearing mice, rapidly inhibited ribonucleotide reductase in tumor cells in a time-dependent manner. MDL 101,731 showed curative activity against mouse L1210 and P388[R] leukemias. The compound showed significant antitumor activity against solid tumors, such as B16 melanoma, Lewis lung carcinoma in mice, human ovarian, and multidrug-resistant epidermoid carcinoma xenografts in athymic nude mice. MDL 101,731 was well tolerated with acute LD_{50} of more than 200 mg/kg (i.p., i.v., and orally) in mice. The antiproliferative and antitumor activities of the

TABLE 14
Synergistic Antitumor Activity by a Combination of MDL 101,731 and Ara-C against Lewis Lung Carcinoma in Mice

Treatment	Tumor Wt (g) (Mean ± S.E.)	% Inhibition	No. of Foci (Mean ± S.E.)	% Inhibition
Control	3.5 ± 0.4		11.3 ± 4	
MDL 101,731	2.8 ± 0.6	20	15.5 ± 6	0
(5 mg/kg, day 1–7)				
Ara-C	3.83 ± 0.8	0	14.6 ± 5	0
(10 mg/kg, day 7–14)				
MDL 101,731 + Ara-C	1.1 ± 0.2	69	0	100

Note: 3LL cells (1×10^5/mouse) were injected s.c on day 0. Compounds were administered i.p., once daily as indicated. Animals were sacrificed on day 15; tumors dissected and weighed. Pulmonary metastatic foci were also counted.

compound were synergistic with S-phase-specific drugs. Based on these observations, MDL 101,731 is currently under development as an antitumor agent and will enter the clinic in both the U.S. and Japan this year.[52]

ACKNOWLEDGMENTS

We thank E. W. Huber for the NMR spectra and subsequent discussions that led to the submission on **43** for an X-ray crystal structure. In addition, we thank our collaborators D. P. Matthews, D. M. Stemerick, P. Bey, B. J. Lippert, S. Mehdi, R. D. Snyder, J. Zwolshen, and M. Lewis. We also thank S. McLean and N. Reddington for their secretarial help.

REFERENCES

1. **Ashley, G.W., Harris, G., and Stubbe, J.,** The mechanism of *Lactobacillus leichmani* ribonucleotide reductase: evidence for 3′ carbon-hydrogen bond cleavage and a unique role for coenzyme B$_{12}$, *J. Biol. Chem.*, 261, 3958, 1986.
2. **Eliasson, R., Fontecave, M., Jornvall, H., Krook, M., Pontis, E., and Reichard, P.,** The anaerobic ribonucleoside triphosphate reductase from *Escherichia coli* requires S-adenosylmethionine as a cofactor, *Proc. Natl. Acad. Sci. U.S.A.*, 87, 3314, 1990.
3. **Lien, E.J.,** Ribonucleotide reductase inhibitors as anticancer and antiviral agents, in *Progress in Drug Research*, Vol. 31, Jucker, E., Ed., Birkhauser Verlag, Basel, 1987, 101.
4. **Wright, J.A., McClarty, G.A., Lewis, W.H., and Srinivasan, P.R.,** Hydroxyurea and related compounds, in *Drug Resistant Mammalian Cells*, Vol. 1, Gupta, R.S., Ed., CRC Press, Boca Raton, FL, 1989, 15.
5. **Stubbe, J.,** Ribonucleotide reductases, *Adv. Enzymol.*, 63, 349, 1990.
6. **Thelander, L. and Reichard, P.,** Reduction of ribonucleotides, *Annu. Rev. Biochem.*, 48, 133, 1979.
7. **Nutter, L.M. and Cheng, Y.C.,** Nature and properties of mammalian ribonucleoside diphosphate reductase, in *Inhibitors of Ribonucleoside Diphosphate Reductase Activity*, Cory, J.G. and Cory, A.H., Eds., Pergamon Press, New York, 1989, chap. 3.
8. **Reichard, P. and Ehrenberg, A.,** Ribonucleotide reductase — a radical enzyme, *Science*, 221, 514, 1983.
9. **Willing, A., Follmann, H., and Auling, G.,** Ribonucleotide reductase of *Brevibacterium ammoniagenes* is a manganese enzyme, *Eur. J. Biochem.*, 170, 603, 1988.
10. **Stubbe, J.,** Ribonucleotide reductases: amazing and confusing, *J. Biol. Chem.*, 265, 5329, 1990.
11. **Larsson, A. and Sjoberg, B.M.,** Identification of the stable free radical tyrosine residue in ribonucleotide reductase, *EMBO J.*, 5, 2037, 1986.
12. **Bollinger, J.M., Jr., Edmondson, D.E., Huynh, B.H., Filley, J., Norton, J.R., and Stubbe, J.,** Mechanism of assembly of the tyrosyl radical-dinuclear iron cluster cofactor of ribonucleotide reductase, *Science*, 253, 292, 1991.
13. **Nordlund, P., Sjoberg, B.M., and Eklund, H.,** Three-dimensional structure at the free radical protein of ribonucleotide reductase, *Nature*, 345, 593, 1990.

14. **Rose, I.A. and Schweigert, B.S.,** Incorporation of ^{14}C totally labeled nucleosides into nucleic acids, *J. Biol. Chem.,* 202, 635, 1953.

15. **Cory, J.G. and Carter, G.L.,** Drug action on ribonucleotide reductase, *Adv. Enzyme Regul.,* 24, 385, 1985.

16. **Mao, S.S., Johnson, M.I., Bollinger, J.M., Baker, C.H., and Stubbe, J.,** Mechanism-based inhibitors and site-directed mutants as probes of the mechanism of ribonucleotide reductase, in *Molecular Mechanisms in Bioorganic Processes,* Bleasdale, C. and Goldman, B.T., Eds., Royal Society of Chemistry, 1990, 305.

17. **Elford, H.L, Freese, M., Passamani, E., and Morris, H.P.,** Ribonucleotide reductase and cell proliferation. I. Variations of ribonucleotide reductase activity with tumor growth rate in a series of rat hepatomas, *J. Biol. Chem.,* 245, 5258, 1970.

18. **Cory, J.G. and Cory, A.H., Eds,** *Inhibitors of Ribonucleotide Diphosphate Reductase Activity,* Pergamon Press, New York, 1988.

19. **Cory, J.G. and Chiba, P.,** Combination chemotherapy directed at the components of nucleoside diphosphate reductase, in *Inhibitors of Ribonucleotide Diphosphate Reductase Activity,* Cory, J.G. and Cory, A.H., Eds., Pergamon Press, New York, 1989, chap. 13.

20. **Fox, R.M.,** Changes in deoxynucleoside triphosphate pools induced by inhibitors and modulators of ribonucleotide reductase, in *Inhibitors of Ribonucleotide Diphosphate Reductase Activity,* Cory, J.G. and Cory, A.H., Eds., Pergamon Press, New York, 1989, chap. 6.

21. **Moore, E.C. and Hurlbert, R.B.,** The inhibition of ribonucleoside diphosphate reductase by hydroxyrea, guanazole, and pyrazoloimidazole, in *Inhibitors of Ribonucleotide Diphosphate Reductase Activity,* Cory, J.G. and Cory, A.H., Eds., Pergamon Press, New York, 1989, chap. 9.

22. **Swain, M.A. and Galloway, D.A.,** Herpes simplex virus specifies two subunits of ribonucleotide reductase encoded by 3'-coterminal transcripts, *J. Virol.,* 57, 802, 1986.

23. **Cohen, E.A., Gaudreau, P., Brazeau, P., and Langelier, Y.,** Specific inhibition of herpes virus ribonucleotide reductase by a nonapeptide derived from the carboxy terminus of subunit 2, *Nature (London),* 321, 441, 1986.

24. **Dutia, B.M., Frame, M.C., Subak-Sharpe, J.H., Clark, W.N., and Marsden, H.S.,** Specific inhibition of herpes virus ribonucleotide reductase by synthetic peptides, *Nature (London),* 321, 439, 1986.

25. For an excellent review on mechanism-based enzyme inhibitors, see **Silverman, R.B.,** *Mechanism-Based Enzyme Inactivation: Chemistry and Enzymology,* Vol. I and II, CRC Press, Boca Raton, FL, 1988.

26. **Thelander, L., Larson, B., Hobbs, J., and Eckstein, F.,** Active site of ribonucleoside diphosphate reductase from *E. coli.* Oxidation-reduction-active disulfides and the B1 subunit, *J. Biol. Chem.,* 251, 1398, 1976.

27. **Ashley, G.W. and Stubbe, J.,** Progress on the chemical mechanism of the ribonucleotide reductases, in *Inhibitors of Ribonucleoside Diphosphate Reductase Activity,* Cory J.G. and Cory, A.H., Eds., Pergamon Press, New York, 1989, chap. 4.

28. **Salowe, S.P., Ator, M.A., and Stubbe, J.,** Products of the inactivation of ribonucleoside diphosphate reductase of *E. coli* with 2'-azido-2'-deoxyuridine 5'-diphosphate, *Biochemistry,* 26, 3408, 1987.

29. **Baker, C.H., Banzon, J., Bollinger, J.M., Stubbe, J., Samano, V., Robins, M.J., Lippert, B., Jarvi, E., and Resvick, R.,** 2-Deoxy-2'-methylenecytidine and 2'-deoxy-2'2-difluorocytidine 5'-diphosphates: potent mechanism-based inhibitors of ribonucleotide reductase, *J. Med. Chem.,* 34, 1879, 1991.

30. **Casper, E.S., Green, M.R., Brown, T.D., Kelsen, D.P., Kresek, T., and Trochanowski, B.,** Phase II trial of gemcitabine (2',2'-difluorodeoxycytidine) in patients with pancreatic cancer, *Proc. Am. Soc. Clin. Oncol.,* 10 (27 Meet.), 143, 1991.

31. **McCarthy, J.R., Matthews, D.P., Stemerick, D.M., Huber, E.W., Bey, P., Lippert, B.J., Snyder, R.D., and Sunkara, P.S.,** Stereospecific methods to (E) and (Z) terminal fluoro olefins and its application to the synthesis of 2'-deoxy-2'-fluoromethylenenucleosides as potential inhibitors of ribonucleoside diphosphate reductase, *J. Am. Chem. Soc.,* 113, 7439, 1991.

32. **Eriksson, S., Kierdaszuk, B., Munch-Petersen, B., Oberg, B., and Johansson, N.G.,** Comparison of the substrate specificities of human thymidine kinase 1 and 2 and deoxycytidine kinase toward antiviral and cytostatic nucleoside analogs, *Biochem. Biophys. Res. Commun.,* 176, 586, 1991.

33. **McCarthy, J.R., Peet, N.P., LeTourneau, M.E., and Inbasekaran, M.,** DAST in organic synthesis. II. The transformation of sulfoxides to α-fluorothioethers, *J. Am. Chem. Soc.,* 107, 735, 1985.

34. **Robins, M.J. and Wnuk, S.F.,** Fluorination at C5' of nucleosides. Synthesis of the new class of 5'-fluoro 5'-S-aryl (alkyl) thionucleosides from adenosine, *Tetrahedron Lett.,* 29, 5729, 1988.

35. **McCarthy, J.R., Matthews, D.P., Edwards, M.L., Stemerick, D.M., and Jarvi, E.T.,** A new route to vinyl fluorides, *Tetrahedron Lett.,* 31, 5449, 1990.

36. **McCarthy, J.R., Matthews, D.P., and Paolini, J.P.,** Reaction of sulfoxides with diethylaminosulfur trifluoride: preparation of fluoromethyl phenyl sulfone, a reagent for the synthesis of fluoro olefins, *Org. Synth.,* 72, 209, 1993.

37. **Inbasekaran, M., Peet, N.P., McCarthy, J.R., and LeTourneau, M.E.,** A novel and efficient synthesis of fluoromethyl phenyl sulfone and its use as a fluoromethyl Wittig equivalent, *J. Chem. Soc. Chem. Commun.,* 678, 1985.

38. **O'Connor, D.E. and Lyness, W.I.,** The effect of methylmercapto, methylsulfonyl, and methylsulfonyl groups on the equilibrium in three-carbon prototropic systems, *J. Am. Chem. Soc.,* 86, 3840, 1964.

39. **Posner, G.H. and Brunelle, D.J.,** Reaction of α,β-ethylenic sulfur compounds with organocopper reagents, *J. Org. Chem.,* 38, 2750, Footnote 31, 1973.

40. **Matsuda, A., Itoh, H., Takenuki, K., Susaki, T., and Ueda, T.,** Alkyl addition reaction of pyrimidine 2′-ketonucleosides: synthesis of 2′-branched-chain sugar pyrimidine nucleosides, *Chem. Pharm. Bull.,* 36, 945, 1988.

41. **Lee, S.H. and Schwartz, J.,** Stereospecific synthesis of alkenyl fluorides (with retention) via organometallic intermediates, *J. Am. Chem. Soc.,* 108, 2445, 1986.

42. **Watanabe, Y., Ueno, Y., Araki, T., Endo, T., and Okawara, M.,** A novel homolytic substitution on vinylic carbon. A new route to vinyl stannane, *Tetrahedron Lett.,* 27, 215, 1986.

43. **Lippert, B.J. and Sunkara, P.S.,** private communication.

44. **Mehdi, S.,** unpublished results.

45. **Sunkara, P.S., Zwolshen, J.H., Lippert, B.J., Snyder, R.D., Matthews, D.P., and McCarthy, J.R.,** Antitumor activity of (E)-2′-fluoromethylene-2′-deoxycytidine (FMDC, MDL 101,731): a novel inhibitor of ribonucleotide reductase, *Proc. Am. Assoc. Cancer Res.,* 32, 2467, 1991.

46. **Sunkara, P.S., Zwolshen, J.H., Matthews, D.P., and McCarthy, J.R.,** Synergistic antitumor activity of (E)-2′-fluoromethylene-2′-deoxycytidine (FMDC, MDL 101,731), an inhibitor of ribonucleotide reductase in combination with S-phase specific drugs, *Proc. Am. Assoc. Cancer Res.,* 33, 3088, 1991.

47. **Sunkara, P.S. et al.,** manuscript in preparation.

48. **Shen, D., Cardarelli, C., Hwang, J., Cornwell, M., Richert, N., Ishii, S., Pastan, I., and Gottesman, M.M.,** *J. Biol. Chem.,* 261, 7762, 1986.

49. **Stubbe, J.,** personal communication.

50. **Matthews, D.P., Persichetti, R.A., and McCarthy, J.R.,** Improved synthesis of fluoromethyl phenyl sulfone, *Org. Prep. Proceed. Int.,* in press.

51. **Matthews, D.P., Gross, R.S., and McCarthy, J.R.,** A new route to 2-fluoro-1-olefins utilizing a synthetic equivalent for the 1-fluoroethene anion, *Tetrahedron Lett.,* 35, 1027, 1994.

52. **Bitonti, A.J., Dumont, J.A., Bush, T.L., Cashman, E.A., Cross-Doersen, D.E., Wright, P.S., Matthew, D.P., McCarthy, J.R., Sunkara, P.S.,** Regression of human breast tumor xenografts in response to (E)-2′-(fluoromethylene cytidine, an inhibitor of ribonucleotide reductase, *Cancer Res.,* 54, 1485, 1994.

Chapter 2

THE DEVELOPMENT OF HUMAN IMMUNODEFICIENCY VIRUS TYPE 1 REVERSE TRANSCRIPTASE INHIBITORS

Andrew M. Stern

CONTENTS

I. INTRODUCTION

Several lines of converging evidence[1-8] have indicated that the causative agent of the acquired immunodeficiency syndrome (AIDS) is the human immunodeficiency virus (HIV).[9] This conclusion has been corroborated, with tragic consequences, by studies demonstrating that transfusion of anti-HIV-1 antibody-positive blood infected 90% of the recipients and the rate of progression to AIDS within this cohort, 38 months after infection, was similar to the rate reported for other groups at risk.[10] Infection with HIV causes chronic, progressive depletion of CD4+ helper/inducer T

This chapter is dedicated to Charlotte Rogow Stern (1927–1991) with the hope that improved therapy can be added to the courage of our terminally ill.

0-8493-7818-4/95/$0.00+$.50
© 1995 by CRC Press Inc.

lymphocytes.[9,11,12] Together with the infection of macrophages and other cells, this depletion creates an immune deficiency that leads to the cancers and opportunistic infections characteristic of AIDS. Infection of the central nervous system can lead to the clinical manifestations that characterize HIV-induced dementia.[13]

The elucidation of the etiology of AIDS and the efficacy of 3′-azido-3′-deoxythymidine (AZT, zidovudine), proven to prolong survival and to reduce morbidity in patients with established AIDS,[14–21] have prompted intensive research directed toward the development of antiretroviral therapy for HIV infection.[22–26] Virus-specific proteins, essential to the life cycle of HIV, are potential therapeutic targets.[22–26] This chapter will focus upon one such protein, HIV-1 reverse transcriptase (RT),[27–31] and the inhibition of its DNA polymerase activity. In addition to the mechanistic aspects of anti-HIV-1 therapy based upon the use of 2′,3′-dideoxynucleosides,[32,33] the development and mode of action of the newly discovered non-nucleoside inhibitors of HIV-1 RT[34–42] will be discussed. From our current clinical experience, the prospects of dose-limiting toxicities and emergence of resistant variants of HIV-1 make it seem unlikely that a single agent acting upon a single target protein will provide the necessary inhibition of viral proliferation to eliminate all virus-infected cells from a patient. Thus, combination therapy involving RT inhibitors will be discussed as well.

Following the specific high-affinity binding of HIV to the CD4 receptor on the host cell surface,[43,44] and subsequent steps involved in virus internalization and uncoating, viral genomic RNA and RT are released into the cytoplasm. In the cytoplasm, the RNA-dependent DNA polymerase activity of RT catalyzes the formation of a first-strand (minus strand) DNA, using the viral genomic RNA as a template, primed by transfer RNA.[28,45] The intrinsic ribonuclease H (RNase H) activity of RT[46] then catalyzes the orderly degradation of the viral RNA from the RNA-DNA hybrid to permit synthesis of a double-stranded viral DNA. The latter in its linear form is subsequently inserted into the genome of the host cell during a reaction catalyzed by HIV integrase.[47,48] Since RT is an essential component of the HIV life cycle,[49] inhibition of its activity should provide an effective means of preventing the spread of HIV infection. 2′,3′-Dideoxynucleosides, comprising the most important class of anti-HIV-1 agents at this time, likely act *in vivo* by inhibiting the DNA polymerase activity of RT *(vide infra)*.

II. 2′,3′-DIDEOXYNUCLEOSIDES

A. *IN VITRO* STUDIES

In 1985 Mitsuya et al.[33] demonstrated that a thymidine analog 3′-azido-3′-deoxythymidine (AZT) originally synthesized by Horwitz et al.[50] as a cancer chemotherapeutic agent and shown by Ostertag et al.[51,52] to be active against a murine retrovirus, inhibited the infectivity and cytopathic effects of HIV upon human T cells *in vitro*. In addition, at the effective concentrations of AZT, the *in vitro* immune functions of normal T cells remained largely intact,[33] suggesting appreciable selectivity against HIV replication. The addition of thymidine reversed the protective effect of AZT as the HIV-infected target cells were lysed in a dose-dependent fashion. When added alone to either infected or uninfected cells thymidine was not significantly cytotoxic. These results indicated that AZT acted as a competitive analog of thymidine in at least one required step involved in its ability to protect cells against the cytopathic effects of HIV infection. These studies were extended;[32,53] with the ribose moiety of the molecule in the 2′,3′-dideoxy configuration, every purine (adenosine, guanosine, and inosine) and pyrimidine (cytidine and thymidine) nucleoside tested suppressed HIV replication *in vitro*. Under these experimental conditions[32] 2′,3′-dideoxycytidine was the most potent and 2′,3′-deoxythymidine the least potent nucleoside analog tested. Furthermore, AZT and related 2′,3′-dideoxynucleosides were shown to effectively inhibit HIV infection in fresh and cultured human peripheral blood monocytes and macrophages.[54,55] These *in vitro* observations suggested a therapeutic potential for 2′,3′-dideoxynucleosides.

B. MODE OF ACTION

In addition to the requirement for a 2′,3′-dideoxy structure, structure-activity studies demonstrated the need for the presence of the 5′-OH group.[32] This result was consistent with the hypothesis[32] that anabolic phosphorylation (catalyzed by cellular kinases) of the dideoxynucleosides to their respective 5′-triphosphate forms, was necessary for their antiviral activity. Evidence supporting this hypothesis has come from the demonstration that 2′,3′-dideoxynucleosides, including AZT, were sequentially phosphorylated in the cytoplasm of appropriate target cells to ultimately yield the corresponding 5′-triphosphates.[56–59] In cell-free systems, these 2′,3′-dideoxynucleoside-5′-triphosphates potently inhibited RT activity by competing with their natural counterparts and acting as alternative substrates.[56,60–70,152] Due to the lack of a 3′-OH group, their incorporation into the growing DNA chain leads to chain termination as subsequent 5′–3′ phosphodiester bonds cannot form.

The AZT-resistance phenotype of clinical isolates of HIV-1 *(vide infra)* has been shown to be conferred by the RT coding region,[71] strongly supporting RT as the *in vivo* target of the 2′,3′-dideoxynucleosides. However, virion-associated RT from these resistant isolates was as susceptible to inhibition by AZT-triphosphate (AZT-TP) as RT derived from sensitive isolates.[72] The K_m value for thymidine triphosphate (TTP) and the K_i value for AZT-TP were also similar when purified RT cloned from a sensitive and resistant isolate pair were compared.[73] In addition, experiments designed to assess the ability of recombinant RT to incorporate AZT-TP into DNA revealed no differences between these same resistant and sensitive isolates.[73] Previous studies have shown that the characteristics of RTs, containing site-directed mutations that alter AZT-TP sensitivity in cell-free systems, cannot reliably predict the phenotype of the virus containing the same mutations.[49] Thus, the precise mechanism of resistance involving RT has not yet been defined. Therefore, the proposed mechanism of the antiviral activity of AZT (and other 2′,3′-dideoxynucleosides), although consistent with the current data, remains unproven. Alternative mechanisms involving the inhibition of the RNase H activity of RT by AZT-monophosphate[74] and perhaps the inhibition of its DNA polymerase activity by reduced (3′-amino) derivatives of AZT-TP,[63,65] not involving chain termination,[65,75] have been suggested but are not strongly supported by the currently available data.

1. Inhibition of HIV-1 RT DNA Polymerase Activity in Cell-Free Systems

The inhibition of HIV-1 RT DNA polymerase activity by AZT-TP and other 2′,3′-dideoxynucleoside-5′-triphosphates has been considered by Reardon, Goody, and colleagues.[66–68,76] Their studies have defined the kinetic parameters, which can be measured in an enzyme assay, that are most relevant to the antiviral activity of AZT (and other 2′,3′-dideoxynucleosides). Optimization of these kinetic parameters could lead to superior antiviral agents within this specific class.

a. Steady-State Kinetic Model

The steady-state[66] and pre-steady-state kinetic[68] constants have been determined for HIV-1 RT-catalyzed incorporation of 2′-deoxynucleotides and 2′,3′-dideoxynucleotides into defined DNA-primed RNA templates and into defined DNA-primed DNA templates. The results of these studies support the following kinetic model consisting of an ordered mechanism for substrate binding[77] proposed for other DNA polymerases:[78,79]

$$E + TP \xrightleftharpoons{K_{TP}} E \cdot TP \xrightleftharpoons[k_2]{k_1} E \cdot TP \cdot d(d)NTP \xrightarrow{k_P} E \cdot TP_{n+1} \xrightarrow{k_4} E + TP_{n+1} \quad (1)$$

where E, TP, and d(d)NTP represent RT, a defined-sequence DNA-primed RNA (or DNA) template, and 2′-deoxynucleoside-5′-triphosphate (or 2′,3′-dideoxynucleoside-5′-triphosphate), respectively. TP_{n+1} represents TP following incorporation of the required ("next") d(d)NMP as defined by the template sequence.

With several template primers, RT has been shown to catalyze a highly processive polymeriza-tion reaction.[77,80] Under processive polymerization conditions, in the presence of all required dNTPs and saturating TP, k_4 is not on the reaction pathway as multiple nucleotide incorporations occur without $E \cdot TP_{n+x}$ dissociation ($x \gg 1$) ($k_p/k_4 \gg 1$). Under these conditions $k_{cat} = k_p$; the rate-determining step follows dNTP binding and is either phosphodiester bond formation (nucleotide incorporation) *per se*, or a rate-limiting conformational change that immediately precedes nucleotide incorporation.[66,68] The Michaelis constant for dNTP is

$$K_m = \frac{k_2 + k_p}{k_1} \qquad (2)$$

reflecting the sum of the rate constants ($k_2 + k_p$) that determines the rate of production of enzyme forms ($E \cdot TP$ or $E \cdot TP_{n+1} \ldots$ or $E \cdot TP_{n+x}$) free to bind dNTP relative to the second-order rate constant, k_1, that determines the rate of production of enzyme forms ($E \cdot TP \cdot dNTP$ or $E \cdot TP_{n+1} \cdot dNTP \ldots$ or $E \cdot TP_{n+x} \cdot dNTP$) to which dNTP is bound under steady-state conditions. Since $k_2 + k_p > k_2$, the lowest value that K_m can have under processive synthesis conditions is equal to k_2/k_1 or K_D, the dissociation constant of dNTP from its complex with $E \cdot TP$, $E \cdot TP_{n+1}$, or $E \cdot TP_{n+x}$. In fact, under processive conditions K_m has been shown to be equal to K_D as $k_2 \gg k_p$.[66,68]

The use of defined-sequence DNA-primed RNA (or DNA) templates makes it possible to study the incorporation of only one nucleotide per template-primer molecule by including only the "next" dNTP as defined by the sequence of TP. This is in contrast to including all dNTPs that ordinarily would be required for processive synthesis (*vide supra*) on a heterotemplate. Thus, the inclusion of *only* the "next" dNTP forces termination of processive polymerization. (Only $E \cdot TP$ can bind dNTP; $E \cdot TP_{n+1}$ cannot.) Under steady-state conditions, therefore, nucleotide incorporation requires the dissociation of $E \cdot TP_{n+1}$ following each catalytic turnover. Thus, k_4 is on the reaction pathway and is the rate-determining step ($k_{cat} = k_4$), since for an intrinsically processive enzyme such as RT, k_4/k_p is $\ll 1$; the quantitative relationship that defines processivity. The Michaelis constant K'_m, under these conditions is

$$K'_m = \frac{k_4}{k_p}\left(\frac{k_2 + k_p}{k_1}\right) = \frac{k_4}{k_p}(K_m) \qquad (3)$$

One important point that will determine the potency of AZT-TP in cell-free systems (*vide infra*) is that the K'_m value for dNTP under conditions of forced termination of processive polymerization is approximately 100 times[66] lower than the K'_m (K_D) value under processive synthesis. The slow rate of dissociation (k_4) relative to the rate of nucleotide incorporation (k_p) gives rise to the steady-state accumulation of the $E \cdot TP_{n+1}$ complex which cannot bind dNTP under forced termination conditions (*vide supra*). The enzyme is effectively "tied up". In the steady state, the concentration of enzyme forms free to bind dNTP is low relative to total enzyme, allowing apparent saturation of the enzyme to occur at relatively low concentrations of dNTP.

Another important point is that forced termination of processive synthesis mimics in many ways the incorporation of a chain terminator (2',3'-dideoxynucleoside-5'-triphosphate) such as AZT-TP. In a cell-free system under steady-state processive conditions, the incorporation of AZT-TP gives rise to chain termination, requiring the dissociation of the $E \cdot TP_{n+1}$ complex before further synthesis can occur. The inhibition of HIV-1 RT by AZT-TP will be discussed using this kinetic model.

b. Substrate and Inhibitory Characteristics of AZT-TP

In the absence of the other three 2'-deoxynucleoside-5'-triphosphates, the time course for incor-poration of TTP into a defined-sequence DNA-primed RNA template[66] was biphasic. A burst of product formation was observed (stoichiometric with enzyme active sites) which was then followed by a slow steady-state rate with a K'_m value of 0.082 μM. AZT-TP incorporation into the same TP as well as into poly(rA)·oligo(dT) generated similar biphasic time patterns and steady-state K'_m and k_{cat} values.[66] These results indicated that AZT-TP is an efficient substrate for HIV-1 RT, corroborating other studies.[64,65,67,152] The biphasic time course is consistent with a single rapid nucleotide incorporation (governed by k_p in the pre-steady state) followed by a rate-limiting dissociation of the $E·TP_{n+1}$ complex (governed by k_4, determining the steady-state rate) under forced termination of processive polymerization conditions *(vide supra)*.

The similar k_{cat} (k_4) values for both TTP and AZT-TP suggest that incorporation of an AZT-monophosphate (AZT-MP) residue onto the 3′ primer terminus does not significantly affect the rate of dissociation of the TP from RT.[66] This result is consistent with studies that have shown that the 3′-azido-terminated TP does not bind to HIV-1 RT more tightly than the normal TP when TP was poly(rA)·oligo(dT)$_{14}$.[81]

In addition to being incorporated into DNA-primed RNA templates, AZT-TP was also incorporated into the homopolymer DNA-primed DNA template, poly(dA)·oligo(dT), and a defined sequence DNA-primed DNA template. The K'_m values of AZT-TP and TTP were similar to each other but were considerably higher (10- to 50-fold) than the K'_m values obtained with DNA-primed RNA templates *(vide supra)*.[66] In addition, no biphasic time course was evident with these primer templates. The absence of biphasic time courses under forced termination conditions suggested that dissociation of the DNA-primed DNA templates from RT was not rate-determining ($k_p \sim k_4$). This was consistent with a more distributive mode of DNA polymerization (on DNA templates) in which dissociation of $E·TP_{n+1}$ occurred relatively frequently and $E·TP_{n+1}$ did not significantly accumulate in the steady state (accounting for the higher K'_m values).

With the defined sequence DNA-primed RNA and DNA templates and poly(dA)·oligo(dT), AZT-TP and TTP were linear competitive inhibitors of each other's incorporation.[66] K_i values for both TTP and AZT-TP were comparable to their K_m values. This is the required relationship between K_i and K_m for competitive alternative substrate inhibitors.[*82] Thus, AZT-TP appears to be a simple competitive alternative substrate inhibitor with respect to TTP.[66] The K_i values with DNA-primed RNA templates were 25 to 50 times lower than with DNA-primed DNA templates, reflecting the correlation between high potency and high intrinsic processivity (K'_m or K_i is lower when k_4/k_p is lower (see Equation 3)). The dissociation constant (K_D) for AZT-TP has been determined to be 11 $\pm$ 1 μM using a sequence-defined DNA-primed RNA template.[68] This value is 50 times higher than its K_i value using the same template primer, indicating that, as an alternative substrate, AZT-TP is significantly more potent than it would be as a simple competitive dead-end inhibitor of TTP. Thus, the steady-state inhibition of RT is not due to the tight binding of AZT-TP to the enzyme·template primer but to the slow dissociation of the nonproductive chain-terminated enzyme·template primer product complexes that accumulate and "tie up" free enzyme.[66–68,76]

Under steady-state conditions where k_4 is on the reaction pathway in both the presence and absence of a chain terminator such as AZT-TP (during the forced termination of processive synthesis or during distributive synthesis when DNA templates are employed), the K_i value of AZT-TP is equal to its K_m value, as discussed above for an alternative substrate inhibitor. However, when poly(rA)·oligo(dT) is used as a template primer, processive synthesis occurs and k_4 is not on the reaction pathway in the absence of AZT-TP *(vide supra)*. Under these conditions, the observed K_i values for AZT-TP are significantly lower than its K_m value (200 nM)[66] ranging between 4 and 130 nM, depending upon the length of the poly(rA) region following the primer in the RNA template.[*81] The more adenosyl residues in the RNA template that are available for processive incorporation of dTMP moieties, the lower is the observed K_i value for AZT-TP.[81] These results have been discussed[81] and indicate that under steady-state conditions the observed K_i values for chain terminators, such as AZT-TP, are highly dependent upon specific assay conditions and may not reflect intrinsic

---

*The K_i value is an experimentally derived value and is equivalent to the concentration of the inhibitor that doubles the slope of the $1/V$ vs. $1/S$ plot. Its physical meaning depends upon the mechanism of inhibition. For a simple nonreactive, dead-end inhibitor, K_i is equal to K_D. For an alternative substrate, K_i is far more complex and is equivalent to the K_m. For an alternative substrate that disrupts processivity (such as AZT-TP), this relationship will hold only when the uninhibited control reaction is run under nonprocessive conditions (see text).

potency. These results have made it difficult to compare the potencies of different chain terminators as well as their relative selectivity toward RT and normal cellular polymerases. Furthermore, although the kinetic parameters that determine the intrinsic potency of chain terminators under steady-state conditions are related to the ones most relevant to the antiviral activity of these agents, they are not the same *(vide infra)*.[67,68,76]

2. Inhibition of HIV-1 RT DNA Polymerase Activity within Infected Cells

Critical differences in reaction conditions between DNA synthesis catalyzed by RT that occurs in a test tube and the replication of the RNA genome that occurs within a cell that has just been penetrated by HIV results in two distinct mechanisms for RT inhibition by chain terminators such as AZT-TP.[76] In the test tube, the molar ratio of RT to template primer is low (a steady-state kinetic condition); thus, for maximal deoxynucleotide incorporation (the measurement of product formation in the test tube) each RT molecule must move between a large number of potential polymerization sites on different template primer molecules once a round of processive synthesis is completed on a given template primer molecule.* In the presence of AZT-TP, the limited supply of RT in the presence of excess template primer is tied up as a consequence of the slow dissociation rate of chain-terminated template primer·RT product complexes that contributes to the low K_i value of AZT-TP relative to its K_D value *(vide supra)*. Thus, in a cell-free system, the kinetic parameters that characterize complexes that form *subsequent* to the incorporation of AZT-MP (product complexes) are critical for determining inhibition. This is not the case in an infected cell where RT is not limiting. In this case, the molar ratio of RT to the tRNA[Lys-3] primed genomic RNA is approximately 1;[83] thus, genome replication and its inhibition are not at steady state. The rate is not directly proportional to RT, as template primer concentration is limiting. Unlike DNA synthesis in the test tube, where activity is measured by deoxynucleotide incorporation, in the infected cell the only measurement of product that is relevant to viral replication is the successful completion of a full-length DNA copy of the viral genome. In the absence of known repair mechanisms to excise the incorporated 2′,3′-dideoxynucleotide from RNA-DNA intermediates,[84] or in the absence of appreciable pyrophosphorolysis (the reversal of synthesis is slow[83a]), the incorporation of a chain terminator at any position during synthesis will lead to inhibition of viral genome replication. In contrast to the cell-free system, under the conditions in a cell, the incorporation of AZT-MP results in substrate (primed genomic viral RNA) and not RT depletion. Thus, within the infected cell, the probability of 2′,3′-dideoxynucleotide incorporation *per se* is the critical parameter that determines RT inhibition.

The kinetic parameters that define the probability of 2′,3′-dideoxynucleotide incorporation into a template primer have been determined.[66–68,76] The pre-steady-state kinetics of single nucleotide incorporation were found to be consistent with the kinetic model:[68]

$$E + TP \underset{\longleftarrow}{\overset{K_{TP}}{\longrightarrow}} E \cdot TP \underset{\longleftarrow}{\overset{K_D}{\longrightarrow}} E \cdot TP \cdot d(d)NTP \overset{k_p}{\longrightarrow} E \cdot TP_{n+1} \qquad (4)$$

In this scheme E, TP, d(d)NTP, and TP_{n+1} represent RT, a defined-sequence DNA-primed RNA template, 2′-deoxy- (or 2′,3′-dideoxy)-nucleoside-5′-triphosphate, and the template primer extended by the required nucleotide or nucleotide analog, respectively. K_{TP} and K_D are the dissociation constants for RT·template primer and for the dissociation of d(d)NTP from this complex. k_p represents the rate constant for either phosphodiester bond formation or for a conformational change that immediately precedes it.

The effective second-order rate constant defining the rate of incorporation of substrate into the template primer under pre-steady-state conditions is k_p/K_D. This is not only analogous to the specificity constant k_{cat}/K_m, but will be shown *(vide infra)* to be equal to it for this kinetic model.

---

*The frequency of RT·TP dissociation depends upon the intrinsic processivity of the reaction with a certain template primer, the length of the template primer, and the specific assay conditions such as pH, temperature, and ionic strength.

By analogy, k_p'/K_D' represents the corresponding second-order rate constant for a competing chain terminator. The pseudo-first-order rate constants for substrate and chain terminator are $k_p(S)/K_D$ and $k_p'(i)/K_D'$ where (S) and (i) represent the intracellular concentrations of the parent 2′-deoxynucleoside-5′-triphosphate and the competing 2′,3′-dideoxy-5′-triphosphate analog, respectively.

The probability, p, that substrate and not chain terminator is incorporated at each step is given by:

$$p = \frac{k_p S / K_D}{k_p S / K_D + k_p'(i)/K_D'} = \frac{1}{1+r} \quad \text{where } r = \frac{k_p'(i)}{K_D' k_p(S)/K_D}$$

$$\text{when } r \ll 1 \text{ then } p = \frac{1}{1+r} = 1 - r \tag{5}$$

It follows that the probability P that a full-length DNA copy is synthesized in the presence of chain terminator is given by: $P = (p)^n$ or $(1/1 + r)^n$, or $(1 - r)^n$, where n represents the number of potential sites for chain terminator incorporation.

Thus $(1 - P)$ is a direct indicator of the inhibition of HIV genome replication by a chain terminator. The HIV genome (HXB-2 strain) is 9.7 kilobases in length and contains 2164 T, 3411 A, 2370 G, and 1773 C residues. Therefore, $n = 3411$ for AZT-TP and 2370 for ddCTP, for example. Since k_p/K_D values for 2′-deoxynucloside-5′triphosphate:2′,3′-dideoxynucleoside-5′-triphosphate pairs have been determined[68] and since in some cases the intracellular nucleotide pools of these pairs have been measured (see references 68 and 76) at therapeutically effective concentrations of the corresponding 2′,3′-dideoxynucleosides, $(1 - P)$ values can be calculated and this has been done.[67,68,76] The results of these calculations suggest that the observed *in vitro* antiviral effects can be accounted for by the chain termination mechanism discussed in this section.[68,76]

For the few examples studied,[68] k_p/K_D values of 2′-deoxynucleoside-5′-triphosphates differ from the k_p'/k_D' values of the corresponding 2′-3′-dideoxynucleoside-5′-triphosphates in the k_p value and not significantly in the K_D value. This indicates that the absence of a 3′-OH group is reflected in a change of rate for phosphodiester bond formation and not a change in the strength of binding of the corresponding nucleoside 5′-triphosphate to the enzyme·template primer complex.

From a practical point of view the k_p/K_D values need not be determined under pre-steady-state conditions. It has been argued that like other DNA polymerases $k_2 \gg k_p$ *(vide supra)* for RT so that the K_m' value under forced termination conditions is equal to:

$$K_m' = (k_4 / k_p)(K_D) \tag{6}$$

Since $k_4 = k_{cat}$, k_{cat}/K_m' is independent of k_4 and is given by k_p/K_D. Since k_p/K_D is equal to k_{cat}/K_m', the k_p/K_D values can be determined from the k_{cat}/K_m' values under steady-state conditions where sequence-defined DNA-primed RNA (DNA) templates are used under forced termination conditions.[68]

3. Summary of HIV-1 RT Inhibition by 2′,3′-Dideoxynucleoside-5′-Triphosphates

1. 2′,3′-Dideoxynucleoside-5′-triphosphates are competitive alternative substrates for HIV-1 RT. Under steady-state conditions in cell-free systems their K_i values are equal to their K_m values when dissociation of the enzyme·template primer product complex is on the reaction pathway of the uninhibited control reaction (i.e., forced termination of processive synthesis on DNA-primed templates or distributive synthesis on DNA-primed DNA templates).

2. With DNA-primed RNA templates these K_i and K_m values are significantly lower than the corresponding K_D values. The latter are equilibrium dissociation constants of ddNTPs dissociating from enzyme·template primer complexes and would ordinarily reflect the potency of

a simple competitive inhibitor. Thus, as an alternative substrate, AZT-TP, for example, is significantly more potent (50- to 200-fold) than it would be as a simple competitive inhibitor of TTP. It follows that the steady-state inhibition of RT is not due to the tight binding of AZT-TP to the enzyme·template primer complexes but to the slow dissociation of the nonproductive chain terminated enzyme·template primer product complexes that accumulate and "tie up" free enzyme.

3. In a cell-free system the observed K_i values for chain terminators, such as AZT-TP, are highly dependent upon specific assay conditions and may not reflect intrinsic potency.

4. Within the infected cell, it is the k_{cat}/K_m values of the dideoxynucleoside-5′ triphosphates (that can be estimated under steady-state conditions in cell-free systems) and not K_i or K_m values *per se*, along with the intracellular levels of the 2′,3′-dideoxynucleoside-5′-triphosphates and the number of potential sites for chain-terminator incorporation within the replicating viral genome, that can account for the antiviral activity of the dideoxynucleosides.

From these analyses it seems possible to optimize some of the properties of 2′,3′-dideoxynucleosides that determine their antiviral activity. However, from the discussion of the clinical results with AZT that follows, it is apparent that dose-limiting toxicity and the emergence of resistant HIV-1 strains pose additional challenges to controlling HIV infection.

C. CLINICAL RESULTS WITH AZT

Following an initial phase I trial of AZT involving 19 patients with AIDS or AIDS-related complex (ARC) that showed improvements in CD4 cell counts, weight gain, a return of delayed-type cutaneous hypersensitivity reactions, and general well-being,[14] a double-blind, randomized, placebo-controlled trial demonstrated that AZT reduced morbidity and mortality when administered to patients with AIDS or ARC.[15] The beneficial clinical effect of AZT therapy was subsequently confirmed and extended in several other studies involving patients with HIV infection and advanced symptomatic disease.[16–21]

More recently two studies that were designed to determine the efficacy of AZT in patients with HIV infection in a relatively early stage who were either mildly symptomatic (ACTG 016)[85] or asymptomatic (ACTG 019)[86] demonstrated a reduction in the progression to AIDS that was statistically significant, although > 80 to 90% of the patients in both studies did not progress. Furthermore, no effect on survival was shown as the development of AIDS, rather than death, was used as the primary outcome measure. Thus, despite the proven efficacy of AZT, the best time to initiate therapy could not be determined from these studies.

To determine the long-term clinical benefits and liabilities of earlier as compared with later therapy, the Veterans Affairs Cooperative Study Group on AIDS Treatment[87,88] randomly assigned patients with symptomatic HIV infection (but not progressed to AIDS) and CD4+ counts between 200 and 500/mm³ to receive either placebo or AZT (1500 mg/d) until an AIDS-defining illness developed or the CD4+ count fell below 200. At that time all patients began to receive AZT. AZT therapy was withheld from the placebo, or late-therapy group for a mean of 14 months and the average duration of follow-up was 28 months. Early AZT therapy reduced the rate of development of clinical AIDS by almost half, which is similar to the results reported in the ACTG 016 and -019 trials.[85,86] Furthermore, within the first 1 to 3 months of AZT therapy, CD4+ counts increased transiently and serum levels of p24 antigen declined. The key point, however, was that in this trial, 2.5-year survival rates were similar in the early-therapy and late-therapy groups — 77 and 83%, respectively. Thus, although AZT therapy slowed the progression to AIDS, it did not improve overall survival. Once AIDS developed in patients receiving early AZT therapy, more of them tended to have multiple AIDS diagnoses, a slightly higher proportion died, and the median survival time was slightly shorter than in similar patients who received late therapy.

In another study,[89] the probabilities of death were compared among men at similar stages of disease who began AZT therapy before diagnosis of AIDS and among those who either did not take AZT at any time or delayed the start of treatment until after a first AIDS-defining illness. Relative risks of death were calculated for each of five initial disease states on the basis of CD4+ counts and clinical symptoms and signs appearing over follow-up periods of 6, 12, 18, and 24 months. Adjustments were also made for the use of prophylaxis against *Pneumocystis carnii* pneumonia (PCP). After these adjustments, AZT alone significantly reduced mortality at 6, 12, and 18 months of follow-up, but not at 24 months. Among AZT users, those who also used PCP prophylaxis before the development of AIDS had significantly lower mortality at 18 and 24 months than those who did not. Thus, although a reduced mortality from AZT therapy was observed in the early follow-up periods, no significant survival benefit was demonstrated at 24 months from AZT alone.

Despite the clear therapeutic benefits of AZT therapy, it is evident from these studies that disease progression is slowed but not halted with this drug. The clinically significant dose-limiting toxicity of AZT *(vide infra)* in both adults and children and the demonstration of *in vitro* resistance to AZT among nearly all persons with advanced HIV infection who have taken the drug for more than a year *(vide infra)* are two factors that could potentially contribute to the apparent loss of drug benefit observed for AZT.

1. Dose-Limiting Toxicity of AZT

Bone marrow suppression is the most common dose-limiting toxicity of AZT, especially in patients with established AIDS.[14,17,90,91] The red cell lineage is most often affected, but leukopenia and thrombocytopenia can also occur.[25] The pathogenesis of AZT-induced bone marrow suppression is not completely understood. AZT, relative to thymidine, has been shown to be a good substrate for thymidine kinase,[56] an enzyme likely to be involved in the anabolic phosphorylation of AZT.[56] AZT monophosphate, however, is a potent alternative substrate inhibitor of thymidylate kinase.[56] This results in the accumulation of AZT monophosphate[56] and the inhibition of TTP synthesis.[56] Thus, one hypothesis is that the elevated AZT monophosphate levels *per se,* or the reduced levels of TTP could be involved in cellular toxicity. Such an effect would be expected to be most severe in rapidly proliferating cells such as bone marrow progenitor cells.

In addition to effects on bone marrow suppression, there also appears to be an increased risk of myopathy in patients taking AZT, especially with prolonged use.[92] The etiology may relate to a depletion of muscle cell mitochondrial DNA with a proliferation of abnormal mitochondria.[93] This effect has been hypothesized to result from the utilization of AZT-5'-triphosphate by DNA polymerase γ.[94] Although AZT-5'triphosphate is selective for RT,[95] it can nevertheless inhibit the activity of the normal cellular DNA polymerases β and γ,[64,69] as [3H]AZT has been shown to be incorporated into cellular DNA.[96,97] This could potentially contribute to the toxicities observed for AZT.

One obvious approach to minimize the toxicities of AZT is to determine the efficacy at lower doses. Current recommendations are to initiate AZT therapy at a dose of 500 to 600 mg/d[98] compared to 1200 to 1500 mg/d in earlier studies.[15] However, at this dose the efficacy for patients with neurologic disease has not been determined.[98] Since the cerebrospinal fluid/plasma AZT ratio is ≤0.6 and may be as low as 0.1 in some patients, theoretically higher levels in blood might be important to achieve efficacious levels in the central nervous system. Further studies are required to determine if 300 mg/d, shown in one trial to have clinical benefits, can be generally recommended.[99] From a recent study, with the caveat that data were collected from a relatively small number of patients, it was evident that AZT given at a dose of 150 mg/d was not as effective as 600 mg/d when each was given in combination with a comparable dose of dideoxycytidine (ddC); yet 600 mg/d of AZT appeared no more toxic than 150 mg/d even when given with ddC.[100,101]

Other approaches to reducing the toxicities associated with AZT are the use of recombinant erythropoietin[102] and granulocyte-macrophage or granulocyte colony-stimulating factors[103,104] to counteract AZT-induced anemia and neutropenia, respectively. Although these approaches may well improve the therapeutic protocols involving AZT, other approaches involving the development of more selective and less toxic antiviral agents are clearly needed.

2. Emergence of AZT-Resistant HIV-1 Strains

The development of drug-resistant HIV-1 strains occurs during the treatment of AIDS and HIV infection. Resistance to AZT is due to the accumulation of specific amino acid substitutions in RT; studies have indicated that all of the resistance phenotype, as assessed *in vitro,* was conferred by the RT coding region of the respective clinical isolate, and that no significant contribution to resistance was found from other viral gene products.[71] The mutation Arg-70 → Lys has been commonly detected first during AZT therapy.[105] This mutation then disappears from the viral population, being replaced by the mutation Thr-215 → Tyr or Phe that requires a two-nucleotide change[105] and is responsible for an approximate 16-fold increase in resistance to AZT.[71] It is not clear whether the apparent disappearance of the codon 70 mutation was due to true reversion or to a shift in the balance of different strains of the virus populations. Studies suggest that the virus mutant at codon 70 alone is less viable than one that is wild-type at 70 (or mutant at codon 215), prevailing only in the presence of AZT where the wild-type virus is selected against.[105] The mutation Met-41 → Leu appears after the change at 215.[71] This genotype has been shown to confer an increase in resistance of approximately fourfold over a virus containing solely Tyr-215. This is likely to cause a selective advantage for the virus containing both Leu-41 and Tyr-215 in the presence of AZT. The polymerase chain reaction (PCR) analyses have also shown that the appearance of Leu-41 precedes the reappearance of Arg-70. Therefore, Leu-41 may also serve as a compensatory mutation allowing the acquisition of Arg-70 in the presence of Tyr-215.[71] The level of resistance of this combination of mutations is as yet unknown.[71] However, it is likely to exhibit only partial resistance with the subsequent appearance of Asp-67 → Asn, and perhaps Lys-219 → Gln conferring high-level resistance (~180-fold *in toto*).[71]

The clinical significance of *in vitro* resistance to AZT is not yet completely understood. The emergence of resistance HIV is not associated with sudden clinical deterioration.[106] Furthermore, progression from the asymptomatic stage of HIV infection to AIDS can occur in AZT-treated patients without the emergence of highly resistant isolates.[105] Thus, while partial resistance may be a requirement for disease progression during AZT therapy, the variation in time of progression after the appearance of partially resistant isolates suggests that additional factors are involved. The application of rapid genetic analysis as a means of assessing viral sensitivity on large populations of treated individuals may help determine the clinical significance.[71] However, for a complete assessment of AZT resistance by genotypic analysis alone, all commonly occurring mutations[107] must be known and monitored; otherwise, highly resistant isolates could be initially overlooked and a possible correlation between disease progression and *in vitro* resistance would not be observed.

Despite the uncertainty about long-term effects (i.e., cumulative drug toxicity and development of resistance) of AZT therapy and doubt about survival benefit with this therapy, many patients now receive AZT at an early stage of their infection to slow progression of the disease. At later stages of HIV infection, both the intolerance to AZT *(vide supra)*[14,17,90,91] and the likelihood that high-level resistance will emerge increase.[108] These are likely to contribute to the loss of benefit from AZT, necessitating the development of alternative therapies.

D. OTHER DIDEOXYNUCLEOSIDES

Several 2′,3′-dideoxynucleosides have shown anti-HIV activity in culture and are currently being evaluated in clinical trials. The status of these compounds has recently and extensively been reviewed.[22,24-26,109,110] All of these dideoxynucleosides are thought to have a mechanism of antiviral activity similar to that of AZT *(vide supra)* where they become anabolically phosphorylated and act upon RT to prevent HIV replication. Since the enzymes responsible for the phosphorylation vary from compound to compound, each dideoxynucleoside displays rather different activity and toxicity profiles.[25,26] For example, ddI (dideoxyinosine, didanosine, Videx) is a purine analog whose likely pathway for phosphorylation involves 5′-nucleotidase-catalyzed monophosphorylation,[111] conversion to 2′,3′-dideoxyadenosine (ddA) monophosphate (ddAMP), catalyzed by adenylosuccinate synthetase and adenylosuccinate lyase, and then phosphorylation to its active form, 2′,3′-dideoxyadenosine-5′-triphosphate (ddATP).[24,71] ddA is also an effective antiviral agent *in vitro*, but in humans ddI, as a prodrug of ddATP, is preferable. ddI is less toxic than AZT for bone marrow progenitor cells and has been associated with relatively little hematologic toxicity even in patients with prior hematologic intolerance to AZT.[112–115] The dose-limiting toxicity of ddI has been a rather painful peripheral neuropathy and pancreatitis. In the fall of 1991, ddI was approved for use in patients who had experienced treatment failure or who were not able to tolerate AZT therapy.

HIV-1 isolates have been obtained from patients with AIDS who had changed therapy to ddI after at least 12 months of treatment with AZT.[116] The *in vitro* sensitivity to ddI was shown to decrease during a 12-month period following ddI initiation. AZT sensitivity, on the other hand, increased during this same period of time. Analysis of the RT coding region revealed a mutation Leu(74) → Val that was shown to confer ddI resistance. When this mutation was present in the same genome as Thr-215 → Tyr, known to confer AZT resistance (and which persisted in the viral population for at least 12 months after therapy was switched from AZT to ddI), the corresponding isolates showed increased sensitivity to AZT. Further studies confirmed that the mutation responsible for ddI resistance suppressed the effect of the AZT resistance mutation. In conjunction with the results indicating nonoverlapping toxicity profiles of AZT and ddI *(vide supra)*, these studies strongly suggest that combination therapy involving AZT and ddI in either alternating or concurrent administration regimens may be more effective than either drug alone *(vide infra)*. Since 2′,3′-dideoxycytidine (ddC) has been shown to exhibit cross-resistance with ddI,[116] as well as have an overlapping toxicity profile involving neuropathy,[117,118] it is unlikely that combination therapy involving ddC and ddI would be beneficial. AZT and ddC combinations are, however, currently being studied *(vide infra)*.

III. NON-NUCLEOSIDE HIV-1-SPECIFIC RT INHIBITORS

A. DISCOVERY AND DEVELOPMENT

Screening, often thought of as the antithesis of the rational approach, continues to play a very productive role in drug discovery.[119] The recent development of non-nucleoside RT inhibitors is one important example.

1. TIBO Derivatives[34,40]

The rational screening strategy that has led to the discovery of the TIBO (4,5,6,7-tetrahydro-5-methylimidazo[4,5,1-jk] [1,4]benzodiazepin-2(1H)-one and thione) derivatives (Table 1) started with the selection of 600 compounds, all prototypes of different chemical series, which were without effect in standard pharmacological assays, and caused only mild toxicity in rodents.[34] The criteria for compounds selected for screening reduced the likelihood of developing agents that, at a later stage, would show undesirable side effects. These compounds were assayed for inhibition of multiple cycle growth, cytopathicity of HIV-1, and for cytoxicity in a sensitive MT-4 cell system.[34] Thus, the cellular-based assay had the advantage of assessing several different potential targets simultaneously, rather than an isolated single specific target protein. Uptake, metabolism, and cytotoxicity were also reflected in its results.

The original lead compound R14458 derived from this screening approach (Table 1) inhibited the cytopathic effect of HIV-1 at a concentration (IC_{50}) one tenth its cytotoxic concentration (CC_{50}). The *S*-(+) enantiomer (5 position) R78305 was active and the isomer with the *R*-(−) configuration R78304 was without effect against the virus (Table 1). Interestingly, the cytotoxic effects (CC_{50}) of the two compounds were nearly equal. The requirement for the *S*-(+) configuration for anti-HIV-1 activity was shown to be common to all the TIBO derivatives synthesized, although some *R* isomers showed some, albeit reduced, activity.[120] A lipophilic group at the 6-position (Table 1) was shown to be mandatory for activity since the unsubstituted compound was inactive. Polarity due to heteroatoms diminished activity and the presence of an olefin enhanced activity. There is an optimum size requirement of this group (compare R78305 with R79882), since activity was lost with increases or decreases in the length or in the breadth of the side chain.[120] The 3-methyl-2-butenyl group appeared to be the best among the compounds synthesized. Replacement of oxygen at position

TABLE 1
Anti-HIV-1 Activity of TIBO Derivatives in MT-4 Cells

Compound	R₁	R₂	R₃	R₄	EC_{50}^{a} (μM)	CC_{50}^{b} (μM)	Selectivity Index[c]	Relative Potency
R14458	O	(±)CH₃	–CH₂–CH=CH₂	—	62	612	10	1
R78304	O	(–) - (R)CH₃	–CH₂–CH=CH₂	—	>500	592	<1	<1
R78305	O	(+) - (S)CH₃	–CH₂–CH=CH₂	—	70	674	10	1
R79882	O	(±)CH₃	–CH₂–CH=C(CH₃)₂	—	13	840	65	5
R82150	S	(+)CH₃	–CH₂–CH=C(CH₃)₂	—	0.029	574	19,793	2,138
R82913	S	(+)CH₃	–CH₂–CH=C(CH₃)₂	9 Cl	0.026	37	1,423	2,385
R86183	S	(+)CH₃	–CH₂–CH=C(CH₃)₂	8 Cl	0.004	77	19,250	15,500

ᵃ EC_{50}, concentration of compound protecting 50% of MT-4 cells against HIV-1-induced cytopathic effects.
ᵇ CC_{50}, concentration of compound causing 50% cytotoxicity.
ᶜ   Selectivity index represents (ratio of CC_{50} to EC_{50}). With permission.

Data from Reference 40. With permission.

C2 with sulfur (R82150) resulted in dramatic increases in potency and selectivity (Table 1).[121] Further increases in potency resulted from the introduction of a chlorine substituent in position 8.[40,122]

The antiviral profile of the TIBO derivatives has been characterized most remarkably by the extreme sensitivity of HIV-1 strains to the TIBO derivatives and relative insensitivity exhibited by HIV-2 strains, and simian immunodeficiency viruses (SIV). This is in contrast to what has been observed with other anti-HIV agents such as 2′,3′dideoxynucleosides which are active against HIV-2 and SIV.

The mechanism of action for the TIBO derivatives has been elucidated in a series of experiments.[34] R81250 (Table 1) was added at various times following exposure of MT-4 cells to HIV-1 in order to identify the drug-sensitive phase under single growth-cycle conditions. These studies indicated that R81250 interacts with a process that coincides with that of reverse transcription. Polymerase chain reaction (PCR) analysis of HIV-1 DNA from these cells corroborated these findings.[123] Furthermore, HIV-1 antigen and progeny formation from persistently infected HUT-78 cells (where virus production is independent of RT) were not affected by 5 d of incubation with 3.5 μM R81250.[34] TIBO derivatives directly inhibited the DNA polymerase activity of virion-derived and recombinant HIV-1 RT.[34,123]

Consistent with the antiviral profile and high selectivity indices, TIBO compounds did not inhibit the polymerase activity catalyzed by HIV-2 RT, and human DNA polymerases α, β, γ.[34,124] The TIBO enantiospecificity was also observed at the RT level and a high-ranking correlation existed between inhibitory effects of these compounds on HIV-1 RT and their anti-HIV-1 activity in cell culture.[34] Taken together, these studies strongly indicate that the antiviral activity of TIBO compounds results from inhibition of RT-catalyzed DNA polymerase activity. The findings that the resistance of certain mutant HIV-1 strains to R81250 maps within the RT gene and that the corresponding molecularly cloned and expressed mutant RT is insensitive to R81250 *(vide infra)* strongly support the proposed antiviral mechanism of TIBO involving RT inhibition.

In addition to the TIBO derivatives, this screening approach has led to the discovery of two other series of specific HIV-1 antiviral agents that likely act at the level of RT. The first of these series, actually discovered prior to the TIBO derivatives, is derived from 1-[(2)-(hydroxyethoxy)methyl]-6-(phenylthio) thymine (HEPT).[37] A very recently discovered series is derived from 2′,5′-bis-*O*-(*tert*-butyldimethylsilyl)-3′-spiro-5″-(4″-amino-1″,2″-oxathiole-2″,2″-dioxide) (TSAO) pyrimidine.[39] Although structurally, these two series are composed of nucleoside analogs, mechanistically, these compounds appear to act as HIV-1-specific non-nucleoside RT inhibitors *(vide infra)*.

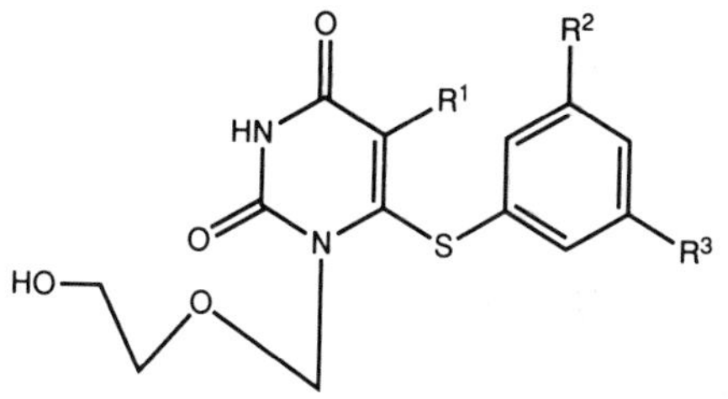

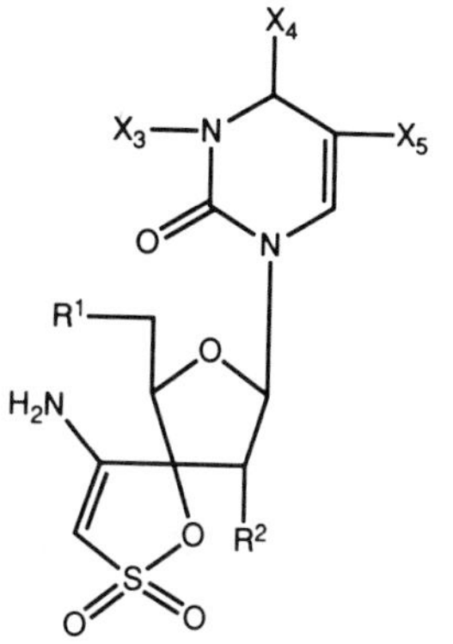

R^1 = methyl, $R^2 = R^3$ = H

X_3 = H, Allyl or methyl
X_4 = OH or NH_2
X_5 = H or methyl
R^1 = O[Si] [Si] = t-butyl-dimethylsilyl
R^2 = O[Si]

HEPT   **TSAO**

A second screening strategy has involved testing for the direct inhibition of the DNA polymerase activity of HIV-1 RT with a structurally diverse group of compounds employing selectivity counterscreens that have included human DNA polymerases α, β, γ. This approach was designed to discover specific RT inhibitors that did not require metabolic activation for antiviral activity and were specific for HIV-1 RT relative to normal cellular polymerases α, β, and γ. Such inhibitors might prove less toxic than the 2′,3′-dideoxynucleosides since anabolic phosphorylation and inhibition of normal cellular polymerases may be important determinants of the dose-limiting toxicities exhibited by the latter class of anti-HIV agents *(vide supra)*. Furthermore, since the kinetic mechanism and stereochemical course of the reaction is similar among the polymerases studied, including HIV-1 RT,[125] it is not likely that a rational approach generating mechanism-based inhibitors would yield specific HIV-1 RT inhibitors. The direct screening of HIV-1 RT has led to the discovery of three structurally distinct series of compounds that, along with the TIBO, HEPT, and TSAO derivatives, constitute a novel class of HIV-1 RT-specific inhibitors.

2. Nevirapine (BI-RG-587)

The first compound discovered utilizing this screening approach was nevirapine (BI-RG-587).[35,41,42,126] This compound satisfied the criteria set for the selection of a preclinical candidate:[126] (1) it possesses IC_{50} values below 100 nM in both the enzyme and cellular-based assays (antiviral activity), (2) it appears nontoxic and specific for RT, showing no significant activity against normal cellular polymerases α, β, γ, δ; (3) it exhibits suitable solubility and bioavailability; and (4) it is relatively stable to metabolic degradation. Furthermore, the selected preclinical candidate should distribute between the brain and plasma and thereby control CNS viral infection. As discussed previously, low CNS/plasma ratios may prevent lower dose protocols for AZT.

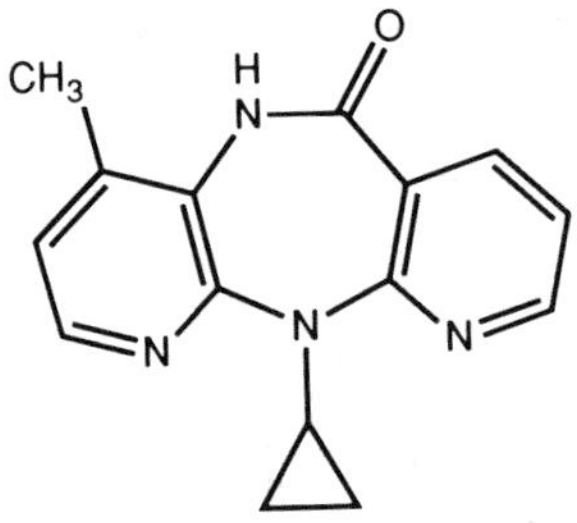

Nevirapine (BI-RG-587)

Among the structurally diverse compounds tested, pyrido[2,3-b][1,4] benzodiazepinone (**I**) derivatives were initially found as weak actives. Since these compounds are structurally related to the M_1-selective, antimuscarinic compound pirenzepine, a large number of pyrido[2,3b][1,4] benzodiazepines (**I**) and pyrido[2,3b][1,5] benzodiazepinones (**II**) and a few dipyrido[3,2-b:2′3′-e] [1,4]diazepinone (**III**) analogs were available for testing.[126] Screening led to (**IV**) as a lead compound and efforts based on initial structure-activity relationships focused on derivatives of **II** and **III**.[126]

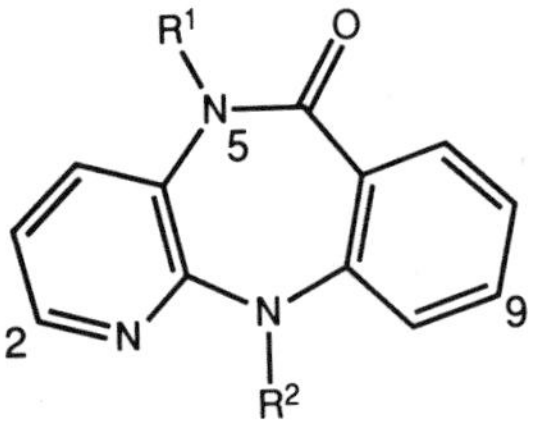

I

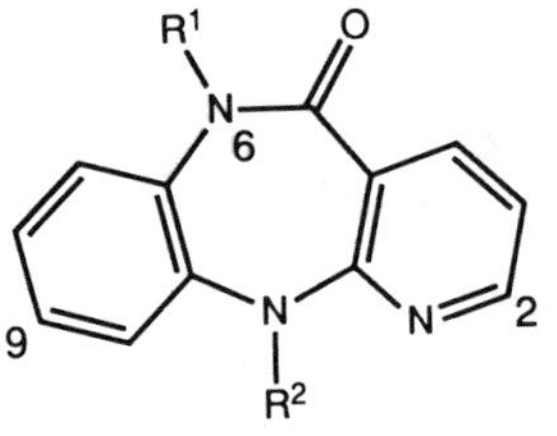

II

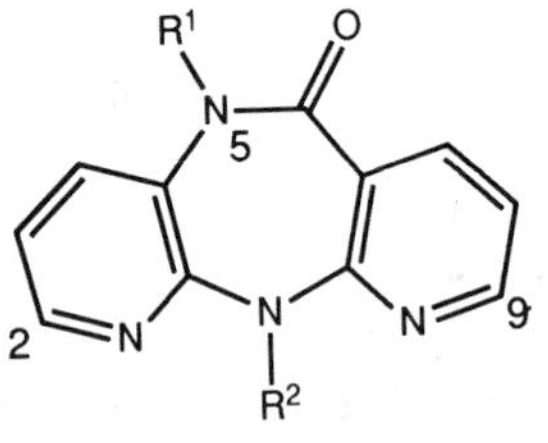

III

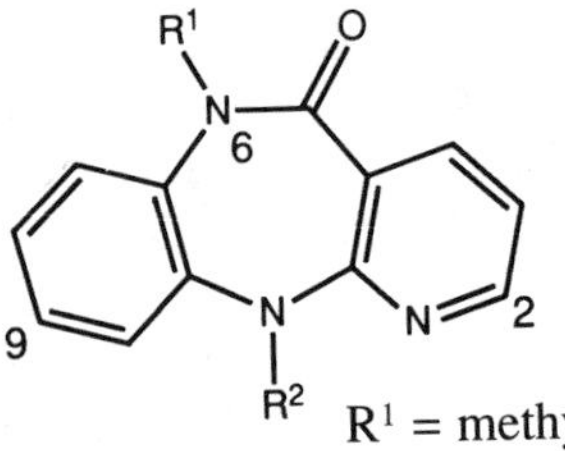

IV R^1 = methyl
    R^2 = ethyl

As a class, the dipyridodiazepinones (**III**) were not only more potent, but were also more water soluble, less cytotoxic, and more resistant to metabolic N-dealkylation than either of the monopyridodiazepinone series (**I** and **II**). The finding that the combination of a 4-methyl substituent and an unsubstituted lactam resulted in maximal activity in the dipyrido **III** series eliminated the dealkylation problem at N-5. Surprisingly, in rats and monkeys the presence of a 4-methyl substituent also significantly reduced dealkylation at N-11. The 11-cyclopropyl compound (nevirapine) exhibited greater bioavailability than the corresponding 11-ethyl derivative. Since nevirapine satisfied all the criteria for the desired properties of a preclinical candidate, it has been selected for clinical evaluation *(vide infra)*.

3. 2-Pyridinones

Like nevirapine, the 2-pyridinone, L-345,516 (Table 2), was discovered in a screening program designed to identify direct and specific inhibitors of HIV-1 RT.[36,127] L-345,516 displayed 30 n*M* potency in the enzyme assay but lacked antiviral activity. This compound, at the oxidation state of formaldehyde was hydrolytically unstable on a time scale that was long for the enzyme assay but short for the cellular-based antiviral assay. Several kinetic experiments indicated that the intrinsic chemical reactivity was not required for HIV-1 RT inhibition. Taken together, these results suggested that stable analogs of L-345,516 might show antiviral activity.

An initial attempt to improve stability involved replacement of the aminomethylene linker of L-345,516 with ethylene to yield L-693,593 (Table 2). Although this compound proved to be hydrolytically stable, it was more than 100-fold weaker than the original lead as an HIV-1 RT inhibitor. Nevertheless, it sufficiently inhibited the spread of HIV-1 infection in H9 human T lymphoid cell culture experiments to indicate that compounds of this structural class could effectively cross cell membranes and produce an antiviral effect[36,127] (Table 2).

An alternate approach, that has led to potent and stable inhibitors in the aminomethylene series, involved replacement of the phthalimide moieties of the original lead with various aromatic and heterocyclic groups. Of the compounds initially explored, the benzoxazole, L-696–040, exhibited relatively good potency in the RT assay with significantly improved antiviral activity. Systematic introduction of methyl groups into the benzoxazole ring of L-696,040 identified the 4' and 7' positions as potency enhancing (Table 2). Disubstitution at these positions with either methyl or chloro yielded the more potent RT inhibitors L-697,639 and L-697,661, both preclinical candidates. Substitution of the phthalimide group in L-693,593 with the benzoxazole moiety resulted in the ethylene-linked L-696,229[127,128] possessing comparable potency to the aminomethyl-linked derivatives L-697,639 and L-697,661.

L-696, 229

These 2-pyridinone derivatives possessed the specificity for HIV-1 RT observed for the original lead L-345,516 and had antiviral activity profiles similar to those of the TIBO derivatives and nevirapine.

Another series of HIV-1 specific RT inhibitors, bis(heteroaryl) piperazines (BHAPs), discovered in a direct HIV-1 RT inhibition screen, has recently been described.[38]

TABLE 2

Structure-Activity Relationships for Inhibition of HIV-1 RT Activity and HIV-1 Infection in Cell Culture by 2-Pyridinone Derivatives

		Potency, nM	
R=	Name	HIV-1 RT IC_{50}[a]	HIV-1 Spread Activity CIC_{95}[b]
	L-345,516	30	c
	L-693,593	3,700	40,000
	L-696,040	210	900
	L-697,639	20	150
	L-697,661	19	100

[a] The IC_{50} is stated as the mean of at least three experiments.
[b] Antiviral activity was determined using HIV-1 strain IIIb in H9 cell culture. Antiviral potencies were expressed as the concentration that inhibited the spread of HIV-1 infection by ≥95% (CIC_{95}).
[c] Hydrolytic instability precluded a reliable determination of this value.

Data from Reference 36. With permission.

$X = H, F$ or OCH_3
$Y = CH_2CH_3$ or $CH(CH_3)_2$

BHAP

B. MODE OF ACTION

Although these non-nucleosides are potent and specific inhibitors of HIV-1 RT, it remains to determine whether this mechanism can account for their antiviral activity. One approach has been to generate non-nucleoside inhibitor-resistant HIV-1 strains[129,130] and determine the mechanism of resistance. Serial passage in the presence of the 2-pyridinone, L-693,593 (Table 2) yielded virus that was approximately 1000-fold resistant to compounds of this class.[129] The RT genes from the L-693,593-resistant proviruses were molecularly cloned by PCR. DNA sequencing of the resistant RT gene revealed single-base-pair changes resulting in conversion of K-103 → N and Y-181 → C. When these specific mutations were introduced into a previously sensitive recombinant HIV-1 strain, L-693,593-resistant virus was generated. These experiments confirmed that these two mutations were responsible for the resistance. Each mutation has been shown to confer partial resistance. Furthermore, bacterially expressed RTs molecularly cloned from L-693,593-resistant virus were sufficiently resistant in enzymatic assays to quantitatively account for the resistance observed in culture, convincingly demonstrating the mechanism of the antiviral activity. Cross-resistance to other non-nucleoside RT inhibitors such as the TIBO derivative R82150 and nevirapine (BI-RG-587) was observed within enzymatic assays as well as in culture. These data suggested that the non-nucleosides constitute a single pharmacologic class of antiviral agent characterized by a common mechanism of RT inhibition. This mechanism appears unique, however, as no cross-resistance with AZT has been observed.[36,129,130,151]

As a class, the non-nucleoside RT inhibitors appear to act reversibly[36,123] and are highly specific for HIV-1 RT.[34–42] In contrast to the 2′,3′-dideoxynucleosides, they are not alternate substrates[36] and do not disrupt processivity.[42,124] In the case of the 2-pyridinones slow binding has been observed.[36] Consistent with the cross resistance among these various non-nucleoside inhibitors *(vide supra),* several studies including direct[36,131] and indirect[132] binding measurements at equilibrium and in the steady state as well as multiple inhibitor kinetic analyses[133] have strongly suggested a common binding site on HIV-1 RT for all of these compounds.[36] Direct binding studies[36,131] and photoaffinity labeling experiments[132] have indicated a single high-affinity binding site for the non-nucleoside inhibitors that has been localized to the p66 subunit of the RT heterodimer.[132] Since the RT heterodimer appears to have a single active site,[134] this stoichiometry is consistent with a Hill slope of 1 obtained from inhibition analyses.

Binding of these non-nucleosides to free enzyme has been measured[131,132] or inferred from kinetic studies.[35,124] However, higher-affinity binding to enzyme·template primer complexes[36,123,133] with certain template primers has been demonstrated. Consistent with these data, in some cases, uncompetitive inhibition with respect to template primer has been observed.[123,133] Although noncompetitive inhibition has been observed,[35,124] purely competitive inhibition patterns with respect to template primer have not been reported for the non-nucleoside RT inhibitors, suggesting that the binding site for these inhibitors does not overlap with the binding site of the template primer.

These non-nucleoside inhibitors appear to be noncompetitive with respect to 2′-deoxynucleoside-5′-triphosphates,[34–42] although apparent competitive inhibition has been observed with respect to TTP for some HEPT derivatives when poly(A)·oligo(dT) was employed[37] and with respect to dATP for TIBO when poly(U)·oligo(dA) was employed.[39] Whether these results represent true competitive inhibition or an effect on K_m mediated by an allosteric interaction has not been resolved. There is some additional evidence suggesting that the non-nucleoside RT inhibitors may bind near or allosterically communicate with the 2′-deoxynucleoside-5′-triphosphate site[131,38] or the pyrophosphate site.[36]

Although the kinetic and binding studies have not provided strong evidence for defining the relationship between the substrate and inhibitor binding sites, several approaches have been used to define these binding sites with respect to the primary sequence of RT. These results have been useful in suggesting a mechanism for inhibition of the RT DNA polymerase activity by this class of compounds.

A photoactive analog of nevirapine, BI-RJ-70,[132,153] has been used to probe the binding site of the non-nucleoside inhibitors on HIV-1 RT. BI-RJ-70, like nevirapine, is a potent and specific inhibitor of HIV-1 RT. Upon UV light illumination, the azido moiety is rapidly transformed into a nitrene intermediate which likely reacts with amino acid side chains at or in close proximity to the inhibitor binding site. The enzyme then becomes covalently labeled and inactivated. Protection from inactivation has been observed in the presence of other non-nucleoside inhibitors. This result, in conjunction with the 1:1 stoichiometry and specific interaction with the p66 subunit, supports the contention that BI-RJ-70 is a specific probe of the non-nucleoside inhibitor binding site.

[³H] BI-RJ-70

The amino acid residues labeled by [³H]BI-RJ-70 at the non-nucleoside binding site were identified by tryptic mapping and peptide sequencing.[153] The primary labeled peptide corresponded to residues 174–199. Within this peptide, the major labeled residue was identified as Y-181. Y-188 also contained significant radioactivity. As previously discussed, a single-base-pair change resulting in a Y-181 → C confers at least partial resistance (200- to 600-fold) to the non-nucleoside RT inhibitors. This mutation results in a corresponding loss of inhibitor binding affinity to the resistant HIV-1 RT.[155] Taken together, these observations strongly suggest that Y-181 and probably Y-188 form part of the binding site on HIV-1 RT for these non-nucleoside inhibitors.

There is strict conservation of tyrosine at positions 181 and 188 of HIV-1 RT. In contrast, the amino acids isoleucine/valine and leucine at positions 181 and 188, respectively, are highly conserved in HIV-2 RT. Since HIV-2 RT is insensitive to these non-nucleoside inhibitors, chimeric constructs involving reciprocal changes at positions 181 and 188 in both HIV-1 and HIV-2 RT backgrounds were generated.[135] The results of these experiments indicated that substitutions at 181 and 188 alone were sufficient to abolish sensitivity of HIV-1 RT but were not sufficient to render complete sensitivity upon HIV-2 RT. Complete sensitivity to the non-nucleoside HIV-1 RT-specific inhibitors was achieved, however, when the entire segment of amino acids 176–190 from HIV-1 RT was exchanged for the corresponding segment in the HIV-2 RT background.[135] Despite some quantitative differences, independent studies have provided similar results.[136,137] The results of these experiments corroborated the conclusions of other studies *(vide supra)* that have defined the binding locus of non-nucleoside inhibitors in relation to the primary sequence of HIV-1 RT.

Tyrosines 181 and 188, critical to non-nucleoside inhibitor binding, flank a conserved sequence motif containing an aspartate-aspartate (asp-asp) doublet (positions 185 and 186 in HIV-1 RT)[138,139] immediately surrounded by nonpolar amino acid residues. This asp-asp motif has been observed in many RNA-dependent RNA and DNA polymerases, and in some DNA-dependent RNA and DNA polymerases.[139] Another conserved region has been identified which also possesses an invariant asp (position 110 in HIV-1 RT).[138,139] The strict conservation of these acidic residues suggests an important role in polymerase function. This hypothesis has been supported by site-directed mutagenesis studies. Mutation of either asp 110, 185, or 186 resulted in complete abolition of RT DNA polymerase activity.[138] Very recent crystallographic studies with HIV-1 RT have confirmed the tenable hypothesis based on analogy with *E. coli* Klenow fragment of DNA polymerase I that asp 110 is spatially close (~5 Å) to asp 185, 186 and is likely to be involved in catalysis.[140] Thus, the binding locus of the non-nucleoside HIV-1 RT inhibitors appears to be proximal to the catalytically essential aspartic acid residues at positions 110, 185, and 186. Binding of the inhibitor could potentially perturb the microenvironment of these residues and result in loss of catalytic activity. Such an effect could be mediated via a direct interaction of the inhibitor with one or more of these residues (perhaps involving a divalent cation) or via an indirect (allosteric) interaction transmitted by one or both tyrosine residues (181, 188). This mechanism of inhibition is in accord with the results of crystallographic studies at 3.5 Å resolution of RT complexed with nevirapine.[140] As the resolution is improved, it would be interesting to determine whether certain regions of the "active site" are accessible from the adjacent non-nucleoside inhibitor binding site. If so, this result could lead to the design of more potent inhibitors while taking advantage of interactions within the "active site" that are theoretically less likely to mutate and lead to viable, resistant virus.

As discussed, several studies,[129,130,135,137,140,153] especially those demonstrating partial resistance,[129,130] have implicated a role for residues 181, 188, and 103 in the binding of non-nucleoside inhibitors to HIV-1 RT, implying that multiple interactions may be required for potent inhibition. Structure-activity relationships among 2-pyridinones suggested that despite a common binding site for these inhibitors the interactions within this binding site may vary considerably from compound to compound.[127] In this regard "resistant" RTs have been shown, either directly[135,137] or have been inferred,[141] to exhibit differential sensitivity toward various non-nucleoside inhibitors. In a striking example, RT [Y(181) → C, K(103) → N] which is resistant to virtually all 2-pyridinones synthesized and cross-resistant to TIBO and nevirapine was almost as sensitive to L-702,019 as was the wild-type RT.[154] In all other respects, L-702,019 behaves as an HIV-1 RT-specific non-nucleoside inhibitor. The significance of this result with respect to overcoming resistance in the clinic (*vide infra*) is uncertain.

L-702,019

C. PRELIMINARY CLINICAL RESULTS

R82913, a potent[142] TIBO derivative (Table 1) which does not inhibit hematopoietic progenitor cell growth *in vitro*[143] was given to 22 patients with AIDS or ARC in a dose escalating pilot study.[144] Doses of 10 to 300 mg administered daily by intravenous infusion were well tolerated for up to 50 weeks with no hematological or biochemical evidence of toxicity. Statistically significant, but small changes in serum p24 antigenemia were observed during the first month of therapy. However, no conclusions could be drawn regarding efficacy in this Phase I study.[144]

Preliminary results of a Phase I study (Protocol 91–1-112, National Institute for Allergy and Infectious Diseases, National Institutes of Health) with the orally bioavailable 2-pyridinone derivative L-697,661 showed an initial decrease of serum p24 antigen expression and an increase in the number of CD4 cells. However, after several weeks of L-697,661 therapy, resistant HIV-1 variants emerged, containing mutations within the RT gene previously observed *in vitro*.[129] The development of resistance was the likely cause for the subsequent increases in p24 antigen expression and decreases in CD4 cell counts. Similar results have been obtained with nevirapine.

These preliminary results, confirming the predictions based upon *in vitro* studies,[129,130] suggest that the utility of the non-nucleoside RT inhibitors as monotherapeutic agents may be precluded by the rapid emergence of resistance. Nevertheless, based upon their relative lack of toxicity and their initial effects upon the surrogate markers, CD4 cell counts and p24 antigenemia, the use of non-nucleoside RT inhibitors in combination with other anti-HIV-1 agents that do not exhibit cross resistance seems warranted. These studies are currently in progress *(vide infra).*

IV. COMBINATION THERAPY INVOLVING RT INHIBITORS

Due to the limitations of the currently available HIV-1 RT inhibitors *(vide supra),* monotherapy could potentially be improved with combinations of two or more agents that are additive or synergistic in their antiviral activities, exhibit nonoverlapping toxicity profiles, and show no significant cross resistance. Ideally, agents should be used whose sole limitation during monotherapy is the inevitable development of resistance and otherwise are not limited by toxicity or by nonoptimal pharmacokinetic properties.

As discussed, the dose-limiting toxicities of AZT are anemia and neutropenia while the most frequently associated toxicity with ddC therapy is a dose-related sensory peripheral neuropathy. The combination of AZT and ddC has additive, and in some assays synergistic *in vitro* inhibitory activity against HIV[145-147] and isolates resistant to AZT *in vitro* remain susceptible to ddC.[116,148] Thus, AZT and ddC possess the characteristics potentially suitable for combination therapy. This approach has been recently evaluated in patients with advanced HIV infection.[100] Although this study involved only 56 patients in a Phase I/II trial, the results appear quite promising. No unusual toxicity was associated with combination therapy and the toxicities that were observed have been well described with either drug alone. Further, the observed toxicity appeared to be neither more frequent nor more severe than that reported with either drug alone, suggesting that the combination did not produce serious additive or synergistic toxicity. The anti-HIV activity of the combination therapy was documented by improvements in weight, CD4 counts, and p24 antigenemia. The reported protocol appeared to produce greater and more persistent effects in patients with advanced HIV infection compared with other study regimens and with the results of previous trials in which AZT monotherapy was used.

Additional trials based upon this pilot study are currently in progress. The efficacy of monotherapy will be compared to combination therapy with AZT and either ddI or ddC. In addition to protocols involving concurrent administration of nucleoside analogs, regimens involving AZT and ddC alternating weekly or monthly[117,149] in the hope of reducing the toxicity of each agent have demonstrated that the total cumulative dose of ddC that could be given without development of neuropathy was at least fivefold greater when ddC was given in an alternating rather than continuous schedule. Further studies are required to determine the efficacy of this approach.

These results involving combination therapy with AZT and ddC and the results from a small trial involving combination therapy with AZT and the putative pyrophosphate analog foscarnet[150] support the concurrent use of multiple RT inhibitors for controlling HIV infection. The HIV-1 RT-specific non-nucleoside inhibitors have been shown to act synergistically with AZT and ddI *in vitro*,[36,151] and display no significant cross resistance[36,129,130,151] or overlapping toxicities with 2′,3′-dideoxy-nucleosides. Clinical studies involving the combination of nucleoside analogs and non-nucleoside RT inhibitors are in progress. As new agents that act upon different steps in the life cycle of HIV become available it is likely that superior combination therapies will result.

ACKNOWLEDGMENTS

I would like to thank Agnes Hendrick for providing valuable literature searches and informational services. Her continuing interest during this project is greatly appreciated. I would also like to thank Jackie Rees and Rose Mazur for their expert assistance in the preparation of this manuscript and my many colleagues at Merck for their valuable contributions and discussions.

REFERENCES

1. **Barre-Sinoussi, F., Chermann, J.C., Rey, F., Nugeyre, T., Chamaret, S., Gruest, J., Dauguet, C., Axler-Blin, C., Brun-Vezinet, F., Rouzioux, C., Rozenbaum, W., and Montagnier, L.,** Isolation of a T-cell lymphotropic virus from a patient at risk for the acquired immunodeficiency syndrome (AIDS), *Science,* 220, 868, 1983.

2. **Popovic, M., Sarngadharan, M.G., Read, E., and Gallo, R.C.,** Detection, isolation, and continuous production of cytopathic retroviruses (HTLV-III) from patients with AIDS and pre-AIDS, *Science,* 224, 497, 1984.

3. **Gallo, R.C., Salahuddin, S.Z., Popovic, M., Shearer, G., Kaplan, M., Haynes, B.F., Palker, T.J., Redfield, R., Oleske, J., Safai, B., White, G., Foster, P., and Markham, P.D.,** Frequent detection and isolation of cytopathic retroviruses (HTLV-III) from patients with AIDS and at risk for AIDS, *Science,* 224, 500, 1984.

4. **Schupbach, J., Popovic, M., Gilden, R.V., Gonda, M.A., Sarngadharan, M.G., and Gallo, R.C.,** Serological analysis of a subgroup of human T-lymphotropic retroviruses (HTLV-III) associated with AIDS, *Science,* 224, 503, 1984.

5. **Sarngadharan, M.G., Popovic, M., Bruch, L., Schupbach, J., and Gallo, R.C.,** Antibodies reactive with human T-lymphotropic retroviruses (HTLV-III) in the serum of patients with AIDS, *Science,* 224, 506, 1984.

6. **Klatzman, D., Barre-Sinoussi, F., Nugeyre, M.T., Dauguet, C., Vilmer, E., Griscelli, C., Brun-Vezinet, F., Rouzioux, C., Gluckman, J.C., Chermann, J.C., and Montagnier, L.,** Selective tropism of lymphadenopathy associated virus (LAV) for helper-inducer T lymphocytes, *Science,* 225, 59, 1984.

7. **Brun-Vezinet, F., Rouzioux, C., Montagnier, L., Chamaret, S., Gruest, J., Barre-Sinoussi, F., Geroldi, D., Chermann, J.C., McCormick, J., Mitchell, S., Piot, P., Taelman, H., Mirlangu, K.B., Wobin, O., Mbendi, N., Mazebo, P., Kalambayi, K., Bridts, C., Desmyter, J., Feinsod, F.M., and Quinn, T.C.,** Prevalence of antibodies to lymphadenopathy-associated retrovirus in African patients with AIDS, *Science,* 226, 453, 1984.

8. **Broder, S. and Gallo, R.C.,** A pathogenic retrovirus (HTLV-III) linked to AIDS, *N. Engl. J. Med.,* 311, 1292, 1984.

9. **Fauci, A.S.,** The human immunodeficiency virus: infectivity and mechanisms of pathogenesis, *Science,* 239, 617, 1988.

10. **Donegan, E., Stuart, M., Niland, J.C., Sacks, H.S., Azen, S.P., Dietrich, S.L., Faucett, C., Fletcher, M.A., Kleinman, S.H., Operskalski, E.A., Perkins, H.A., Pindyck, J., Schiff, E.R., Stites, D.P., Tomasulo, P.A., Mosley, J.W.,** The Transfusion Safety Group, Infection with human immunodeficiency virus type I (HIV-1) among recipients of antibody-positive blood donations, *Ann. Intern. Med.,* 113, 733, 1990.

11. **Bowen, D.L., Lane, H.C., and Fauci, A.S.,** Immunopathogenesis of the acquired immunodeficiency syndrome, *Ann. Intern. Med.,* 103, 704, 1985.

12. **Ho, D.D., Pomerantz, R.J., and Kaplan, J.C.,** Pathogenesis of infection with human immunodeficiency virus, *N. Engl. J. Med.,* 317, 278, 1987.

13. **Price, R.W., Brew, B., Sidtis, J., Rosenblum, M., Scheck, A.C., and Cleary, P.,** The brain in AIDS: central nervous system HIV-1 infection and AIDS dementia complex, *Science,* 239, 586, 1988.

14. **Yarchoan, R., Weinhold, K.J., Lyerly, H.K., Gelmann, E., Blum, R.M., Shearer, G.M., Mitsuya, H., Collins, J.M., Myers, C.E., Klecker, R.W., Markham, P.D., Durack, D.T., Nusinoff-Lehrman, S., Barry, D.W., Fischl, M.A., Gallo, R.C., Bolognesi, D.P., and Broder, S.,** Administration of 3′azido-3′deoxythymidine, an inhibitor of HTLV-III/LAV replication, to patients with AIDS or AIDS-related complex, *Lancet,* i, 575, 1986.

15. **Fischl, M.A., Richman, D.D., Grieco, M.H., Gottlieb, M.S., Volberding, P.A., Laskin, O.L., Leedom, S.M., Groopman, J.E., Mildvan, D., Schooley, R.T., Jackson, G.G., Durack, D.T., King, D.,** and the AZT Collaborative Working Group, The efficacy of azidothymidine (AZT) in the treatment of patients with AIDS and AIDS-related complex: a double-blind, placebo-controlled trial, *N. Engl. J. Med.,* 317, 185, 1987.

16. **Creagh-Kirk, T., Doi, P., Andrews, E., Nusinoff-Lehrman, S., Tilson, H., Hoth, D., and Barry, D.W.,** Survival experience among patients with AIDS receiving zidovudine. Follow-up of patients in a compassionate plea program, *JAMA,* 260, 3009, 1988.

17. **Dournon, E., Rozenbaum, W., Michon, C., Perrone, C., DeTruchis, P., Bouvet, E., Levacher, M., Matheron, S., Gharakhanian, S., Girard, P.M., Salmon, S., Leport, C., Dazza, M.C., Regnier, B.,** and the Claude Bernard Hospital AZT Study Group, Effects of Zidovudine in 365 consecutive patients with AIDS or AIDS-related complex, *Lancet,* ii, 1297, 1988.

18. **Fischl, M.A., Richman, D.D., Causey, D.M., Grieco, M.H., Bryson, Y., Mildvan D., Laskin, O.L., Groopman, J.E., Volberding, P.A., Schooley, R.T., Jackson, G.G., Durack, D.T., Andrews, J.C., Nusinoff-Lehrman, S., Barry, D.W.,** and the AZT Collaborative Working Group, Prolonged zidovudine therapy in patients with AIDS and advanced AIDS-related complex, *JAMA,* 262, 2405, 1989.

19. **Richman, D.D. and Andrews, J.,** Results of continued monitoring of participants in the placebo-controlled trial of zidovudine for serious human immunodeficiency virus infection, *Am. J. Med.,* 85 (Suppl. 2A), 208, 1988.

20. **Stambuk, D., Youle, M., Hawkins, D., Farthing, C., Shanson, D., Farmer, R., Lawrence, A., and Gazzard, B.,** The efficacy and toxicity of azidothymidine (AZT) in the treatment of patients with AIDS and AIDS-related complex (ARC): an open uncontrolled treatment study, *Q. J. Med.,* 70, 161, 1989.

21. **Swanson, C.E. and Cooper, D.A.** (for the Australian Zidovudine Study Group), Factors influencing outcome of treatment with zidovudine of patients with AIDS in Australia, *AIDS,* 4, 749, 1990.

22. **Connolly, K.J. and Hammer, S.M.,** Antiretroviral therapy: reverse transcriptase inhibition, *Antimicrob. Agents Chemother.,* 36, 245, 1992.

23. **Connolly, K.J. and Hammer, S.M.,** Antiretroviral therapy: strategies beyond single-agent reverse transcriptase inhibition, *Antimicrob. Agents Chemother.,* 36, 509, 1992.

24. **Sachs, M.K.,** Antiretroviral chemotherapy of human immunodeficiency virus infections other than with azidothymidine, *Arch. Intern. Med.,* 152, 485, 1992.

25. **Yarchoan, R., Pluda, J.M., Perno, C.-F., Mitsuya, H., and Broder, S.,** Anti-retroviral therapy of human immunodeficiency virus infection: current strategies and challenges for the future, *Blood,* 78, 859, 1991.

26. **Mitsuya, H., Yarchoan, R., and Broder, S.,** Molecular targets for AIDS therapy, *Science,* 249, 1533, 1990.

27. **Jacobo-Molina, A. and Arnold, E.,** HIV reverse transcriptase structure-function relationships, *Biochemistry,* 30, 6351, 1991.

28. **Goff, S.P.,** Retroviral reverse transcriptase: synthesis, structure, and function, *J. AIDS,* 3, 817, 1990.

29. **Rey, M.A., Spire, B., Dormont, D., Barre-Sinoussi, F., Montagnier, L., and Chermann, J.C.,** Characterization of the RNA dependent DNA polymerase of a new human T-lymphotropic retrovirus (lymphadenopathy associated virus), *Biochem. Biophys. Res. Commun.,* 121, 126, 1984.

30. **Hoffman, A.D., Banapour, B., and Levy, J.A.,** Characterization of the AIDS-associated retrovirus reverse transcriptase and optimal conditions for its detection in virions, *Virology,* 147, 326, 1985.

31. **di Marzo-Veronese, F., Copeland, T.D., DeVico, A.L., Rahman, R., Oroszlan, S., Gallo, R.C., and Sarngadharan, M.G.,** Characterization of highly immunogenic p66/p51 as the reverse transcriptase of HTLV-III/LAV, *Science,* 231, 1289, 1986.

32. **Mitsuya, H. and Broder, S.,** Inhibition of the *in vitro* infectivity and cytopathic effect of human T-lymphotropic virus type III/lymphadenopathy-associated virus (HTLV-III/LAV) by 2′,3′dideoxynucleosides, *Proc. Natl. Acad. Sci. U.S.A.,* 83, 1911, 1986.

33. **Mitsuya, H., Weinhold, K.J., Furman, P.A., St. Clair, M.H., Nusinoff-Lehrman, S., Gallo, R.C., Bolognesi, D., Barry, D.W., and Broder, S.,** 3′azido-3′deoxythymidine (BW A509U): an antiviral agent that inhibits the infectivity and cytopathic effect of human T-lymphotropic virus type III/lymphadenopathy-associated virus *in vitro, Proc. Natl. Acad. Sci. U.S.A.,* 82, 7096, 1985.

34. **Pauwels, R., Andries, K., Desmyter, J., Schols, D., Kukla, M.S., Breslin, H.J., Raeymaeckers, A., Van Gelder, J., Woestenborghs, R., Heykants, J., Schellekens, K., Janssen, M.A.C., DeClercq, E., and Jenssen, P.A.J.,** Potent and selective inhibition of HIV-1 replication *in vitro* by a novel series of TIBO derivatives, *Nature (London),* 343, 470, 1990.

35. **Merluzzi, V.J., Hargrave, K.D., Labadia, M., Grozinger, K., Skoog, M., Wu, J.C., Shih, C.K., Eckner, K., Hattox, S., Adams, J., Rosenthal, A.S., Jaanes, R., Eckner, R.J., Koup, R.A., and Sullivan, J.L.,** Inhibition of HIV-1 replication by a nonnucleoside reverse transcriptase inhibitor, *Science,* 250, 1411, 1990.

36. **Goldman, M.E., Nunberg, J.H., O'Brien, J.A., Quintero, J.C., Schleif, W.A., Freund, K.F., Gaul, S.L., Saari, W.S., Wai, J.S., Hoffman, J.M., Anderson, P.S., Hupe, D.J., Emini, E.A., and Stern, A.M.,** Pyridinone derivatives: specific human immuno deficiency virus type I reverse transcriptase inhibitors with antiviral activity, *Proc. Natl. Acad. Sci. U.S.A.,* 88, 6863, 1991.

37. **Baba, M., Shigeta, S., Tanaka, H., Miyasaka, T., Ubasawa, M., Umezu, K., Walker, R.T., Pauwels, R., and DeClercq, E.,** Highly potent and selective inhibition of HIV-1 replication by 6-phenylthiouracil derivatives, *Antiviral Res.,* 17, 245, 1992.

38. **Romero, D.L., Busso, M., Tan, C.K., Reusser, F., Palmer, J.R., Poppe, S.M., Aristoff, P.A., Downey, K.M., So, A.G., Resnick, L., and Tarpley, W.G.,** Nonnucleoside reverse transcriptase inhibitors that potently and specifically block human immunodeficiency virus type 1 replication, *Proc. Natl. Acad. Sci. U.S.A.,* 88, 8806, 1991.

39. **Balzarini, J., Perez-Perez, M.J., San-Felix, A., Schols, D., Perno, C.-F., Vandamme, A.-M., Camarasa, M.-J., and DeClercq, E.,** 2′,5′-Bis-o-(tert-butyldimethylsilyl)-3′-spiro-5″-(4″-amino-1″,2″-oxathiole-2″,2″-dioxide)pyrimidine (TSAO) nucleoside analogs: highly selective inhibitors of human immunodeficiency virus type 1 that are targeted at the viral reverse transcriptase, *Proc. Natl. Acad. Sci. U.S.A.,* 89, 4392, 1992.

40. **Pauwels, R., Andries, K., Debyser, Z., Kukla, M., Schols, D., Desmyter, J., DeClercq, E., and Janssen, P.A.J.,** TIBO derivatives: a new class of highly potent and specific inhibitors of HIV-1 replication, *Biochem. Soc. Trans.,* 20, 509, 1992.

41. **Grob, P.M., Wu, J.C., Cohen, K.A., Ingraham, R.H., Shih, C.-K., Hargrave, K.D., McTague, T.L., and Merluzzi, V.J.,** Nonnucleoside inhibitors of HIV-1 reverse transcriptase: nevirapine as a prototype drug, *AIDS Res. Hum. Retroviruses,* 8, 145, 1992.

42. **Skoog, M.T., Hargrave, K.D., Miglietta, J.J., Kopp, E.B., and Merluzzi, V.J.,** Inhibition of HIV-1 reverse transcriptase and virus replication by a non-nucleoside dipyrido-diazepinone BI-RG-587 (Nevirapine), *Med. Res. Rev.,* 12, 27, 1992.

43. **Dalgleish, A.G., Beverly, P.C., Clapham, P.R., Crawford, D.H., Greaves, M.F., and Weiss, R.A.,** The CD4(T4) antigen is an essential component of the receptor for the AIDS retrovirus, *Nature (London),* 312, 763, 1984.

44. **Klatzmann, D., Champagne, E., Chamaret, S., Gruest, J., Guetard, D., Hercend, T., Gluckman, J.C., and Montagnier, L.,** T-lymphocyte T4 molecule behaves as the receptor for human retrovirus LAV, *Nature (London),* 312, 767, 1984.

45. **Barat, C., Lullien, V., Schatz, O., Keith, G., Nugeyre, M.T., Gruninger-Leitch, F., Barre-Sinoussi, F., LeGrice, S.F.J., and Darlix, J.L.,** HIV-1 reverse transcriptase specifically interacts with the anticodon domain of its cognate primer tRNA, *EMBO J.*, 8, 3279, 1989.

46. **Hansen, J., Schulze, T., Mellert, W., and Moelling, K.,** Identification and characterization of HIV-specific RNase H by monoclonal antibody, *EMBO J.*, 7, 239, 1988.

47. **Brown, P.O., Bowerman, B., Varmus, H.E., and Bishop, J.M.,** Retroviral integration: structure of the initial covalent product and its precursor, and a role for the viral IN protein, *Proc. Natl. Acad. Sci. U.S.A.*, 86, 2525, 1989.

48. **Lightfoote, M.M., Coligan, J.E., Folks, T.M., Fauci, A.S., Martin, M.A., and Venkatesan, S.,** Structural characterization of reverse transcriptase and endonuclease polypeptides of the acquired immunodeficiency syndrome retrovirus, *J. Virol.*, 60, 771, 1986.

49. **Larder, B.A., Kemp, S.D., and Purifoy, D.J.M.,** Infectious potential of human immunodeficiency virus type I reverse transcriptase mutants with altered inhibitor sensitivity, *Proc. Natl. Acad. Sci. U.S.A.*, 86, 4803, 1989.

50. **Horwitz, J.P., Chua, J., and Noel, M.,** Nucleosides v. the monomesylates of 1-(2'-deoxy-β-D-lyxofuranosyl) thymine, *J. Org. Chem.*, 29, 2076, 1964.

51. **Ostertag, W., Roesler, G., Krieg, C.J., Kind, J., Cole, T., Crozier, T., Gaedicke, G., Stinheider, G., Klug, N., and Dube, S.,** Induction of endogenous virus and of thymidine kinase by bromodeoxyuridine in cell cultures transformed by friend virus, *Proc. Natl. Acad. Sci. U.S.A.*, 71, 4948, 1974.

52. **Krieg, C.J., Ostertag, W., Clauss, U., Pragnell, I.B., Swetly, P., Roesler, G., and Weimann, B.J.,** Increase in intracisternal A-type particles in friend cells during inhibition of friend virus (SFFV) release by interferon or azidothymidine, *Exp. Cell Res.*, 116, 21, 1978.

53. **Mitsuya, H., Jarrett, R.F., Matsukura, M., diMarzo-Veronese, F., deVico, A.L., Sarngadharan, M.G., Johns, D.G., Reitz, M.S., and Broder, S.,** Long-term inhibition of human T-lymphotropic virus type III/lymphadenopathy-associated virus (human immunodeficiency virus) DNA synthesis and RNA expression in T cells protected by 2',3'dideoxynucleosides *in vitro, Proc. Natl. Acad. Sci. U.S.A.*, 84, 2033, 1987.

54. **Perno, C.F., Yarchoan, R., Cooney, D.A., Hartman, N.R., Gartner, S., Popovic, M., Hao, Z., Gerrard, T.L., Wilson, Y.A., Johns, D. G., and Broder, S.,** Inhibition of human immunodeficiency virus (HIV-1/HTLV-III Ba-L) replication in fresh and cultured human peripheral blood monocytes/macrophages by azidothymidine and related 2',3'-dideoxynucleosides, *J. Exp. Med.*, 168, 1111, 1988.

55. **Skinner, M.A., Matthews, T.J., Greenwall, T.K., Bolognesi, D.P., and Hebdon, M.,** AZT inhibits HIV-1 replication in monocytes, *J. AIDS*, 1, 162, 1988.

56. **Furman, P.A., Fyfe, J.A., St. Clair, M., Weinhold, K., Rideout, J.L., Freeman, G.A., Nusinoff-Lehrman, S., Bolognesi, D.P., Broder, S., Mitsuya, H., and Barry, D.W.,** Phosphorylation of 3'-azido-3'-deoxythymidine and selective interaction of the 5'-triphosphate with human immunodeficiency virus reverse transcriptase, *Proc. Natl. Acad. Sci. U.S.A.*, 83, 8333, 1986.

57. **Cooney, D.A., Dalal, M., Mitsuya, H., McMahon, S.B., Nadkarni, M., Balzarini, J., Broder, S., and Johns, D.G.,** Initial studies on the cellular pharmacology of 2',3'-dideoxycytidine, an inhibitor of HTLV-III infectivity, *Biochem. Pharmacol.*, 35, 2065, 1986.

58. **Cooney, D.A., Ahluwalia, G., Mitsuya, H., Fridland, A., Johnson, M., Hao, Z., Dalal, M., Balzarini, J., Broder, S., and Johns, D.G.,** Initial studies on the cellular pharmacology of 2',3'-dideoxyadenosine, an inhibitor of HTLV-III infectivity, *Biochem. Pharmacol.*, 36, 1765, 1987.

59. **Ahluwalia, G., Cooney, D.A., Mitsuya, H., Fridland, A., Flora, K.P., Hao, Z., Dalal, M., Broder, S., and Johns, D.G.,** Initial studies on the cellular pharmacology of 2',3'-dideoxyinosine, an inhibitor of HIV infectivity, *Biochem. Pharmacol.*, 36, 3797, 1987.

60. **Hao, Z., Cooney, D.A., Hartman, N.R., Perno, C.F., Fridland, A., deVico, A., Sarngadharan, M.G., Broder, S., and Johns, D.G.,** Factors determining the activity of 2',3'-dideoxynucleosides in suppressing human immunodeficiency virus *in vitro, Mol. Pharmacol.*, 34, 431, 1988.

61. **Starnes, M.C. and Cheng, Y-C.,** Cellular metabolism of 2',3'-dideoxycytidine, a compound active against human immunodeficiency virus *in vitro, J. Biol. Chem.*, 262, 988, 1987.

62. **Balzarini, J., Kang, G.-J., Dalal, M., Herdewijn, P., DeClercq, E., Broder, S., and Johns, D.G.,** The anti-HTLV-III (anti-HIV) and cytotoxic activity of 2',3'-didehydro-2'3'-dideoxynucleosides: a comparison with their parental 2',3'-dideoxynucleosides, *Mol. Pharmacol.*, 32, 162, 1987.

63. **Cheng, Y.-C., Dutschman, G.E., Bastow, K.F., Sarngadharan, M.G., and Ting, R.Y.C.,** Human immunodeficiency virus reverse transcriptase. General properties and its interaction with nucleoside triphosphate analogs, *J. Biol. Chem.*, 262, 2187, 1987.

64. **St. Clair, M.H., Richards, C.A., Spector, T., Weinhold, K.J., Miller, W.H., Langlois, A.J., and Furman, P.A.,** 3'-azido-3'-deoxythymidine triphosphate as an inhibitor and substrate of purified human immunodeficiency virus reverse transcriptase, *Antimicrob. Agents Chemother.*, 31, 1972, 1987.

65. **Kedar, P.S., Abbotts, J., Kovacs, T., Lesiak, K., Torrence, P., and Wilson, S.H.,** Mechanism of HIV reverse-transcriptase-enzyme primer interaction as revealed through studies of a dNTP analog, 3'-azido-dTTP, *Biochemistry*, 29, 3603, 1990.

66. **Reardon, J.E. and Miller, W.H.,** Human immunodeficiency virus reverse transcriptase: substrate and inhibitor kinetics with thymidine 5'-triphosphate and 3'-azido-3'-deoxythymidine 5'-triphosphate, *J. Biol. Chem.*, 265, 20302, 1990.

67. **Muller, B., Restle, T., Reinstein, J., and Goody, R.S.,** Interaction of fluorescently labeled dideoxynucleotides with HIV-1 reverse transcriptase, *Biochemistry,* 30, 3709, 1991.

68. **Reardon, J.E.,** Human immunodeficiency virus reverse transcriptase: steady-state and pre-steady-state kinetics of nucleotide incorporation, *Biochemistry,* 31, 4473, 1992.

69. **White, E.L., Parker, W.B., Macy, L.J., Shaddix, S.C., McCaleb, G., Secrist, J.A., III, Vince, R., and Shannon, W.M.,** Comparison of the effect of carbovir, AZT, and dideoxynucleoside triphosphates on the activity of human immunodeficiency virus reverse transcriptase and selected human polymerases, *Biochem. Biophys. Res. Commun.,* 161, 393, 1989.

70. **Balzarini, J., Hao, Z., Herdewijn, P., Johns, D.G., and DeClercq, E.,** Intracellular metabolism and mechanism of antiretrovirus action of 9-(2-phosphonylmethoxyethyl) adenine, a potent anti-human immunodeficiency virus compound, *Proc. Natl. Acad. Sci. U.S.A.,* 88, 1499, 1991.

71. **Kellam, P., Boucher, C.A.B., and Larder, B.A.,** Fifth mutation in human immunodeficiency virus type 1 reverse transcriptase contributes to the development of high-level resistance to zidovudine, *Proc. Natl. Acad. Sci. U.S.A.,* 89, 1934, 1992.

72. **Larder, B.A., Darby, G., and Richman, D.D.,** HIV with reduced sensitivity to zidovudine (AZT) isolated during prolonged therapy, *Science,* 243, 1731, 1989.

73. **Larder, B.A. and Kemp, S.D.,** Multiple mutations in HIV-1 reverse transcriptase confer high-level resistance to zidovudine (AZT), *Science,* 246, 1155, 1989.

74. **Tan, C.K., Civil, R., Mian, A.M., So, A.G., and Downey, K.M.,** Inhibition of the RNase H activity of HIV reverse transcriptase by azidothymidylate, *Biochemistry,* 30, 4831, 1991.

75. **Kovacs, T., Parkanyi, L., Pelczer, I., Cervantes-Lee, F., Pannell, K.H., and Torrence, P.F.,** Solid-state and solution conformation of 3'-amino-3' deoxythymidine, precursor to a noncompetitive inhibitor of HIV-1 reverse transcriptase, *J. Med. Chem.,* 34, 2595, 1991.

76. **Goody, R.S., Muller, B., and Restle, T.,** Factors contributing to the inhibition of HIV reverse transcriptase by chain-terminating nucleotides *in vitro* and *in vivo, FEBS Lett.,* 291, 1, 1991.

77. **Majumdar, C., Abbotts, J., Broder, S., and Wilson, S.H.,** Studies on the mechanism of human immunodeficiency virus reverse transcriptase, *J. Biol. Chem.,* 263, 15657, 1988.

78. **Kuchta, R.D., Mizrahi, V., Benkovic, P.A., Johnson, K.A., and Benkovic, S.J.,** Kinetic mechanism of DNA polymerase I (Klenow), *Biochemistry,* 26, 8410, 1987.

79. **Patel, S.S., Wong, I., and Johnson, K.A.,** Pre-steady-state kinetic analysis of processive DNA replication including complete characterization of an exonuclease-deficient mutant, *Biochemistry,* 30, 511, 1991.

80. **Huber, H.E., McCoy, J.M., Seehra, J.S., and Richardson, C.C.,** Human immunodeficiency virus 1 reverse transcriptase, *J. Biol. Chem.,* 264, 4669, 1989.

81. **Ma, Q.-F., Bathurst, I.C., Barr, P.J., and Kenyon, G.L.,** The observed inhibitory potency of 3'-azido-3'-deoxythymidine 5'triphosphate for HIV-1 reverse transcriptase depends on the length of the poly(rA) region of the template, *Biochemistry,* 31, 1375, 1992.

82. **Segel, I.H.,** *Enzyme Kinetics: Behavior and Analysis of Rapid Equilibrium and Steady-State Enzyme Systems,* John Wiley & Sons, New York, 1975, 113.

83. **Varmus, H. and Swanstrom, R.,** Replication of retroviruses, in *RNA Tumor Viruses,* 2nd ed., Weis, R., Teich, N., Varmus, H., and Coffin, J., Eds., Cold Spring Harbor Laboratory Press, Cold Spring Harbor, NY, 1984, 369.

83a. **Reardon, J.,** personal communication.

84. **Ji, J. and Loeb, L.A.,** Fidelity of HIV-1 reverse transcriptase copying RNA *in vitro, Biochemistry,* 31, 954, 1992.

85. **Fischl, M.A., Richman, D.D., Hansen, N., Collier, A.C., Carey, J.T., Para, M.F., Hardy, W.D., Dolin, R., Powderly, W.G., Allan, J.D., Wong, B., Merigan, T.C., McAuliffe, V.J., Hyslop, N.E., Rhame, F.S., Balfour, H.H., Jr., Spector, S.A., Volberding, P., Pettinelli, C., Anderson, J.,** and the AIDS Clinical Trials Group, The safety and efficacy of zidovudine (AZT) in the treatment of subjects with mildly symptomatic human immunodeficiency virus type 1 (HIV) infection: a double-blind, placebo-controlled trial, *Ann. Intern. Med.,* 112, 727, 1990.

86. **Volberding, P.A., Lagakos, S.W., Koch, M.A., Pettinelli, C., Myers, M.W., Booth, D.K., Balfour, H.H., Jr., Reichman, R.C., Bartlett, J.A., Hirsch, M.S., Murphy, R.L., Hardy, W.D., Soeiro, R., Fischl, M.A., Bartlett, J.G., Merigan, T.C., Hyslop, N.E., Richmann, D.D., Valentine, F.T., Corey, L.,** and the AIDS Clinical Trials Group of the National Institute of Allergy and Infectious Diseases, Zidovudine in asymptomatic human immunodeficiency virus infection. A controlled trial in persons with fewer than 500 CD4 positive cells per cubic millimeter, *N. Engl. J. Med.,* 322, 941, 1990.

87. **Hamilton, J.D., Hartigan, P.M., Simberkoff, M.S., Day, P.L., Diamond, G.R., Dickinson, G.M., Drusano, G.L., Egorin, M.J., George, W.L., Gordin, F.M., Hawkes, C.A., Jensen, P.C., Klimas, N.G., Labriola, A.M., Lahart, C.J., O'Brien, W.A., Oster, C.N., Weinhold, K.J., Wray, N.P., Zolla-Pazner, S.B.,** and the Veterans Affairs Cooperative Study Group on AIDS Treatment, A controlled trial of early versus late treatment with zidovudine in symptomatic human immunodeficiency virus infection: results of the Veterans Affairs Cooperative Study, *N. Engl. J. Med.,* 326, 437, 1992.

88. **Corey, L. and Fleming, T.R.,** Treatment of HIV infection-progress in perspective, *N. Engl. J. Med.,* 326, 484, 1992.

89. **Graham, N.M., Zeger, S.L., Park, L.P., Vermund, S.H., Detels, R., Rinaldo, C.R., and Phair, J.P.,** The effects on survival of early treatment of human immunodeficiency virus infection, *N. Engl. J. Med.,* 326, 1037, 1992.

90. **Richman, D.D., Fischl, M.A., Grieco, M.H., Gottlieb, M.S., Volberding, P.A., Laskin, O.L., Leedom, J.M., Groopman, J.E., Mildvan, D., Hirsch, M.S., Jackson, G.G., Durack, D.T., and Nusinoff-Lehrman, S.,** Group TACW: the toxicity of azidothymidine (AZT) in the treatment of patients with AIDS and AIDS-related complex: a double-blind, placebo-controlled trial, *N. Engl. J. Med.,* 317, 192, 1987.

91. **Gill, P.S., Rarick, M., Brynes, R.K., Causey, D., Laureiro, C., and Levine, A.M.,** Azidothymidine associated with bone marrow failure in the acquired immunodeficiency syndrome (AIDS), *Ann. Intern. Med.,* 107, 502, 1987.

92. **Dalakas, M.C., Illa, I., Pezeshkpour, G.H., Laukaitis, J.P., Cohen, B., and Griffin, J.L.,** Mitochondrial myopathy caused by long-term zidovudine therapy, *N. Engl. J. Med.,* 322, 1098, 1990.

93. **Arnaudo, E., Dalakas, M., Shanske, S., Moraes, C.T., DiMauro, S., and Schon, E.A.,** Depletion of muscle mitochondrial DNA in AIDS patients with zidovudine-induced myopathy, *Lancet,* 337, 508, 1991.

94. **Simpson, M.V., Chin, C.D., Keilbaugh, S.A., Lin, T.-S., and Prusoff, W.H.,** Studies on the inhibition of mitochondrial DNA replication by 3′-azido-3′-deoxythymidine and other dideoxynucleoside analogs which inhibit HIV-1 replication, *Biochem. Pharmacol.,* 38, 1033, 1989.

95. **Huang, P., Farquhar, D., and Plunkett, W.,** Selective action of 3′-azido-3′-deoxythymidine 5′-triphosphate on viral reverse transcriptases and human DNA polymerases, *J. Biol. Chem.,* 265, 11914, 1990.

96. **Sommadossi, J.-P., Carlisle, R., and Zhou, Z.,** Cellular pharmacology of 3′-azido-3′-deoxythymidine with evidence of incorporation into DNA of human bone marrow cells, *Mol. Pharmacol.,* 36, 9, 1989.

97. **Avramis, V.I., Markson, W., Jackson, R.L., and Gomperts, E.,** Biochemical pharmacology of zidovudine in human T-lymphoblastoid cells (CEM), *AIDS,* 3, 417, 1989.

98. **Fischl, M.A., Parker, C.B., Pettinelli, C., Wulfshon, M., Hirsch, M.S., Collier, A.C., Antoniskis, D., Ho, M., Richman, D.D., Fuchs, E., Merigan, T.C., Reichman, R.C., Gold, J., Steigbigel, N., Leoung, G.S., Rasheed, S., Tsiatis, A.,** and the AIDS Clinical Trials Group, A randomized controlled trial of a reduced daily dose of zidovudine in patients with the acquired immunodeficiency syndrome, *N. Engl. J. Med.,* 323, 1009, 1990.

99. **Collier, A.C., Bozzette, S., Coombs, R.W., Causey, D.M., Schoenfeld, D.A., Spector, S.A., Pettinelli, C.B., Davies, G., Richman, D.D., Leedom, J.M., Kidd, P., and Corey, L.,** A pilot study of low-dose zidovudine in human immunodeficiency virus infection, *N. Engl. J. Med.,* 323, 1015, 1990.

100. **Meng, T.-C., Fischl, M.A., Boota, A.M., Spector, S.A., Bennett, D., Bassiakos, Y., Lai, S., Wright, B., and Richman, D.D.,** Combination therapy with zidovudine and dideoxycytidine in patients with advanced human immunodeficiency virus infection: a phase I/II study, *Ann. Intern. Med.,* 116, 13, 1992.

101. **Fauci, A.S.,** Combination therapy for HIV infection: getting closer, *Ann. Intern. Med.,* 116, 85, 1992.

102. **Fischl, M., Galpin, J.E., Levine, J.D., Groopman, J.E., Henry, D.H., Kennedy, P., Miles, S., Robbins, W., Starrett, B., Zalusky, R., Abels, R.I., Tsai, H.C., and Rudnick, S.A.,** Recombinant human erythropoietin for patients with AIDS treated with zidovudine, *N. Engl. J. Med.,* 322, 1488, 1990.

103. **Groopman, J.E.,** Granulocyte-macrophage colony-stimulating factor in human immunodeficiency virus disease, *Semin. Hematol.,* 27 (Suppl. 3), 8, 14, 1990.

104. **Pluda, J.M., Yarchoan, R., Smith, P.D., McAtee, N., Shay, L.E., Oette, D., Maha, M., Wahl, S.M., Myers, C.E., and Broder, S.,** Subcutaneous recombinant granulocyte-macrophage colony-stimulating factor used as a single agent and in an alternating regimen with azidothymidine in leukopenic patients with severe human immunodeficiency virus infection, *Blood,* 76, 463, 1990.

105. **Boucher, C.A.B., O'Sullivan, E., Mulder, J.W., Ramautarsing, C., Kellam, P., Darby, G., Lange, J.M.A., Goudsmit, J., and Larder, B.A.,** Ordered appearance of zidovudine resistance mutations during treatment of 18 human immunodeficiency virus-positive subjects, *J. Infect. Dis.,* 165, 105, 1992.

106. **Boucher, C.A.B., Tersmette, M., Lange, J.M.A., Kellam, P., DeGoede, R.E.Y., Mulder, J.W., Darby, G., Goudsmit, J., and Larder, B.A.,** Zidovudine sensitivity of human immunodeficiency viruses from high-risk, symptom-free individuals during therapy, *Lancet,* 336, 585, 1990.

107. **Japour, A.J., Chatis, P.A., Eigenrauch, H.A., and Crumpacker, C.S.,** Detection of human immunodeficiency virus type 1 clinical isolates with reduced sensitivity to zidovudine and dideoxyinosine by RNA-RNA hybridization, *Proc. Natl. Acad. Sci. U.S.A.,* 88, 3092, 1991.

108. **Richman, D.D., Grimes, J.M., and Lagakos, S.W.,** Effect of stage of disease and drug dose on zidovudine susceptibilities of isolates of human immunodeficiency virus, *J. AIDS,* 3, 743, 1990.

109. **DeClercq, E.,** HIV inhibitors targeted at the reverse transcriptase, *AIDS Res. Hum. Retroviruses,* 8, 119, 1992.

110. **Matthews, S.J., Cersosimo, R.J., and Spivack, M.L.,** Zidovudine and other reverse transcriptase inhibitors in the management of human immunodeficiency virus-related disease, *Pharmacotherapy,* 11, 419, 1991.

111. **Johnson, M.A. and Fridland, A.,** Phosphorylation of 2′-3′dideoxyinosine by cytosolic 5′-nucleotidase of human lymphoid cells, *Mol. Pharmacol.,* 36, 291, 1989.

112. **Connolly, K.J., Allan, J.D., Fitch, H., Jackson-Pope, L., McLaren, C., Canetta, R., and Groopman, J.,** Phase I study of 2′,3′dideoxyinosine administered orally twice daily to patients with AIDS or AIDS-related complex and hematologic intolerance to zidovudine, *Am. J. Med.,* 91, 471, 1991.

113. **Yarchoan, R., Pluda, J.M., Thomas, R.V., Mitsuya, H., Brouwers, P., Wyvill, K.M., Hartman, N., Johns, D.G., and Broder, S.,** Long-term toxicity/activity profile of 2′,3′-dideoxyinosine in AIDS or AIDS-related complex, *Lancet,* 336, 526, 1990.

114. **Cooley, T.P., Kunches, L.M., Saunders, C.A., Ritter, J.K., Perkins, C.J., McLaren, C., McCaffrey, R.P., and Liebman, H.A.,** Once-daily administration of 2′,3′dideoxyinosine (ddI) in patients with the acquired immunodeficiency syndrome or AIDS-related complex, *N. Engl. J. Med.,* 322, 1340, 1990.

115. **Lambert, J.S., Seidlin, M., Reichman, R.C., Plank, C.S., Laverty, M., Morse, G.D., Knupp, C., McLaren, C., Pettinelli, C., Valentine, F.T., and Dolin, R.,** 2′,3′-dideoxyinosine (ddI) in patients with the acquired immunodeficiency syndrome or AIDS-related complex: a phase I trial, *N. Engl. J. Med.,* 322, 1333, 1990.

116. **St. Clair, M.H., Martin, J.L., Tudor-Williams, G., Bach, M.C., Vavro, C.L., King, D.M., Kellam, P., Kemp, S.D., and Larder, B.A.,** Resistance to ddI and sensitivity to AZT induced by a mutation in HIV-1 reverse transcriptase, *Science,* 253, 1557, 1991.

117. **Yarchoan, R., Thomas, R.V., Allain, J.P., McAtee, N., Dubinsky, R., Mitsuya, H., Lawley, T.J., Safai, B., Myers, C., Perno, C., Klecker, R., Wills, R., Fischl, M., McNeely, M., Pluda, J., Leuther, M., Collins, J., and Broder, S.,** Phase I studies of 2′,3′-dideoxycytidine in severe human immunodeficiency virus infection as a single agent and alternating with zidovudine, *Lancet,* I, 76, 1988.

118. **Merigan, T.C., Skowron, G., Bozzette, S.A., Richman, D., Uttamchandani, R., Fischl, M., Schooley, R., Hirsch, M., Soo, W., Pettinelli, C., Schaumburg, H.,** and the ddC Study Group of the AIDS Clinical Trials Group, Circulating p24 antigen levels and responses to dideoxycytidine in human immunodeficiency virus (HIV) infections. A phase I and II study, *Ann. Intern. Med.,* 110, 189, 1989.

119. **Hirschmann, R.,** Medicinal chemistry in the golden age of biology: lessons from steroid and peptide research, *Angew. Chem. Int. Ed. Engl.,* 30, 1278, 1991.

120. **Kukla, M.J., Breslin, H.J., Pauwels, R., Fedde, C.L., Miranda, M., Scott, M.K., Sherrill, R.G., Raeymaekers, A., Van Gelder, J., Andries, K., Janssen, M.A.C., DeClercq, E., and Janssen, P.A.J.,** Synthesis and anti-HIV-1 activity of 4,5,6,7-tetrahydro-5-methylimidazo[4,5,1-jk]-[1,4]-benzodiazepin-2-(1H)-one (TIBO) derivatives, *J. Med. Chem.,* 34, 746, 1991.

121. **Kukla, M.J., Breslin, H.J., Diamond, C.J., Grous, P.P., Ho, C.Y., Miranda, M., Rodgers, J.D., Sherrill, R.G., DeClercq, E., Pauwels, R., Andries, K., Moens, L.J., Janssen, M.A.C., and Janssen, P.A.J.,** Synthesis and anti-HIV-1 activity of 4,5,6,7-tetrahydro-5-methylimidazo[4,5,1-jk]-[1,4]-benzodiazepin-2-(1H)-one (TIBO) derivatives. II, *J. Med. Chem.,* 34, 3187, 1991.

122. **Parker, K.A. and Coburn, C.A.,** Regioselectivity in intramolecular nucleophilic aromatic substitution. Synthesis of the potent anti HIV-1 8 halo TIBO analogs, *J. Org. Chem.,* 57, 97, 1992.

123. **Debyser, Z., Pauwels, R., Andries, K., Desmyter, J., Kukla, M.J., Janssen, P.A.J., and DeClercq, E.,** An antiviral target on reverse transcriptase of human immunodeficiency virus type 1 revealed by tetrahydroimidazo-[4,5,1-jk]-[1,4]-benzodiazepin-2-(1H)-one and -thione derivatives, *Proc. Natl. Acad. Sci. U.S.A.,* 88, 1451, 1991.

124. **Frank, K.B., Noll, G.J., Connell, E.V., and Sim, I.S.,** Kinetic interaction of human immunodeficiency virus type-1 reverse transcriptase with the antiviral tetrahydroimidazo [4,5,1-jk]-[1,4]-benzodiazepine-2-(1H)-thione compound, R82150, *J. Biol. Chem.,* 266, 14232, 1991.

125. **Hopkins, S., Furman, P.A., and Painter, G.R.,** Investigation of the stereochemical course of DNA synthesis catalyzed by human immunodeficiency virus type-1 reverse transcriptase, *Biophys. Biochem. Res. Commun.,* 163, 106, 1989.

126. **Hargrave, K.D., Proudfoot, J.R., Grozinger, K.G., Cullen, E., Kapadia, J.R., Patel, U.R., Fuchs, V.U., Mauldin, S.C., Vitous, J., Behnke, M.L., Klunder, J.M., Pal, K., Skiles, J.W., McNeil, D.W., Rose, J.M., Chow, G.C., Skoog, M.T., Wu, J.C., Schmidt, G., Engel, W., Eberlein, W.G., Saboe, T.D., Campbell, S.J., Rosenthal, A.S., and Adams, J.,** Novel non-nucleoside inhibitors of HIV-1 reverse transcriptase. I. Tricyclic pyridobenzo- and dipyridodiazepinones, *J. Med. Chem.,* 34, 2231, 1991.

127. **Saari, W.S., Hoffman, J.M., Wai, J.S., Fisher, T.E., Rooney, C.S., Smith, A.M., Thomas, C.M., Goldman, M.E., O'Brien, J.A., Nunberg, J.H., Quintero, J.C., Schleif, W.A., Emini, E.A., Stern, A.M., and Anderson, P.S.,** 2-pyridinone derivatives: a new class of nonnucleoside, HIV-1 specific reverse transcriptase inhibitors, *J. Med. Chem.,* 34, 2922, 1991.

128. **Goldman, M.E., O'Brien, J.A., Ruffing, T.L., Nunberg, J.H., Schleif, W.A., Quintero, J.C., Siegl, P.K.S., Hoffman, J.M., Smith, A.M., and Emini, E.A.,** L-696,229 specifically inhibits human immunodeficiency virus type 1 reverse transcriptase and possesses antiviral activity *in vitro, Antimicrob. Agents Chemother.,* 36, 1019, 1992.

129. **Nunberg, J.H., Schleif, W.A., Boots, E.J., O'Brien, J.A., Quintero, J.C., Hoffman, J.M., Emini, E.A., and Goldman, M.E.,** Viral resistance to human immunodeficiency virus type 1-specific pyridinone reverse transcriptase inhibitors, *J. Virol.,* 65, 4887, 1991.

130. **Richman, D., Shih, C.-K., Lowy, I., Rose, J., Prodanovich, P., Goff, S., and Griffin, J.,** HIV-1 mutants resistant to non-nucleoside inhibitors of reverse transcriptase arise in tissue culture, *Proc. Natl. Acad. Sci. U.S.A.,* 88, 11241, 1991.

131. **Dueweke, T.J., Kezdy, F., Waszak, G.A., Deibel, M.R., Jr., and Tarpley, W.G.,** The binding of a novel bishetero-arylpiperazine mediates inhibition of human immunodeficiency virus type-1 reverse transcriptase, *J. Biol. Chem.,* 267, 27, 1992.

132. **Wu, J.C., Warren, T.C., Adams, J., Proudfoot, J., Skiles, J., Raghavan, P., Perry, C., Potoki, I., Farina, P.R., and Grob, P.M.,** A novel dipyridodiazepinone inhibitor of HIV-1 reverse transcriptase acts through a nonsubstrate binding site, *Biochemistry,* 30, 2022, 1991.

133. **Tramontano, E. and Cheng, Y.-C.,** HIV-1 reverse transcriptase inhibition by a dipyridodiazepinone derivative: BI-RG-587, *Biochem. Pharmacol.,* 43, 1371, 1992.

134. **LeGrice, S.F.J., Naas, T., Wohlgensinger, B., and Schatz, O.,** Subunit-selective mutagenesis indicates minimal polymerase activity in heterodimer-associated p51 HIV-1 reverse transcriptase, *EMBO J.,* 10, 3905, 1991.

135. **Shih, C.-K., Rose, J.M., Hansen, G.L., Wu, J.C., Bacolla, A., and Griffin, J.A.,** Chimeric human immunodeficiency virus type 1/type 2 reverse transcriptases display reversed sensitivity to nonnucleoside analog inhibitors, *Proc. Natl. Acad. Sci. U.S.A.,* 88, 9878, 1991.

136. **Shaharabany, M. and Hizi, A.,** The catalytic functions of chimeric reverse transcriptases of human immunodeficiency viruses type 1 and type 2, *J. Biol. Chem.,* 267, 3674, 1992.

137. **Condra, J.H., Emini, E.A., Gotlib, L., Graham, D.J., Schlabach, A.J., Wolfgang, J.A., Colonno, R.J., and Sardana, V.V.,** Identification of the human immunodeficiency virus reverse transcriptase residues that contribute to the activity of diverse nonnucleoside inhibitors, *Antimicrob. Agents Chemother.,* 36, 1441, 1992.

138. **Lowe, D.M., Parmar, V., Kemp, S.D., and Larder, B.A.,** Mutational analysis of two conserved sequence motifs in HIV-1 reverse transcriptase, *FEBS Lett.,* 282, 231, 1991.

139. **Delarue, M., Poch, O., Tordo, N., Moras, D., and Argos, P.,** An attempt to unify the structures of polymerases, *Protein Eng.,* 3, 461, 1990.

140. **Kohlstaedt, L.A., Wang, J., Friedman, J.M., Rice, P.A., and Steitz, T.A.,** Crystal structure at 3.5 Å resolution of HIV-1 reverse transcriptase complexed with an inhibitor, *Science,* 256, 1783, 1992.

141. **Balzarini, J., Perez-Perez, M.-J., San-Felix, A., Velazquez, S., Camarasa, M.-J., and DeClercq, E.,** [2′,5′-Bis-O-(tert-butyldimethylsilyl)]3′-spiro-5″-(4″ amino-1″,2″-oxathiole-2″,2″-dioxide) (TSAO) derivatives of purine and pyrimidine nucleosides as potent and selective inhibitors of human immunodeficiency virus type 1, *Antimicrob. Agents Chemother.,* 36, 1073, 1992.

142. **White, E.L., Buckheit, R.W., Jr., Ross, L.J., Germany, J.M., Andries, K., Pauwels, R., Janssen, P.A.J., Shannon, W.M., and Chirigos, M.A.,** A TIBO derivative, R82913, is a potent inhibitor of HIV-1 reverse transcriptase with hetero-polymer templates, *Antiviral Res.,* 16, 257, 1991.

143. **Geissler, R.G., Ganser, A., Ottman, O.G., Kojouharoff, G., Reutzel, P., Andries, K., Schellekens, K., and Hoelzer, D.,** TIBO R82913 A new HIV-1 inhibiting agent does not inhibit hematopoietic progenitor cells, *AIDS Res. Hum. Retroviruses,* 7, 1021, 1991.

144. **Pialoux, G., Youle, M., DuPont, B., Gazzard, B., Cauwenbergh, G.F.M.J., Stoffels, P.A.M., Davies, S., De Saint Martin, J., and Janssen, P.A.J.,** Pharmacokinetics of R82913 in patients with AIDS or AIDS-related complex, *Lancet,* 338, 140, 1991.

145. **Baba, M., Pauwels, R., Balzarini, J., Herdewijn, P., DeClercq, E., and Desmyter, J.,** Ribavirin antagonizes inhibitory effects of pyrimidine 2′,3′-dideoxynucleosides but enhances inhibitory effects of purine 2′,3′-dideoxynucleosides on replication of human immunodeficiency virus *in vitro, Antimicrob. Agents Chemother.,* 31, 1613, 1987.

146. **Tornevik, Y. and Eriksson, S.,** 2′,3′-dideoxycytidine toxicity in cultured human CEM T lymphoblasts: effects of combination with 3′-azido-3′deoxythymidine and thymidine, *Mol. Pharmacol.,* 38, 237, 1990.

147. **Eron, J.J., Hirsch, M.S., Merrill, D.P., Chou, T.-C., and Johnson, V.A.,** Synergistic inhibition of HIV-1 by the combination of zidovudine (AZT) and 2′,3′-dideoxycytidine (ddc) *in vitro,* Seventh Int. Conf. on Aids. Florence, Italy, June 16–21, 1991 (Abstract W.B., 2116).

148. **Larder, B.A., Chesebro, B., and Richman, D.D.,** Susceptibilities of zidovudine-susceptible and -resistant human immunodeficiency virus isolates to antiviral agents determined by using a quantitative plaque reduction assay, *Antimicrob. Agents Chemother.,* 34, 436, 1990.

149. **Skowron, G. and Merigan, T.C.,** Alternating and intermittent regimens of zidovudine (3′-azido-3′ deoxythymidine) and dideoxycytidine (2′,3′-dideoxycytidine) in the treatment of patients with acquired immunodeficiency syndrome (AIDS) and AIDS-related complex, *Am. J. Med.,* 88 (Suppl. 5B), 20S, 1990.

150. **Jacobson, M.A., van der Horst, C., Causey, D.M., Dehlinger, M., Hafner, R., and Mills, J.,** *In vivo* additive antiretroviral effect of combined zidovudine and foscarnet therapy for human immunodeficiency virus infection (ACTG protocol 053), *J. Infect. Dis.,* 163, 1219, 1991.

151. **Richman, D.D., Rosenthal, A.S., Skoog, M., Eckner, R.J., Chou, T.C., Sabo, J.P., and Merluzzi, V.J.,** BI-RG-587 is active against zidovudine-resistant human immunodeficiency virus type 1 and synergistic with zidovudine, *Antimicrob. Agents Chemother.,* 35, 305, 1991.

152. **Heidenreich, O., Kruhoffer, M., Grosse, F., and Eckstein, F.,** Inhibition of human immunodeficiency virus 1 reverse transcriptase by 3′-azidothymidine triphosphate, *Eur. J. Biochem.,* 192, 621, 1990.

153. **Cohen, K.A., Hopkins, J., Ingraham, R.H., Pargellis, C., Wu, J.C., Palladino, D.E.H., Kinkade, P., Warren, T.C., Rogers, S., Adams, J., Farina, P.R., and Grob, P.M.,** Characterization of the binding-site for nevirapine (BI-RG-587) a nonnucleoside inhibitor of human immunodeficiency virus type 1 reverse transcriptase, *J. Biol. Chem.,* 266, 14670, 1991.

154. **Goldman, M.E., O'Brien, J.A., Ruffing, T.L., Schleif, W.A., Sardana, V.V., Condra, J.H., Hoffman, J.M., and Emini, E.A.,** in preparation.

155. **Freund, K. and Stern, A.,** unpublished observations.

Chapter 3

THE DEVELOPMENT OF POTENT ANGIOTENSIN II RECEPTOR ANTAGONISTS USING A NOVEL PEPTIDE PHARMACOPHORE MODEL

Richard M. Keenan, Joseph Weinstock, Judith C. Hempel, James M. Samanen, David T. Hill, Nambi Aiyar, Eliot H. Ohlstein, and Richard M. Edwards

CONTENTS

I. INTRODUCTION

A. THE RENIN-ANGIOTENSIN SYSTEM

The renin-angiotensin system (Figure 1) plays a key role in the regulation of blood pressure, fluid and electrolyte balance, and blood volume. In the initial step of the biochemical cascade, the

angiotensinogen

| RENIN

angiotensin I

| ACE

angiotensin II

FIGURE 1.   The renin-angiotensin system.

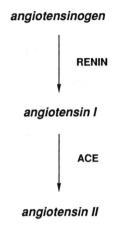

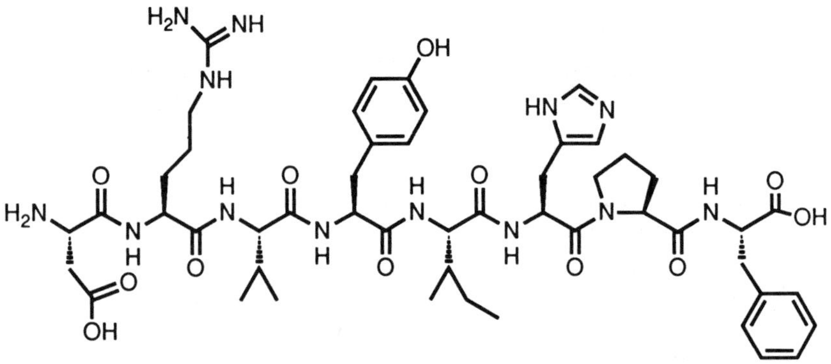

FIGURE 2.   Structure of the octapeptide angiotensin II (H-Asp[1]-Arg[2]-Val[3]-Tyr[4]-Ile[5]-His[6]-Pro[7]-Phe[8]-OH).

proteolytic enzyme renin catalyzes the breakdown of the high-molecular-weight peptide, angiotensinogen, to angiotensin I. A second enzyme, angiotensin-converting enzyme or ACE, causes the further breakdown of angiotensin I into angiotensin II (AII), which is the pressor hormone responsible for much of the cardiovascular effects of the renin-angiotensin system. The pharmacology of AII has recently been reviewed in depth.[1] In short, AII can cause and sustain hypertension by a variety of mechanisms, including acting directly on the blood vessels to cause vasoconstriction, exerting an effect on the kidney leading to salt and water retention, and stimulating the synthesis of the potent mineralocorticoid aldosterone in the adrenal gland. In addition, AII also possesses significant activity in the central nervous system.

The octapeptide AII (Figure 2) is the most prominent active member of a series of angiotensin peptides. Its decapeptide precursor angiotensin I (AI), containing the additional two amino acids His and Leu at the C-terminus, is relatively inactive. The desaminoterminal aspartic acid heptapeptide angiotensin III (AIII) has activity similar to that of AII in many areas of the body. Further truncation of the amino-terminus leads to the hexapeptide AII(3–8), or AIV, which has recently been reported to be an endogenous renal vasodilator[2] and have psychoactive properties.[3] The des-carboxyterminal phenylalanine analog AII(1–7) may play a major role in the brain and heart. It has been reported to stimulate prostaglandin synthesis and vasopressin release and cause neuronal excitation.[4]

AII exerts its effects via interaction with specific cellular receptors. Two major AII receptor subtypes have been identified: AT-1 which appears to be responsible for most of the cardiovascular effects of AII, and AT-2, which is found in the brain and reproductive system. Little is known about the consequences of AT-2 stimulation at this time.[5] By the use of molecular biological techniques, two distinct AT-1 receptors have been isolated from the rat, which are now designated AT-1a and AT-1b. These receptors appear to have seven helical transmembrane domains typical of other G-protein-linked receptors.[6] The heptapeptide AIII also interacts at the AT-1 and AT-2 receptors, whereas the actions of AIV appear to be mediated via a distinct cellular receptor.

B. TARGETS FOR THERAPEUTIC INTERVENTION IN THE RENIN-ANGIOTENSIN SYSTEM

A number of approaches have been investigated to inhibit the renin-angiotensin system as a means to treat hypertension and other cardiovascular disorders. ACE inhibitor therapy, which causes a decrease in the biosynthesis of AII, has become a widely used strategy to control hypertension. Two ACE inhibitors, captopril and enalapril, are the leading antihypertensive drugs used currently in clinical practice. When used as a single agent, only about half the patients respond satisfactorily; but when used in conjunction with a diuretic, patient response increases to over 80%.[1] ACE inhibitors have also been shown to be useful in congestive heart failure, and their utility is also being investigated in renal failure. In general, the ACE inhibitors are very effective drugs with relatively few side effects. However, the nonspecific hydrolase activity of ACE presents a potential drawback to ACE inhibitor therapy. For example, ACE participates in the degradation of the peptide analgesic enkephalins as well as the vasodepressor and inflammatory nonapeptide bradykinin. Elevated levels of bradykinin may cause some of the unwanted side effects such as dry cough and angioedema associated with the use of ACE inhibitors.[7]

Inhibition of renin, an enzyme with more stringent requirements for substrate recognition, has also been studied as a strategy to interfere with the renin-angiotensin system.[8] Although considerable progress has been made in the development of **potent renin inhibitors**, these agents generally suffer from low oral bioavailability and short duration of action. Both ACE inhibitors and renin inhibitors exert their beneficial antihypertensive effect by lowering the plasma concentration of AII.

Ideally, the most direct approach to interfere with the renin-angiotensin system would be to inhibit the binding of the effector hormone AII to its receptor. Competitive AII receptor antagonists would be expected to have similar therapeutic effects and indications as the ACE inhibitors, but might lack the unwanted side effects associated with inhibition of other ACE-mediated pathways. The synthesis of numerous peptide analogs of AII led to the discovery of potent peptide antagonists.[1] The single replacement of the aromatic L-phenylalanine at position eight of AII by an aliphatic L-alanine or L-isoleucine led to antagonist activity. Further replacement of Asp[1] by sarcosine (*N*-methyl-glycine) caused an enhancement of binding and imparted resistance to aminopeptidases responsible for the degradation of AII. [Sar[1], Ala[8]]AII is the classical AII antagonist saralasin, which displays potent and selective AII receptor antagonist activity both *in vitro* and *in vivo*. However, peptide analogs such as saralasin had limited therapeutic utility due to partial agonist activity, short duration of action, and lack of appreciable oral bioavailability.

The design of nonpeptide AII receptor antagonists to overcome the inherent liabilities of peptide therapeutics has presented a significant challenge to researchers in the field. Only recently, almost 20 years after the initial discovery of saralasin, have potent nonpeptide AII antagonists with significant oral activity been developed. A weakly potent nonpeptide antagonist, CV-2947 (**1**), an imidazole-acetic acid derived from similar compounds being studied as diuretics, was reported in the patent literature.[9] Research on improving the activity of **1** led to more potent AII receptor antagonists. Prototypical of the newer agents are SK&F 108566 (**2**)[10] and losartan (DuP 753) (**3**),[11] which retain the basic imidazole core found in the structure of **1**. Herein, we elaborate on the development of imidazole-5-acrylic acids such as SK&F 108566 as a novel class of highly potent and specific nonpeptide AII receptor antagonists. This research included the use of an overlay hypothesis of **1** on a unique pharmacophore model of AII to assist in small molecule optimization. The success of this endeavor demonstrates how such a comparison of a native peptide with nonpeptide ligands can be a fruitful component of the peptidomimetic drug design process.

1   2   3

II. THE OVERLAY HYPOTHESIS

A. OVERLAY HYPOTHESIS STRATEGY

In using an overlay hypothesis in peptidomimetic drug design, it is assumed that a small molecule will bind to the receptor with an affinity that is dependent on the closeness of the mimicry with strategic portions of the peptide which define a receptor pharmacophore. Thus, improvements in affinity in a series of small molecules are derived from increasing the overall similarity to a three-dimensional array of key binding elements on the peptide. As a general rule, attempting to mimic groups on the peptide such as charged groups or aromatic rings prone to be involved in strong binding interactions, and introducing these groups into the small molecule in a conformationally defined manner will be an effective strategy to rapidly evaluate an overlay hypothesis and potentially increase affinity.

In practice, the process of employing an overlay hypothesis strategy in drug design consists of deriving a pharmacophore model of the peptide ligand, identifying a small molecule lead through screening or *de novo* design and synthesis, equating groups in the small molecule with groups in the peptide ligand, and visualizing an overlay by molecular modeling. Based on the principles outlined above, new small molecules designed to overlay the peptide pharmacophore model are synthesized and tested. Any observed increases in affinity tend to lend support to the original overlay, whereas decreased affinity requires modification of the hypothesis. At any point, it may be necessary to consider alternate pharmacophore models of the peptide to rationalize increases in small molecule affinity and aid in the generation of novel overlay hypotheses. The iterative process ideally could drive the evolution of more potent compounds to the point where affinity or potency is no longer the issue, but where practical drug factors such as oral bioavailability, duration of action, toxicity, and ease of synthesis become limiting.

Thus, in order to use an overlay hypothesis, a useful pharmacophore model of the peptide ligand must be available. However, in peptides containing more than a few amino acid residues, an enormous number of low-energy conformations are accessible in solution. In the peptide-receptor complex, the receptor may induce a conformation not particularly favorable for the peptide in solution, adding further complexity to the problem. The synthesis of peptide analogs, especially those containing conformational constraints, is an invaluable component of the peptidomimetic drug design process. These analogs not only provide important structure-activity relationships (SAR) in identifying critical binding groups on the peptide, but also may help determine peptide backbone angles involved in a possible receptor-bound conformation. As opposed to peptides, simple small molecules generally have more limited conformational freedom. Overlaying a relatively rigid small molecule on a larger peptide can also serve to narrow down the number of possible useful conformations of the peptide. Therefore, even a weakly potent small molecule proves extremely valuable as a starting point when using an overlay hypothesis in drug design. The report of the

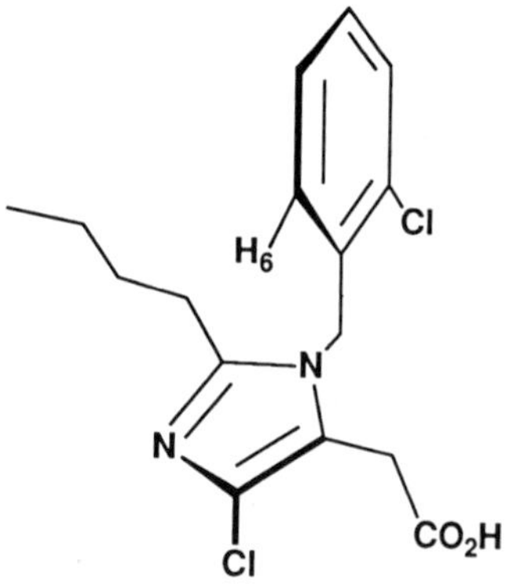

FIGURE 3. Benzylimidazole **1** depicting the orthogonal relationship of the phenyl and imidazole rings.

benzylimidazole **1** sparked our interest in the application of an overlay hypothesis strategy to aid in the design of nonpeptide angiotensin II antagonists.

B. PRELIMINARY COMPARISON OF THE NONPEPTIDE WITH AII

1. Analysis of the Initial Nonpeptide CV 2947

Pharmacological evaluation of the benzylimidazole CV 2947 (**1**) showed it to be a specific, competitive AII receptor antagonist which displayed modest levels of both *in vitro* and *in vivo* activity.[10] Our strategy to generate small molecules with improved activity involved modification of **1** to more closely resemble AII based on an overlay hypothesis. Therefore, **1** was studied to determine a conformation which would be useful for overlay comparisons and which might help generate a pharmacophore model of AII. A striking feature of the nuclear magnetic resonance (NMR) spectrum of **1** was a signal for the 6-hydrogen of the phenyl ring at 6.5 ppm, which is shifted upfield from the other aromatic hydrogens at 7.25 ppm by the electron cloud of the imidazole ring. Thus, the phenyl and imidazole rings appeared to align orthogonally in a low-energy solution conformation of **1,** as shown in Figure 3. Energy minimization studies[12] on **1** supported this conformation and suggested that the phenyl is in a plane approximately coplanar with the dipole of the imidazole ring defined by the N-C-N bonds. The conformationally restricted compounds **4** and **5,** in which a methyl group was placed systematically on the benzyl and acetic acid methylenes, were synthesized to explore the importance of the orthogonal relationship for AII binding.[13] In **4** (NMR data determined on the ethyl ester), the peak for the 6-hydrogen of the phenyl ring appeared at 6.93 ppm which is 0.5 ppm downfield from the corresponding peak for **6**, suggesting that it is less influenced by the electron cloud of the imidazole. Conversely, the peak for the 6-hydrogen of the phenyl in **5** appeared at 6.4 ppm, suggesting that the phenyl-imidazole relationship is similar to that of **6** and, thus, **1**. The binding IC_{50} for **4, 5**, and **6** are 880, 54, and 38 μM, respectively, indicating that the orthogonal relationship of the phenyl and imidazole rings might be associated with enhanced binding potency. On the other hand, the low affinity of **4** could be due to a negative steric interaction of the methyl group with the receptor.

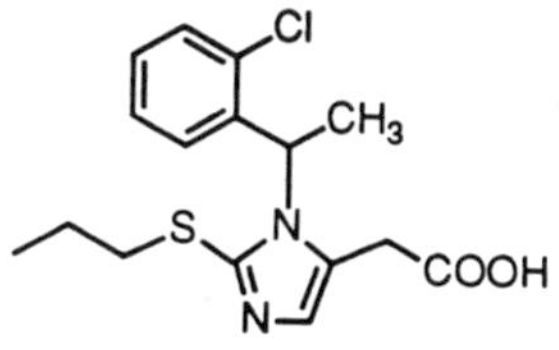

4 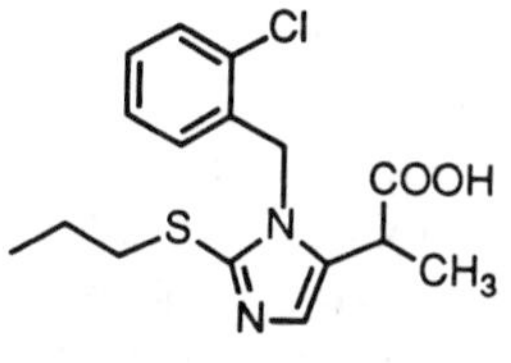   **5** 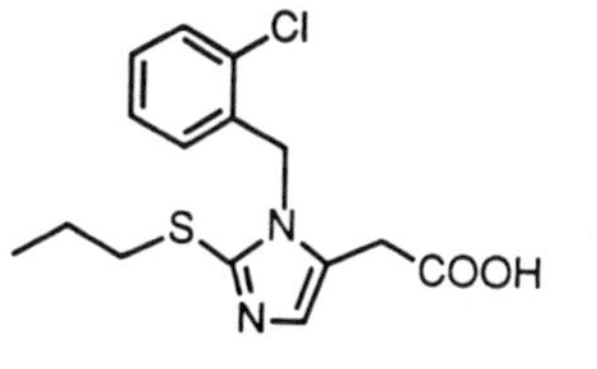   **6**

2. Initial Peptide Models of Angiotensin II

A pharmacophore model of AII was required in order to use an overlay hypothesis strategy in the design of more potent small molecule receptor antagonists. Among the many available in the literature, two were utilized in our initial modeling studies. The first, a statistical model derived from NMR coupling constants,[14] was modified to be consistent with constrained AII analog SAR. Thus, the SAR of numerous peptide analogs containing conformationally constrained replacements for key amino acid residues in the octapeptide sequence had been reinvestigated.[15] Examination of those constrained analogs which maintained activity generated a set of allowed phi/psi angles for important backbone residues in AII which were then incorporated into the literature model.[16] In our initial work, we also utilized a solution conformation of AII described by Smeby and Fermandjian[17a] which had been derived from physical chemical data and peptide SAR.

3. Preliminary Comparison of the Nonpeptide with AII

A preliminary comparison of **1** and AII revealed that the benzylimidazole and the octapeptide shared a number of common structural elements for potential overlap. For example, each molecule contained an imidazole ring: the core imidazole ring in **1** and the imidazole ring found at His[6] in AII. Also, the carboxylic acid of **1** could align with either Asp[1] or the C-terminal carboxylate of AII and the *N*-benzyl ring of **1** could in theory mimic any of three aromatic rings in AII: Tyr[4], His[6], or Phe[8]. Likewise, the lipophilic butyl sidechain of **1** might overlay any of three hydrophobic residues in AII: Val[3], Ile[5], or Pro[7]. Finally, **1** could either mimic a conformation of octapeptide agonists such as AII, but lack the necessary size or critical structural features to elicit agonist activity, or overlay structurally related octapeptide receptor antagonists such as saralasin which may bind to the receptor in a different conformation. Thus, a large number of overlay hypotheses could in theory be generated for the small molecule **1** and AII.

Careful examination of existing AII peptide SAR helped refine our overlay hypothesis strategy and focus our synthetic efforts. The literature suggested that Arg[2], Tyr[4], His[6], and Phe[8] appeared to be the functionally important amino acids, whereas the hydrophobic residues Val[3], Ile[5], and Pro[7] played more of a role in affecting the conformation of the peptide than in binding.[17] At the six position in AII, there was a marked preference for sidechains with unsubstituted five-membered heterocyclic rings, discounting a potential overlay of either the tetra-substituted imidazole ring or *N*-benzyl of **1** at this position. In addition, the retention of activity observed with residue 2–8 heptapeptides (i.e., AIII) and Sar[1]-substituted analogs of AII,[17] both of which lack the N-terminal Asp residue, favored the C-terminal carboxylate as the more likely region of potential overlay for the carboxylic acid of **1**. In fact, the C-terminal region of AII had been identified as crucial for both receptor recognition and activation.[17,18] For example, the single replacement of L-Phe at position eight with D-Phe or aliphatic amino acids converted potent agonists to potent antagonists. Bearing this peptide SAR in mind, we concentrated our synthetic efforts on the further examination of a particularly intriguing initial overlay hypotheses involving the overlay of the *N*-benzyl ring of **1** with the Phe[8] aromatic ring of AII.

C. INITIAL OVERLAY HYPOTHESES

1. The Phe[8] Overlay Hypothesis

In addition to the peptide data, other evidence pointed to a possible overlay of the small molecule with position eight of AII. For example, there were similarities in the SAR for the C-terminal carboxylate of AII and the acetic acid in **1**. Thus, esterification or reduction of this acid in AII caused a loss of activity,[19] as was observed for the acetic acid in the imidazole series.[9b,20] According to a Phe[8] overlay hypothesis,[21] the acid and aromatic ring of the benzylimidazole would align with the corresponding elements of Phe[8]. In addition, the butyl chain could align with the hydrophobic Pro[7] region and the imidazole ring of the small molecule may mimic the Pro[7]-Phe[8] peptide bond. An overlay of **1** on the C-terminal region of AII is depicted in Figure 4. Consideration of the Phe[8]

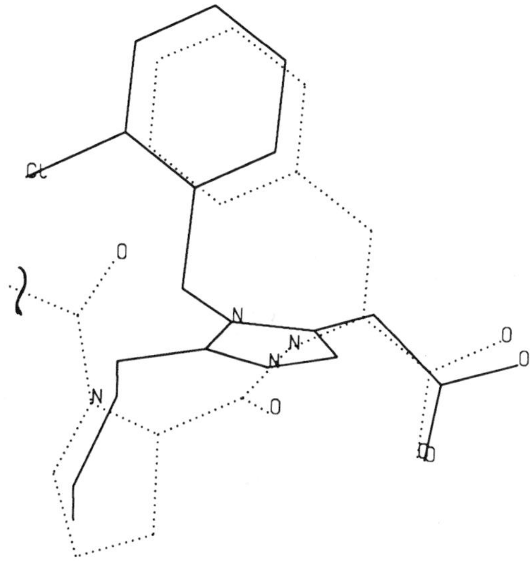

FIGURE 4. Overlay of **1** (solid) on the Pro[7]-Phe[8] C-terminal region of AII (dotted).

overlay theory led to the design of compounds such as **7** and **8**, in which the benzyl group was moved over onto the acetic acid sidechain to more closely resemble the phenylalanine residue. Unfortunately, these compounds proved to be substantially less active than **1**,[22] and an alternate overlay hypothesis for the phenyl ring of **1** was considered.

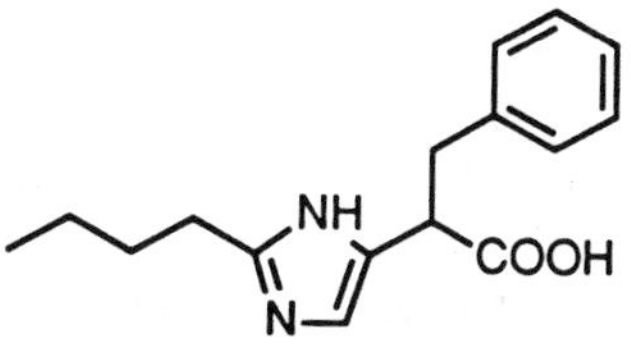

2. The Tyr[4] Overlay Hypothesis

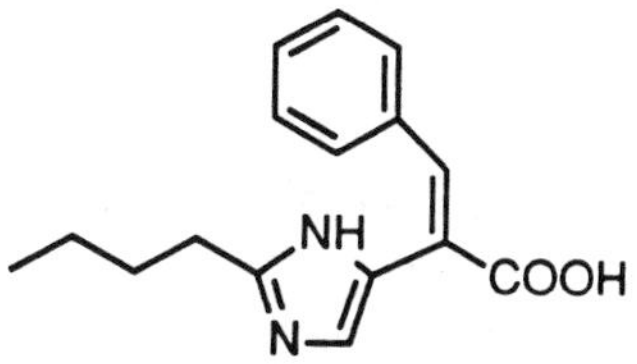

The aromatic ring at Tyr[4] of AII represented a second possible area of overlap for the *N*-benzyl group of **1**. In a solution conformation,[17a] the tyrosine aromatic ring is positioned over and pseudo-orthogonal to the peptide backbone, an alignment reminiscent of the orthogonal relationship between the benzyl and imidazole rings observed in the nuclear Overhauser effect (NOE) studies of **1**. Therefore, in another overlay hypothesis, the *N*-benzyl of **1** was aligned with the tyrosine aromatic ring of AII (Figure 5).[21] Such an overlay allows the 2-butyl of **1** to mimic the adjacent hydrophobic region of Ile[5], and the acetic acid to lay in the peptide backbone. The imidazole ring in the small molecule could be serving as a template which holds the critical binding elements in the proper orientation.[23] With the Tyr[4] overlay theory in mind, a number of analogs were made in which the chain length and degree of branching in the imidazole 2-substituent were altered. Also, functional groups such as aromatic and heteroaromatic rings were introduced at the 2-position of the imidazole nucleus in attempts to mimic other important binding groups of AII.[24] For synthetic ease, these modifications were examined in the thioalkyl series exemplified by **6**. However, no increases in potency were seen, and the Tyr[4] overlay hypothesis was put aside.

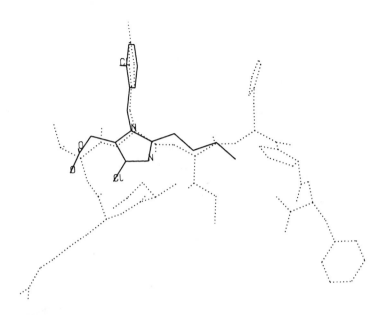

FIGURE 5.   Overlay of **1** (solid) with a solution conformation[17a] of AII (dotted) depicting the Tyr[4] overlay hypothesis.

3. Tyr[4]-Phe[8] Overlay Hypothesis

Thus, two separate overlay hypotheses concerning the resemblance of **1** and the native peptide had been proposed, but no improvement in potency was observed for any of the analogs designed based on either of these proposals. Despite the lack of success to date, there were elements of each hypothesis which remained attractive. The similarity in SAR still pointed to the carboxylate at the C-terminus of AII as a reasonable candidate for overlay by the acetic acid of the nonpeptide, although the data suggested that the *N*-benzyl ring of **1** did not appear to be overlaying the phenylalanine aromatic ring. The other reasonable candidate for the phenyl group of **1** to overlay remained the Tyr[4] aromatic ring. As with position eight, position four of AII had been identified as an important region for receptor binding as well as a key determinant of agonist vs. antagonist activity in the octapeptide series. For example, aromatic analogs which lacked the hydrophilic phenol of tyrosine at position four lost agonist activity, but were potent AII antagonists.[17] However, proposing a Tyr[4] overlay using a literature model for the conformation of AII left the acetic acid of **1** distant from the C-terminal carboxylate of AII. We wanted to investigate the possibility that the phenyl group of **1** may be able to mimic the Tyr[4] aromatic ring while the carboxyl of **1** maintained a good overlap with the C-terminal carboxylate of AII. Thus, a Tyr[4]-Phe[8] overlay hypothesis became the focus of our molecular modeling efforts.[16]

D. MODELING OF TYR[4]-PHE[8] HYPOTHESIS
1. Proposal of a Novel Model for the Conformation of AII

The existing literature models for the conformation of AII[25] could not adequately support the hypothesis that the phenyl ring of **1** overlaid the Tyr[4] aromatic region while the carboxyl maintained its alignment with the C-terminus of AII. Therefore, in order to effectively compare the nonpeptide with AII, it was necessary to develop a novel model for the conformation of AII which would be consistent with both octapeptide SAR and the structural requirements of the small molecule. Thus, the model which incorporated the set of allowed phi/psi angles deduced from a study of constrained amino acid SAR was first folded around **1** to bring Tyr[4] in close proximity to the C-terminal carboxylate of AII to more closely fit the proposed overlay with the small molecule. A flexible fit algorithm[26] was then used to match the Tyr[4] aromatic ring, hydrophobic Ile[5] sidechain, and C-terminal carboxylate of AII with the *N*-benzyl ring, butyl sidechain, and carboxylic acid of **1**. The resultant model, Model II (Figure 6), contained a loop encompassing the 4–8 residues of AII to bring the Tyr[4] aromatic ring into the vicinity of the C-terminal carboxylic acid. Model II, a product of the

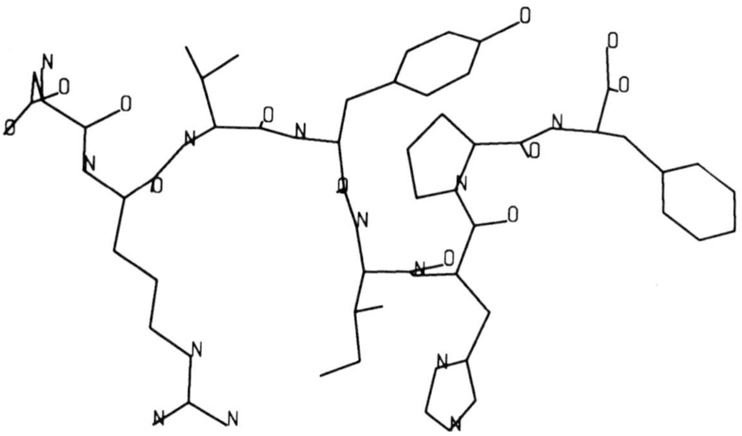

FIGURE 6. Model II, a working model of AII which positions Tyr[4] near the C-terminal carboxylate.

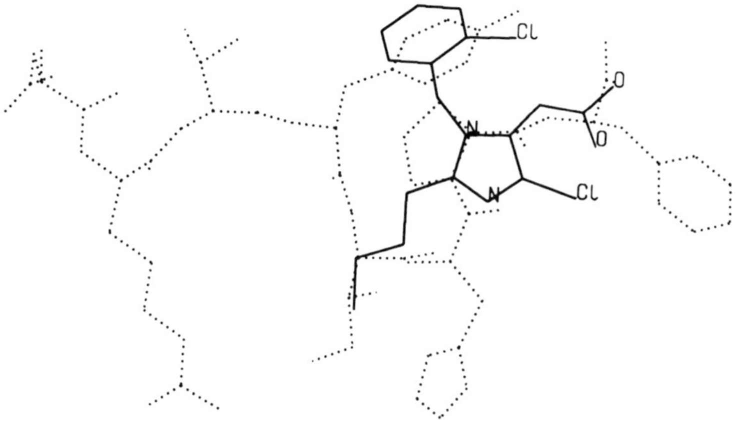

FIGURE 7. Benzylimidazole **1** (solid) overlaid on Model II (dotted).

combination of a study of peptide SAR, NMR studies, and molecular modeling, has served as a valuable template guiding the design of a series of potent small molecule receptor antagonists.[16]

2. Proposed Structural Modifications to Benzylimidazole Nucleus

In the proposed overlay of **1** on Model II (Figure 7), the phenyl ring of **1** aligns with the tyrosine phenolic ring and the carboxylic acid of **1** points in the same direction as the C-terminal carboxylic acid of AII. In addition, the butyl sidechain on the imidazole ring in **1** can extend into the lipophilic region near the hydrophobic Ile[5] of AII. According to this overlay hypothesis, the small molecule **1** covered the Tyr[4] aromatic ring, Ile[5] sidechain, and the Phe[8] carboxylate of AII but failed to reach other regions of the octapeptide known to be important for affinity, such as the Arg[2] and His[6] sidechains and Phe[8] aromatic ring. Envisioning that the addition of at least one more key binding element would be necessary for improving the affinity of a small molecule antagonist, the region around Phe[8] appeared most accessible for structural modifications of **1** to increase its resemblance to AII and thus possibly enhance potency. From the overlay (Figure 7), it was apparent that the chain connecting the acid to the imidazole ring of **1** could be lengthened to better approximate the

TABLE 1
ACID SIDECHAIN STUDY

no.	R	IC_{50} $(\mu M)^a$	K_b $(\mu M)^b$
1^c	CO_2H	43	2.7
9	CO_2H	12	1.8
10	CO_2H	16.5	2.8
11	CO_2H	124	14
12	CO_2H	8.9	0.81
13	CO_2H	>10	15
14	CO_2H	10.6	17
15	CO_2H	23.5	3.5
16	CO_2H	33.9	4.8
17	CO_2H	>100	17.2

[a] Inhibition of $[^{125}I]$AII specific binding to rat mesenteric arteries, n = 3-5. [b] Inhibition of AII-induced vasoconstriction of the rabbit aorta, n = 3-5. [c] Imidazole ring substituted with a 4-Cl.

TABLE 2
Substituted (E) Acrylic Acids

no.	R$_1$	R$_2$	IC$_{50}$ (μM) [a]	K$_b$ (μM) [b]
12	H	H	8.9	0.81
18	H	Me	22.3	0.97
19	H	nBu	6.9	0.16
20	H	Ph	20.1	0.60
21	H	CH$_2$Ph	2.6	0.064
21c[c]	H	CH$_2$Ph	38.0	8.7
22	H	CH$_2$CH$_2$Ph	2.9	0.42
23	Me	H	3.6	0.66
24	nBu	H	12.0	0.73
25	CH$_2$Ph	H	9.8	0.33
26	Me	CH$_2$Ph	28.4	0.76

[a,b] See Table 1. [c] Z isomer.

separation between the aromatic ring of Tyr[4] and the C-terminal carboxylate. Subsequently, attaching a substituent to such an extended acid sidechain to interact with the binding pocket for the octapeptide's C-terminal sidechain could lead to a desired increase in affinity. For example, another aryl group would mimic either an agonist ([L-Phe[8]]-AII) or antagonist ([D-Phe[8]]-AII) at position eight. Alternatively, a new aliphatic substituent would more closely resemble potent octapeptide antagonists such as saralasin or [Sar[1], Ile[8]]-AII. Our research on modifying the acetic acid sidechain of **1** to better fit the proposed pharmacophore model was the first step in the evolution of substituted imidazole-5-acrylic acids as a new class of potent small molecule AII receptor antagonists.[10,27] As potency has increased in the small molecules, the original model of AII has necessarily undergone some minor revisions. Nevertheless, the original premise involving the concomitant overlay of both the Tyr[4] and Phe[8] regions of AII by the small molecule imidazole antagonists has remained intact.

III. RESULTS OF STRUCTURAL MODIFICATIONS

A. BIOLOGICAL ASSAYS

The compounds which were synthesized as part of our research on nonpeptide angiotensin II receptor antagonists were evaluated for activity using two different *in vitro* screens: competitive binding vs. radiolabeled AII in a rat mesenteric artery receptor preparation (this tissue has since been shown to consist largely of the AT-1 receptor subtype);[28] and inhibition of AII induced vasoconstriction in isolated rabbit aorta strips. The competitive binding assay measured the intrinsic affinity of a particular compound for the AII receptor while the rabbit aorta assay verified the results obtained in the receptor binding screen as well as tested the compound's ability to counteract the effects of AII in a physiologically relevant system. As we were interested in both a compound's intrinsic affinity and its ability to functionally antagonize AII, both assays were monitored for improvement in potency in the SAR study. Although the relative order of activity for the two *in vitro* screens did not always agree, the most potent compounds exhibited good activity in both assays. Currently, there is no good explanation for the discrepancy between the two assays, although a number of factors, such as different receptor subtype populations, variations in tissue distribution, or species difference may account for some of the observed lack of correlation.

B. EXTENDED ACID SIDECHAINS

There were few data available on different carboxylic acid chain lengths in the small molecule imidazole series at the time we initiated our research.[9b] As a preliminary investigation of our overlay hypothesis, we examined longer tethers connecting the imidazole ring and the carboxylic acid, introducing rigidity into the chain in addition to varying its length.[27] Since preliminary experiments had shown that the chloro group at C-4 in **1** was not essential for activity,[10] this atom was replaced with hydrogen for synthetic ease. The other substituents on the imidazole ring were held constant to determine the effects of alterations only in the C-5 acid sidechain. The data presented in Table 1 indicated that although extended acid sidechains of varying length could be accommodated by the receptor, the best activity was observed with the *(E)* acrylic acid **12**. This analog was more active than the corresponding *(Z)* acrylic acid **11** or the saturated derivative **10**. In the aryl series (**15–17**), the *para* and *meta* isomers displayed more potency than the *ortho* analog. Taken together, these results suggest that the carboxylate prefers to extend away from the imidazole core of the molecule when interacting with the receptor, as suggested by the molecular modeling. The rigid *(E)* acrylic acid, which lengthened the acid sidechain in a conformationally restricted manner while maintaining potency, represented a promising lead for further exploration.

C. ADDITION OF ARYL RING TO MIMIC PHE[8] SIDECHAIN

The retention of activity observed upon extending the acetic acid sidechain encouraged investigation of the second aspect of the overlay hypothesis: attachment of a substituent to interact with the binding pocket for the C-terminal sidechain.[27] The results are presented in Table 2. Addition of an arylmethyl appendage *alpha* to the carboxylic acid to mimic Phe[8] (**21**) led to a notable increase in potency, as predicted by the overlay hypothesis. The α-benzyl analog proved superior to either the α-phenyl (**20**) or α-phenethyl (**22**) analogs, especially in the rabbit aorta assay. The pattern observed in the unsubstituted series continued as the *(E)* α-benzyl acrylic acid was more active than its *(Z)* isomer. In contrast to the *alpha* position, the β-methyl-substituted analog **23** exhibited comparable activity to the β-benzyl compound **25**. However, the combination of α-benzyl with β-methyl lowered affinity. Although the benzyl group was introduced *alpha* to the carboxylate on a number of extended acid sidechains, this modification proved most effective in the *(E)* acrylic acid series (data not shown).

The overlay hypothesis had correctly pointed to functionalization of the acid sidechain as a means to increase potency. Although attachment of a simple benzyl group did increase potency, it was left to a more traditional approach — synthesis and evaluation of a variety of aryl analogs — to discover even more potent AII receptor antagonists. A number of substituents on the newly introduced phenyl ring were examined in the α-benzyl acrylic acid series.[27] A condensed SAR is presented in Table 3. Two general trends were observed. First, addition of electron-rich groups such as hydroxy or methoxy, especially at the *para* position, led to enhanced affinity. Second, there was a general intolerance for increased bulkiness by the receptor. For example, the 3,4-methylenedioxy derivative **28** had more affinity than the more sterically demanding 3,4-dimethoxy compound **29**. Similarly, the primary amine or free phenolic compounds displayed better binding than their larger methylated counterparts.

The general preference for electron-rich groups as well as the intolerance for increased bulk was also observed in a series of acrylic acid antagonists with heterocyclic rings at this position (Table 3).[27]

TABLE 3
α-Arylmethyl Acrylic Acid Analogs

no.	Ar	IC$_{50}$ (μM)[a]	K$_b$ (μM)[b]
21	C$_6$H$_5$	2.6	0.064
27	4-OMeC$_6$H$_4$	0.81	0.12
28	3,4-(OCH$_2$O)C$_6$H$_3$	0.47	0.13
29	3,4-(OMe)$_2$C$_6$H$_3$	6.8	0.62
30	4-OHC$_6$H$_4$	0.12	0.47
31	3,4-(OH)$_2$C$_6$H$_3$	0.058	0.14
32	4-NH$_2$C$_6$H$_4$	3.0	0.32
33	4-N(Me)$_2$C$_6$H$_4$	14.6	0.59
34	3,4-Cl$_2$C$_6$H$_3$	25.6	13.7
35	2-furyl	1.2	0.05
35c[c]	2-furyl	>100	14.4
36	3-furyl	0.36	0.03
37	2-thienyl	0.44	0.05
38	3-thienyl	0.33	0.17
39	CH$_2$-2-thienyl	0.86	0.14
40	2-pyridyl	0.82	2.9
41	3-pyridyl	0.30	0.18
42	4-pyridyl	0.15	0.17
43	4(5)-imidazoyl	0.25	0.31

[a,b] See Table 1. [c] Z isomer.

Comparison of the α-furfuryl analogs **35** and **35c** revealed that as potency has increased, the difference in activity between the *(Z)* and *(E)* acrylic acid isomers has become more pronounced. In the pyridyl series (**40–42**), a stepwise increase in activity was observed going from the 2- to 3- to 4-substituted analogs, a trend similar to that seen with the analogous series of methoxy-substituted phenyl rings. As in the benzyl example, one carbon homologation of the connecting chain to furnish the thiophenethyl analog did not improve activity. In addition to having submicromolar affinity for

TABLE 4
Acid Extensions

no.	R	IC$_{50}$ (μM)^a	K$_b$ (μM)^b
37	CO$_2$H	0.44	0.05
44	CH$_2$OH	0.52	0.87
45	CONH$_2$	1.55	1.60
46	CH=CHCO$_2$H (E)	0.70	0.54
47	CH$_2$OCH$_2$CO$_2$H	0.43	0.05
48	CO-NHCH$_2$CO$_2$H	0.21	0.04
49	CO-D-Phe	1.61	0.63
50	CO-L-Phe	1.55	0.11

^{a,b} See Table 1.

the receptor, the heterocyclic compounds exhibited oral activity and increased potency in the *in vivo* rat model. For example, the α-thienylmethyl compound **37** inhibited the pressor effects of exogenously administered AII in a normotensive rat with an ID$_{50}$ of 3.6 mg/kg i.v. and 4.8 mg/kg i.d.[27]

D. REEVALUATION OF THE MODELING HYPOTHESIS

The enhanced activity observed on replacing the phenyl ring attached to the acrylic acid sidechain with more electron-rich rings such as 4-hydroxyphenyl, 2-thienyl, 4(5)-imidazoyl, and 4-pyridyl pointed to an alternate possible overlay of this group with either Tyr[4] or His[6], the two more electron-rich aryl residues of AII. The similarity in binding potency for the acid and alcohol analogs of the 2-thienylmethyl compound (Table 4) suggested the carboxylic acid may not be involved in a strong ionic or hydrogen-bonding interaction with the receptor but simply be overlaying the peptide backbone of AII. If this were the case, further extension of the acrylic acid sidechain may lead to better overlay of the acid with the C-terminal carboxylate of AII. A small group of analogs were synthesized to investigate this hypothesis (Table 4). Although some exhibited potency in the *in vitro* assays, none was more active than the simple carboxylic acid. Also, the imidazole analog (**43**, Table 3), which may have been expected to show enhanced activity, was not more potent than the other heterocyclic ring analogs. Thus, our best hypothesis for a potential overlay of the thienyl ring of the acrylic acid imidazole antagonists remained the phenyl ring at Phe[8]. As further support

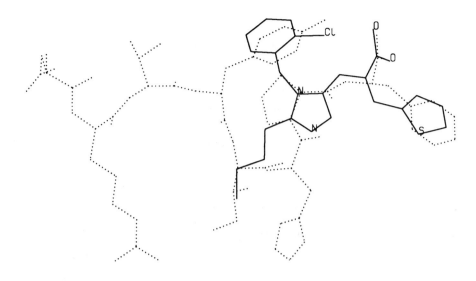

FIGURE 8. Imidazole-5-acrylic acid **37** (solid) overlaid on Model II (dotted).

for this theory, [Sar¹, Thi⁸]AII was subsequently synthesized and found to exhibit better binding affinity than the simple phenyl analog [Sar¹]AII.[10]

A proposed overlay of the α-thienylmethyl acrylic acid analog on the proposed pharmacophore model of AII, Model II, is depicted in Figure 8. The appended thiophene ring and acrylic acid of the imidazole antagonist align with the phenyl ring and carboxylate of the C-terminal phenylalanine of AII. The butyl chain attached at C-2 of the imidazole and the 2-chlorophenyl group in the small molecule maintain their respective overlays with the hydrophobic region near Ile⁵ and the Tyr⁴ aromatic ring in AII. Interestingly, the aryl-substituted acrylic acids cannot mimic the agonist conformation at the C-terminus of AII proposed by Marshall,[29] where the position eight phenylalanine aromatic ring lies over the plane of the Pro⁷-Phe⁸ amide bond. To accommodate the rigid thienylmethyl-substituted acrylic acid, the phenylalanine aromatic ring of AII is instead swung away from the adjacent Pro⁷-Phe⁸ peptide bond as shown. Thus, the peptide model was adjusted to better fit a nonpeptide with increased potency. This conformation at the C-terminus may represent an antagonist conformation reminiscent of AII peptide antagonists containing a D-aryl amino acid at position eight.[18] The weaker affinity of the simple alkyl-substituted acrylic acid analogs as compared to the benzyl-substituted compounds argues against a proposed overlay on AII peptide antagonists containing alkyl amino acid residues at position eight.

E. ALTERATION OF *N*-BENZYL RING SUBSTITUTION TO MIMIC TYR⁴

As predicted by the modeling, extending the acetic acid chain to an *(E)* acrylic acid and then attaching a benzyl group to mimic Phe⁸ gave increased potency.[27] Our overlay hypothesis also proposed a simultaneous overlay of the Tyr⁴ region of AII by the *N*-benzyl ring of the imidazole antagonists. Examination of Figure 8 suggested that adding *N*-benzyl substituents which mimic the phenolic group — by either donating or accepting a hydrogen bond — may generate improved activity. As was the case with the aromatic ring on the acid sidechain, altering the substitution pattern on the *N*-benzyl aromatic ring proved to be a fruitful area of research.[10,30] Among the many different analogs synthesized, the hydroxy, nitro, and the carboxylic acid analogs provided the most interesting data (Table 5). At the *ortho* position, substituting the hydroxyl group **(52)** for the chlorine led to an order of magnitude increase in binding affinity. Electron-withdrawing groups at the *ortho* position, such as nitro, trifluoromethyl, and cyano **(55–57)**, gave similar results. On the other hand, substituting at the *ortho* position with a carboxylic acid **(60)**, which is isoelectronic with the nitro group but has an overall

TABLE 5
N-Benzyl Substitution

no.	X	IC_{50} (nM)[a]	K_b (nM)[b]
37	2-Cl	440	51
51	H	4530	268
52	2-OH	34	90
53	3-Me, 4-OH	21	33
54	3-Me, 4-OMe	66	70
55	$2\text{-}CF_3$	260	51
56	2-CN	22	20
57	$2\text{-}NO_2$	31	57
58	$3\text{-}NO_2$	210	47
59	$4\text{-}NO_2$	620	50
60	$2\text{-}CO_2H$	6000	2250
61	$3\text{-}CO_2H$	3.3	11
2	$4\text{-}CO_2H$ (SK&F 108566)	1.0	0.21
62	$4\text{-}CONH_2$	0.56	12
63	$4\text{-}CN_4H$	0.4	0.25
64	$4\text{-}CO_2Me$	39	250
65	2-Cl, $4\text{-}CO_2H$	1.45	0.02
66	$2\text{-}NO_2$, $4\text{-}CO_2H$	0.15	0.85
67	$4\text{-}(2'\text{-}CO_2H\text{-}C_6H_4)$	6.58	1.56
68	$4\text{-}CH_2CO_2H$	59.6	4.77

[a,b] See Table 1.

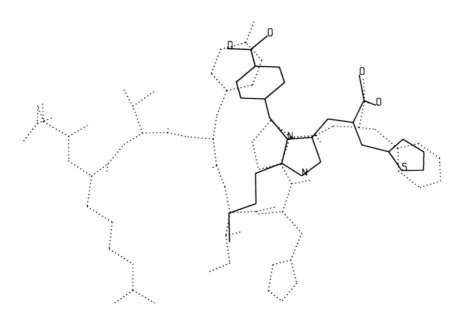

FIGURE 9. SK&F 108566 (**2**, solid) overlaid on a modified Model II (dotted).

negative charge, resulted in a comparatively inactive compound. At the *para* position, a different pattern emerged. Whereas the hydroxy analog **53** was still potent, the nitro analog **59** had poor activity. However, substituting the carboxylic acid at the *para* position led to a significant increase in activity.[10,30] The 4-CO$_2$H analog (**62**, SK&F 108566) displayed remarkable affinity for the receptor — equivalent to angiotensin II. Importantly, this dramatic increase in activity carried over to the *in vivo* models, where SK&F 108566 showed both enhanced potency and increased duration of action.[10,28] It is interesting to note that in the carboxy-substituted series, activity decreased going from 4- to 3- to 2-substitution, whereas the opposite trend is observed in the nitro-substituted analogs.

Some further modifications to the diacid nucleus are also presented in Table 5.[30] Since an increase in affinity on going from 2-chloro to 4-carboxy was also observed by the DuPont chemists in the parent acetic acid series.[11] We attached the DuPont biphenyl-carboxylic acid group in hopes of seeing a further improvement. Activity dropped off for this analog (**67**), indicating that the biphenylimidazole class of nonpeptide antagonists may bind differently to the receptor. Also, extending the acid chain by one carbon (**68**) led to a loss of activity, and conversion to the tetrazole (**63**), a modification which was very effective in the DuPont series, did not generate any increase in *in vivo* potency or duration (data not shown) although high binding affinity was retained.[30] The 4-carboxamide analog **62** also exhibited tight binding, but the corresponding methyl ester **64** lost affinity, suggesting a polar group is crucial at the *para* position for high affinity. Finally, the combination of the 4-carboxy with a 2-chloro or 2-nitro retained both *in vitro* and *in vivo* potency.[10] Since in the imidazol-5-acetic acid series, the 2-chloro group on the *N*-benzyl ring caused the benzyl ring to favor an orthogonal relationship with the imidazole ring,[12] the potent imidazole-5-acrylic acids such as SK&F 108566 may adopt a similar conformation at the receptor.

IV. OVERLAY COMPARISONS OF SK&F 108566

A. OVERLAY ON AN UPDATED PHARMACOPHORE MODEL OF AII

The potent small molecule diacid SK&F 108566 can be positioned to overlay the Tyr[4] and Phe[8] residues of AII (Figure 9), consistent with the fundamental premise of our modeling hypothesis. By slightly opening the χ^1 angle of Tyr[4] in Model II,[16] the tyrosine phenol in AII can align with the *para* carboxylic acid of SK&F 108566, which suggests that the carboxy group of SKF 108566 may in fact be interacting with the same hydrogen bond-accepting residue on the receptor as the Tyr[4] phenol. As before

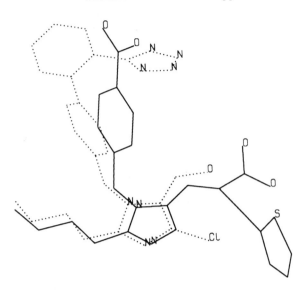

FIGURE 10. Overlay of butylimidazole portions of SK&F 108566 (**2**, solid) and DuP 753 (**3**, dotted).

with the C-terminal aryl region, this refinement of the peptide model at Tyr[4] was done to rationalize increased potency in the nonpeptide series. The proposed overlay at the C-terminus is not affected, as the acid and thiophene ring of SK&F 108566 can maintain their alignment with the corresponding elements of Phe[8]. The butyl chain lies in the hydrophobic region near Ile[5] and the *(E)* double bond may in fact mimic a peptide bond in this region of the AII backbone. Thus, there are a number of possible areas of overlap which may account for the observed potency of the small molecule.

Recently, a potent peptide analog of AII was reported where Val[3] and Ile[5] were replaced by a homoCys[3]-homoCys[5] disulfide bridge, creating a larger hydrophobic surface with these two amino acid sidechains occupying the same face of the peptide.[31] Since we have proposed an overlay of the small molecule imidazoles with the C-terminal region of AII, this change in the peptide's conformation can be incorporated into our model without affecting the overlay hypothesis.[32] The butyl chain of the small molecule can maintain access to the hydrophobic region which now encompasses both Ile[5] and Val[3]. The rest of the areas of overlap remain unaffected.

B. OVERLAY COMPARISON WITH ANOTHER NONPEPTIDE ANTAGONIST

In addition to overlaying SK&F 108566 on the native peptide AII, it is also interesting to compare it to other potent nonpeptide AII antagonists such as DuP 753 (Losartan).[11] Of the many possible overlay comparisons of SK&F 108566 with DuP 753, two will be presented here. The first overlay (Figure 10) is generated by lining up the structural feature in each molecule leftover from their common ancestor **1**, the butylimidazole portion. In this overlay, the benzoic acid of SK&F 108566 and the tetrazole of DuP 753 can be superimposed and the acrylic acid of SK&F 108566 and the hydroxymethyl of DuP 753 point in the same general direction. Taking into account the known *in vivo* oxidation of the hydroxymethyl group of DuP 753 to the corresponding carboxylic acid to yield an active metabolite responsible for much of the pharmacology[11] increases the resemblance to SK&F 108566. However, in this overlay comparison DuP 753 lacks functionality in the vicinity of the thiophene ring of SK&F 108566, and SK&F 108566 likewise does not overlay the outer phenyl ring of DuP 753.

The second overlay comparison emphasizes the similarities in the gross structural modifications done to **1** in the development of the two distinct classes of AII receptor antagonists, in spite of clearly different design strategies by the two research groups.[33] It is apparent that in each molecule an extended acid chain containing another aryl ring has been appended onto the imidazole nucleus. Overlaying the two extended acid sidechains, the *N*-biphenyltetrazole of DuP 753 and the thienylmethyl-substituted acrylic acid of SK&F 108566, furnishes the overlay depicted in Figure 11. According to this overlay, the benzoic acid portion of SK&F 108566 can also be superimposed on the substituted imidazole ring

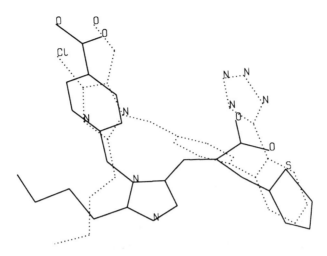

FIGURE 11. Overlay of extended acid sidechains of SK&F 108566 (**2**, solid) and DuP 753 (**3**, dotted).

of DuP 753. This overlay comparison suffers because it does not allow the alignment of the common butyl sidechains of each molecule. Still, this hydrophobic portion of each molecule may interact with a large hydrophobic pocket in the AII receptor, such as the one which provides the space for the hydrophobic region of AII spanning Val^3, Ile^5, and Pro^7. We feel this overlay may better explain the similar activity of the two structurally distinct nonpeptide antagonists.

If DuP 753 aligns with SK&F 108566 as in Figure 11, by analogy, it could also overlay our model of the peptide AII at the C-terminus. The tetrazole and the phenyl ring to which it is attached may mimic the phenylalanine carboxylic acid and phenyl ring, and the latent carboxylic acid on the imidazole can align with the Tyr^4 phenol. The many other nonpeptide antagonists which incorporate DuPont-like biphenyltetrazoles attached to a heterocyclic core may align similarly with SK&F 108566 and AII. Conversely, it is always possible that even related molecules may bind to a receptor in quite different ways.

V. CONCLUSION

In conclusion, starting from a weak nonpeptide antagonist, we have been able to develop a series of potent and specific small molecule AII receptor antagonists: the imidazole-5-acrylic acids. These compounds were designed by overlaying the small molecule **1** on the octapeptide AII. After overlay hypotheses based on literature models of AII led to no improvement in potency, a novel pharmacophore model based on the hypothesis that the benzylimidazole-acetic acid antagonist **1** may be overlaying both the Tyr^4 and Phe^8 regions of AII was proposed. The overlay correctly predicted that extension of the acid sidechain and attachment of a second aryl residue to mimic the C-terminal phenylalanine region of AII would lead to increased activity. Next, alteration of the *N*-benzyl ring substitution to more closely resemble Tyr^4, the second basic tenet of our initial modeling hypothesis, resulted in the discovery of SK&F 108566, an extremely potent nonpeptide angiotensin II receptor antagonist. Thus, the iterative use of overlay hypothesis strategy played a key role in the design of a potent small molecule therapeutic candidate.

Finally, it is important to recognize the limitations of the peptide pharmacophore modeling. Although the prediction by the modeling that an additional aryl ring to mimic the C-terminal phenylalanine would enhance activity proved accurate, the modeling could not predict which aryl ring system would prove most effective. It was left to synthesis of a variety of aryl ring systems to discover that the thienylmethyl-substituted acrylic acid provided the best combination of *in vitro* activity and *in vivo* potency. Similarly, the discovery of the *para* carboxylic acid on the *N*-benzyl ring was a result of an intensive analog effort around *N*-benzyl ring substituents under the general goal to mimic the tyrosine phenol. Also, we have no physical evidence to support the hypothesis that the nonpeptide and the peptide interact with the same surface region of the receptor or that any of the models utilized represents a true bioactive conformation of AII. Nevertheless, the use of peptide pharmacophore modeling correctly pointed to the acid sidechain and *N*-benzyl ring as promising places to alter the original molecule. Thus, in spite of the acknowledged limitations, thinking about the relationship of the peptide and various nonpeptides in terms of overlay hypotheses has proven very helpful in accomplishing the objective of designing potent nonpeptide receptor antagonists. Similar analysis of other peptides may furnish working models, or templates, such as Model II, to aid in the design of other nonpeptide receptor antagonists. Therefore, our work on the discovery of potent angiotensin II receptor antagonists may serve as a model for future drug design efforts.

REFERENCES

1. **Garrison, J.C. and Peach, M.C.,** Renin and angiotensin, in *The Pharmacological Basis of Therapeutics,* 8th ed., Gilman, A.G., Rall, T.W., Nies, A.S., and Taylor, P., Eds., Pergamon Press, New York, 1990, 749.

2. **Coleman, J.K.M., Ong, B., Sardinia, M., Harding, J., and Wright, J.W.,** Changes in renal blood flow due to infusions of angiotensin II (3–8) [AIV] in normotensive rats, *FASEB J.,* 6, A981, 1992.

3. **Braszko, J.J., Wlasienko, J., Koziolkiewicz, W., Janecka, A., and Wisausewski, K.,** The 3–7 fragment of angiotensin II is probably responsible for its psychoactive properties, *Brain Res.,* 542, 49, 1991.

4. **Ferrario, C.M., Brosnihan, K.B., Diz, D.I., Jaiswal, N., Khosla, M.C., Mitsted, A., and Tallant, E.A.,** Angiotensin (1–7): a new hormone of the angiotensin system, *Hypertension,* 18 (Suppl. III), 126, 1991.

5. **Balla, T., Baukal, A.J., Eng, S., and Catt, K.J.,** Angiotensin II receptor subtypes and biological responses in the adrenal cortex and medulla, *Mol. Pharmacol.,* 40, 401, 1991.

6. **Mauzy, C.A., Egloff, A.M., Wu, L.-H., Zhu, G., Oxender, D.L., and Chung, F.-Z.,** Cloning of a new subtype of rat angiotensin II type 1 receptor gene, *FASEB J.,* 6, A1578, 1992.

7. **Coulter, D.M. and Edwards, I.R.,** Cough associated with captopril and enalapril, *Br. Med. J.,* 294, 1521, 1987.

8. **Greenlee, W.J.,** Renin inhibitors, *Med. Res. Rev.,* 10, 173, 1990.

9. (a) **Furukawa, Y., Kishimoto, S., and Nishikawa, K.,** Hypotensive imidazole-5-acetic acid derivatives, U.S. Patent 4,355,040, 1992. (b) **Furukawa, Y., Kishimoto, S., and Nishikawa, K.,** Hypotensive imidazole derivatives, U.S. Patent 4,340,598, 1982.

10. **Weinstock, J., Keenan, R.M., Samanen, J., Hempel, J., Finkelstein, J.A., Franz, R.G., Gaitanopoulos, D.E., Girard, G.R., Gleason, J.G., Hill, D.T., Morgan, T.M., Peishoff, C.E., Aiyar, N., Brooks, D.P., Fredrickson, A.C., Ohlstein, E.H., Ruffolo, R.R., Stack, E.J., Sulpizio, A.C., Weidley, E.F., and Edwards, R.M.,** 1-(Carboxybenzyl)-imidazole-5-acrylic acids: potent and selective angiotensin II receptor antagonists, *J. Med. Chem.,* 34, 1514, 1991.

11. **Duncia, J.V., Carini, D.J., Chiu, A.T., Johnson, A.L., Price, W.A., Wong, P.C., and Timmermans, P.B.M.W.M.,** The discovery of DuP 753, a potent, orally active nonpeptide angiotensin II receptor antagonist, *Med. Res. Rev.,* 12, 149, 1992.

12. **Hempel, J.C.,** unpublished results. Calculations were carried out with the use of MAXIMIN in SYBYL Version 3.1 (Tripos Associates).

13. **Weinstock, J. and Gaitanopoulos, G.R.,** unpublished results.

14. **Premilat, S. and Maigret, B.,** Statistical molecular models for angiotensin II and enkephalin related to NMR coupling constants, *J. Phys. Chem.,* 84, 293, 1980.

15. **Samanen, J.M., Cash, T., Narindray, D., Brandeis, E., Adams, W., Weideman, H., and Yellin, T.,** An investigation of angiotensin II agonist and antagonist analogs with 5,5-dimethylthiazolidine-4-carboxylic acid and other constrained amino acids, *J. Med. Chem.,* 34, 3036, 1991.

16. **Samanen, J.M., Weinstock, J., Hempel, J.C., Keenan, R.M., Hill, D.T., Ohlstein, E.H., Weidley, E.F., Aiyar, N., and Edwards, R.,** A molecular model of angiotensin II for rational design of small antagonists with enhanced potency, in *Peptides, Chemistry and Biology,* Smith, J.A. and Rivier, J.E., Eds., ESCOM, Leiden, 1992, 386.

17. (a) **Smeby, R.R. and Fermandjian, S.,** Conformation of angiotensin II, in *Chemistry and Biochemistry of Amino Acids, Peptides and Proteins,* Vol. 5, Weinstein, B., Ed., Marcel Dekker, New York, 1978, 117. (b) **Peach, M.J.,** Structural features of angiotensin II which are important for biologic activity, *Kidney Int.,* 15, S-3, 1979. (c) **Bumpus, F.M. and Koshla, M.C.,** Angiotensin analogs as determinants of the physiologic role of angiotensin and its metabolites, in *Hypertension: Physiopathology and Treatment;* Genest, J., Koiw, E., and Kuchel, O., Eds., McGraw-Hill, New York, 1977, 183. (d) **Samanen, J., Cash, T., Narindray, D., Brandeis, E., Yellin, T., and Regoli, D.,** The role of position four in angiotensin II antagonism: a structure-activity study, *J. Med. Chem.,* 32, 1366, 1989. (e) **Samanen, J., Narindray, D., Cash, T., Brandeis, E., Adams, W., Yellin, T., Eggleston, D., DeBrosse, C., and Regoli, D.,** Potent angiotensin II antagonists with non-β-branched amino acids in position 5, *J. Med. Chem.,* 32, 466, 1989. (f) **Samanen, J., Brandeis, E., Narindray, D., Adams, W., Cash, T., Yellin, T., and Regoli, D.,** The importance of residues two (arginine) and six (histidine) in high affinity angiotensin II antagonists, *J. Med. Chem.,* 31, 737, 1988.

18. (a) **Hsieh, K., LaHann, T.R., and Speth, R.C.,** Topographic probes of angiotensin and receptor: potent angiotensin II agonist containing diphenylalanine and long-acting antagonists containing biphenylalanine and 2-indan amino acid in position 8, *J. Med. Chem.,* 32, 898, 1989. (b) **Samanen, J., Narindray, D., Adams, W., Cash, T., Yellin, T., and Regoli, D.,** Effects of D-amino acid substitution on antagonist activities of angiotensin II analogues, *J. Med. Chem.,* 31, 510, 1988.

19. **Hsieh, K. and Marshall, G.R.,** Role of C-terminal carboxylate in angiotensin II activity: alcohol, ketone, and ester analogues of angiotensin II, *J. Med. Chem.,* 29, 1968, 1986.

20. **Weinstock, J., Hill, D.T., and Girard, G.R.,** unpublished results.

21. **Samanen, J.M., Hempel, J.C., Peishoff, C.E., Weinstock, J., and Keenan, R.M.,** Molecular models of angiotensin II in the design of potent nonpeptide antagonists, in *Angiotensin II Receptors: Molecular Biology, Biochemistry, and Pharmacology,* Ruffolo, R.R., Ed., CRC Press, Boca Raton, FL, in press, 1994.

22. **Keenan, R.M., Hempel, J.C., Weinstock, J., Samanen, J.M., and Edwards, R.M.,** unpublished results.

23. An analogous role has been proposed for the central diazapine ring in the nonpeptide cholecystokinin receptor antagonist. See: **Pincus, R.M., Carty, R.P., Chen, J., Lubowshy, J., Avitable, M., Shan, D., Scheraga, H.A., and Murphy, R.B.,** On the biologically active structures of cholecystokinin, little gastrin, and enkephalin in the gastrointestinal system, *Proc. Natl. Acad. Sci. U.S.A.,* 84, 4821, 1987.

24. **Hill, D.T., Weinstock, J., and Girard, G.R.,** unpublished results.

25. For a comprehensive bibliography of proposed models of AII, see: **Dunica, J.V., Chiu, A.T., Carini, D.J., Gregory, G.B., Johnson, A.L., Price, W.A., Wells, G.J., Wong, P.C., Calabrese, J.C., and Timmermans, P.B.M.W.M.,** The discovery of potent nonpeptide angiotensin II receptor antagonists: a new class of potent antihypertensives, *J. Med. Chem.,* 33, 1312, 1990. SYBYL (Tripos Associates); see Reference 16.

26. SYBYL (Tripos Associates); see Reference 16.

27. **Keenan, R.M., Weinstock, J., Finkelstein, J.A., Franz, R.G., Gaitanopoulos, D.E., Girard, G.R., Hill, D.T., Morgan, T.M., Samanen, J.M., Hempel, J.C., Eggleston, D., Aiyar, N., Griffin, E., Ohlstein, E.H., Stack, E.J., Weidley, E.F., and Edwards, R.M.,** Imidazole-5-acrylic acids: potent nonpeptide angiotensin II receptor antagonists designed using a novel peptide pharmacophore model, *J. Med. Chem.,* 35, 3858, 1992.

28. **Edwards, R.M., Aiyar, N., Ohlstein, E.H., Weidley, E.F., Griffin, E., Ezekiel, M., Keenan, R.M., Ruffolo, R.R., and Weinstock, J.,** Pharmacological characterization of the nonpeptide angiotensin II receptor antagonist, SK&F 108566, *J. Pharmacol. Exp. Ther.,* 260, 175, 1992.

29. **Marshall, G.R.,** Determination of the receptor bound conformation of angiotensin, *Dtsch. Apoth. Ztg.,* 85, 2783, 1986.

30. **Keenan, R.M., Weinstock, J., Finkelstein, J.A., Franz, R.G., Gaitanopoulos, D.E., Girard, G.R., Hill, D.T., Morgan, T.M., Samanen, J., Aiyar, N., Griffin, E., Ohlstein, E.H., Stack, E.J., Weidley, E.F., and Edwards, R.M.,** Potent nonpeptide angiotensin II receptor antagonists. 2. 1-(Carboxybenzyl)imidazole-5-acrylic acids. *J. Med. Chem.,* 36, 1880, 1993.

31. (a) **Spear, K.L., Brown, M.S., Reinhard, E.J., McMahon, E.G., Olins, G.M., Paloma, M.A., and Patton, D.R.,** Conformational restriction of angiotensin II: cyclic analogues having high potency, *J. Med. Chem.,* 33, 1935, 1990. (b) **Sugg, E.F., Dolan, C.A., Patchett, A.A., Chang, R.S.L., Faust, K.A., and Lotti, V.J.,** Cyclic disulfide analogues of [Sar^2, Ile^8]-angiotensin II, in *Peptides: Chemistry, Structure, and Biology,* Rivier, J.E. and Marshall G.R., Eds., ESCOM, Leiden, 1990, 305.

32. **Samanen, J.M., Peishoff, C.E., Keenan, R.M., and Weinstock, J.M.,** Refinement of a molecular model of angiotensin II (AII) employed in the discovery of potent nonpeptide antagonists, *Bioorg. Med. Chem. Lett.,* 3, 909, 1993.

33. See References 11 and 25 for a discussion of the strategy employed in the design of DuP 753.

Chapter 4

ENDOTHELIN STRUCTURE AND DEVELOPMENT OF RECEPTOR ANTAGONISTS

Annette M. Doherty

CONTENTS

I. INTRODUCTION

Endothelin (ET) is a potent vasoconstrictor peptide originally isolated from the culture supernatant of porcine aortic endothelial cells.[1,2] Since the initial identification of porcine ET which is identical to human ET (termed ET-1), three further members of this family of peptides have been reported, endothelin-2 (ET-2), endothelin-3 (ET-3), and mouse vasoactive intestinal contractor (VIC) (Figure 1).[2-7] Although VIC was originally thought to be a fourth isoform, a recent report suggests that VIC might be a murine (and rat) form of ET-2 because of the identical sizes of the mRNAs for the precursor form of ET-2. ET-1, -2, and -3 all appear to be distinct gene products. The

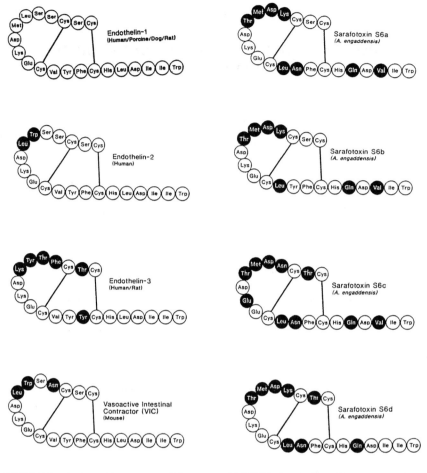

FIGURE 1. Amino acid sequences of the endothelin peptide family. Filled circles, residues different from those in ET-1.

FIGURE 2. Amino acid sequences of the sarafotoxin peptide family. Filled circles, residues different from those in ET-1.

sarafotoxins (SRTXs) isolated from the venom of the Israeli burrowing asp, *Atractaspis engaddensis,* show a remarkable sequence homology with the endothelin peptides (Figure 2).[8–10] ET-1 is some tenfold more potent than the vasoconstrictor angiotensin II, and has extremely long-lasting pressor effects.[6,11–13] It is currently not known whether ET-1 acts in a paracrine or autocrine manner or what role it may play in normal physiology and or pathophysiology of human disease.

II. BIOLOGICAL ACTIONS

Numerous reports have described the many biological actions of the ET peptide family (Table 1).[11–63] The peptides have effects on hemodynamics, renal, cardiovascular, and endocrine function. Intravenous infusion of endothelin to anesthetized dogs increased arterial pressure by increasing peripheral vascular resistance, coronary vascular resistance, and decreasing cardiac output. Renal blood flow and glomerular filtration rate were markedly reduced, accompanied by a reduction in natriuresis and increases in plasma renin activity.[46]

In contrast, low-dose infusions of ET-1 and -3 have elicited only a vasodilatory action.[19,20] There have been several reports describing the initial transient but potent vasodilation of the ETs that appears to be selective for certain arterial beds.[20,21] The effect has been observed to occur *in vivo* in ganglionic-blocked animals and, therefore, cannot be due to a reflex response to the vasoconstrictor component.[20] Low doses of ET-3 have been reported to cause continuous vasodilation of mesenteric

TABLE 1
Biological Actions of Endothelin

Effect	Ref.
Potent and long-lasting constriction of isolated vascular/nonvascular smooth muscle	16–18
Coronary arterial vasoconstriction, increased perfusion pressure	17–19
Vasodilation in certain tissues at low doses	20–22
Positive inotropic and chronotropic activity	18, 23–25
Reduction of cardiac output	26
Proarrythmogenic activity	27
Mitogenic actions	28–33
Induction of hypertrophy of cardiomyocytes	34
Inhibition of platelet aggregation *in vivo* (not *in vitro*) via release of EDRF or prostacyclin	35,36
Chemotactic activity	37,38
Potent contractile agonist activity in human isolated bronchus	39–42
Inhibition of renin release	43
Reduction of renal blood flow and glomerular filtration rate	44–46
Inhibition of arginine vasopressin release	47
Stimulation of substance P release	48
Stimulation of catecholamine release	49–51
Stimulation of EDRF release	52
Stimulation of ANP secretion	53,54
Increased release of prostacyclin	55
Stimulation of gastrointestinal hemorrhagic and necrotic damage in rat mucosa	56,57
Inhibition of prolactin secretion	58
Stimulation of luteinizing hormone, follicle stimulating hormone, and thyroid stimulating hormone release from primary monolayer cultures of rat anterior pituitary cells	58
Stimulation of pituitary gonadotropin release	59,60
Stimulation of glycogenolysis in rat heptocytes	61,62
Stimulation of PDGF secretion in cultured human mesangial cells	63

arteries preconstricted with norepinephrine and to be accompanied by elevation of cyclic nucleotides.[22] It is thought that the vasodilation elicited by the ETs is mediated via a single receptor subtype[64] in certain tissue beds. Interestingly, low threshold concentrations of ET-1, that exhibit no significant contraction, potentiate that response to other vasoconstrictor hormones such as norepinephrine and serotonin.[65,66]

Circulating ET-1 is eliminated rapidly from the circulation with a half-life of 7 min; however, blood pressure following i.v. administration of exogenous ET-1 is sustained at a high level for a long period of time, suggesting a prolonged tight binding to its receptors.[67]

III. BIOSYNTHESIS — ENDOTHELIN PROCESSING

Endothelin-1 is generated from a 203-amino acid peptide known as preproendothelin, by an unknown dibasic endopeptidase. This enzyme cleaves the prepropeptide to a 38- (human) or 39- (porcine) amino acid peptide known as big endothelin or proendothelin. Big ET is then cleaved by an elusive enzyme, known as endothelin-converting enzyme or ECE, to afford the biologically active molecule ET-1 (Figure 3).[2] Big ET is only 1% as potent as ET-1 in inducing contractile activity in vascular strips but it is equally potent *in vivo* at raising blood pressure, presumably by rapid conversion to ET-1.[68] The true identity and cellular location of ECE is still somewhat unclear, although there have been numerous reports describing possible proteases in both the cytoplasm and membrane-bound cellular fractions.[69–78] Many groups have chosen to isolate ECE from endothelial cells of various species, since endothelin is known to be synthesized and secreted by this cell type. It was initially reported that two types of protease activity were present in porcine or bovine

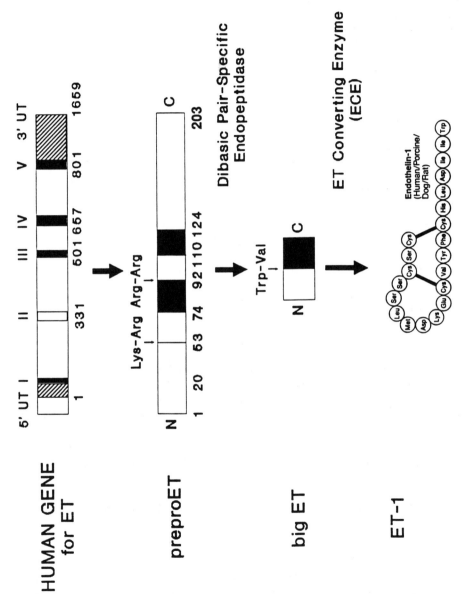

FIGURE 3. Biosynthesis of endothelin.

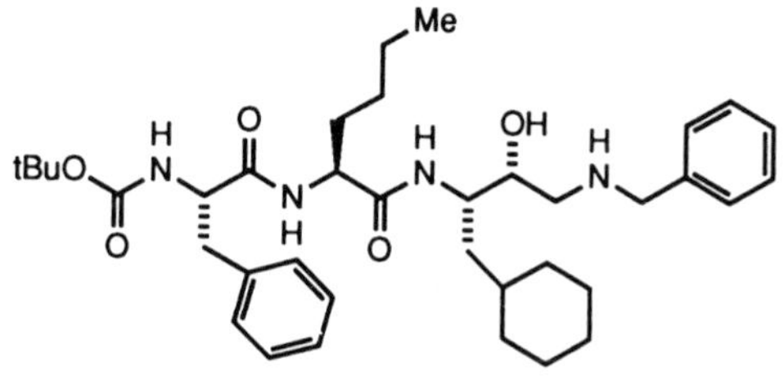

FIGURE 4. Structure of the cathepsin E inhibitor SQ 32,056.

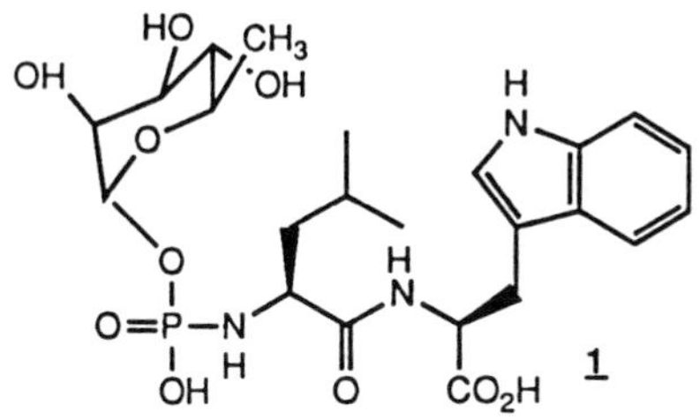

FIGURE 5. Structure of phosphoramidon.

endothelial cells that could cause conversion of big ET to ET *in vitro*.[69,70,75,77,78] However, it was subsequently found that the aspartic protease activity from porcine endothelial cells, thought to be predominantly cathepsin D, also caused further degradation of ET-1 and was therefore unlikely to be the true ECE.[71] Moreover, human cathepsin D also causes rapid degradation of ET-1.[72]

Human cathepsin E has been shown to convert Big ET to ET *in vitro*.[72] There has been an *in vivo* study where several potent cathepsin E inhibitors were administered to conscious rats prior to a pressor challenge with big ET-1. One of the inhibitors, SQ 32,056 (3 mg/kg i.v.), (Figure 4) blocked both the big ET-1 and ET-1 pressor responses.[73] It is not known whether this inhibitor is selective for cathepsin E over closely related aspartic proteinases and whether the effects observed may indicate that cathepsin E is not a physiologically relevant ECE.[73]

There has been one study showing that the intracellular accumulation of pepstatin, an aspartic protease inhibitor, did not inhibit ET-1 production in cultured bovine aortic endothelial cells.[74] Stronger evidence that ECE is in fact a neutral metalloprotease has begun to appear,[75-78] although there are no known specific metalloprotease ECE inhibitors to confirm these findings *in vivo*. However, the nonspecific metalloproteinase inhibitor phosphoramidon has been shown to inhibit the intracellular conversion of big ET-1 to ET-1 in cultured vascular endothelial cells and smooth muscle cells.[76]

ET-converting activity has been detected in both the membranous and cytosolic fractions of cultured porcine, bovine, and human endothelial cells.[75] Micromolar concentrations of phosphoramidon (Figure 5) have been shown to block the pressor response of big ET both *in vitro* and *in vivo*.[77,79-81] Phosphoramidon, being a potent neutral endopeptidase (NEP) inhibitor (EC 3.4.24.11), would be expected to potentiate the actions of a number of other vasoactive peptides including atrial natriuretic peptide (ANP). However, kelatorphan, a specific NEP inhibitor, did not inhibit the production of ET, indicating that this enzyme is not ECE.[81] Phosphoramidon has been shown to potentiate the ET-1-induced bronchopulmonary response in guinea pigs, indicating the involvement of endopeptidase-like enzymes, present in airway tissue, in the metabolism of ET-1.[82] It has also been reported that phosphoramidon is able to inhibit vasoconstrictor effects evoked by intravenous injections of big ET-1 in anesthetized pigs, but did not have any effect on the plasma ET-1 level.[83] It should be noted that phosphoramidon, being a rather general metalloproteinase inhibitor, may be acting via a number of mechanisms to cause these *in vivo* effects. Clearly the

TABLE 2
Cleavage of the Trp21-Val22 Bond of Big
ET and Fragments

ECE	Substrate	ECE Converting Activity (Relative)
Microsomal	Big ET-1	100
	Big ET-2	5
	Big ET-3	0
	Big ET[1–37]	115
	Big ET[16–37]	371
	Big ET[1–31]	1
	Big ET[17–26]	0
Cytosolic	Big ET-1	100
	Big ET-2	0
	Big ET-3	0
	Big ET[1–37]	98
	Big ET[16–37]	439
	Big ET[1–31]	8
	Big ET[17–26]	0

Data from Reference 84.

discovery of specific ECE inhibitors is awaited to elucidate the true identity of ECE. However, it is not clear at present that there is a single enzyme responsible for this cleavage.

Investigations of the structural requirements for the metalloproteinase ECE have indicated that the residues 32–37 of Big ET-1 are important for conversion (Table 2), while the amino-terminal loop structure appears to interfere with the access of ECE to big ET.[84] The ability of the metalloproteinase ECE derived from bovine aortic endothelial cells to cleave various big ET analogs is shown in Table 2.[84] Transition-state substrate analogs of big ET may be one possible approach to the development of specific ECE inhibitors.

IV. TISSUE DISTRIBUTION OF ET-1 AND RELATED PEPTIDES

The name "endothelin" obviously derives from the endothelial cells from which this peptide was originally isolated. However, it has since been found that the ETs are produced in many other cell types, including airway epithelial cells,[85] renal cell lines,[86,87] macrophages,[88] and several cancer cell lines.[89] Immunoreactive ET is widely distributed throughout many vascular and nonvascular tissues.[90] ET-1 has also been detected in cerebral cortex, cerebellum, brain stem, basal ganglia, and hypothalamus of the human brain, with a much lower abundance in the pituitary gland.[91,92] Colocalization studies in the cortex have shown that ET-1 mRNA and immunoreactivity are present in cells that express NPY mRNA and immunoreactivity. ET peptides and their precursors have also been detected in the human cerebrospinal fluid.[93] Micro-autoradiography has revealed high densities of iodinated ET-1, -2, and -3 in the human uterus localized to glandular epithelial cells and blood vessels.[94] Endothelin immunoreactivity has also been detected in human placenta, chorion, and amnion.[95,96]

V. RECEPTOR SUBTYPES

Cloning and expression of two ET subtypes, termed ET_A and ET_B, from bovine and rat lung, respectively, has been reported.[97,98] The receptors have been classified according to the relative affinities of the agonists from the endothelin and sarafotoxin peptide families (Table 3).[99] Autoradiographic studies using radiolabeled ligands have been reported to study the receptor subtype

TABLE 3
Receptor Subtypes

	ET_A	ET_B
Agonist affinity	ET-1 = ET-1 >> ET-3 >>> SRTX-6c	ET-1 = ET-2 = ET-3 = SRTX-6c
Source of cloned receptors	Bovine lung[97] Rat A10 cell[107] Human placenta[101–103]	Rat lung[98] Human jejunum/liver[100,106] Human placenta[105] Porcine cerebellum[108]
Localization	Vascular smooth muscle,[118] heart,[118] lung,[118] intestine,[115] brain[116]	Brain,[116,117] endothelial cells, lung, kidney

TABLE 4
Distribution of Human Binding Sites

Location	Ref.
Cardiac tissue: nerves, atria, ventricles, coronary arteries	118,119
Renal tissue: glomeruli > papillae	120
Adrenal glands: zona glomerulosa > medulla	121
Central nervous system	116,117
Lung, pulmonary artery, bronchus, pulmonary vein	118,122
Gastrointestinal tract	115
Fetal placental vessels	123
Human umbilical cord	124

distribution in both animal and in human tissues. More recently the ET_A and ET_B human receptors have also been cloned from a variety of sources (Table 3).[100–105]

Both receptor subtypes are rhodopsin-like in structure and are coupled to G-proteins. The ET_A receptor clearly mediates vasoconstriction and is widely localized in vascular smooth muscle of cardiovascular tissue origin, in the lung and in certain regions of the brain (Table 4).[97,106] The ET_B or nonselective receptor, recognizing the ET isopeptides with equal affinity, was originally known as the nonvascular smooth muscle receptor (Figure 6).[98,107] This receptor is localized in endothelial cells and in certain regions of the brain and has been associated with vasodilator activity, perhaps through the release of the endothelium derived relaxing factor (EDRF).[64] However, it has been reported that the rabbit pulmonary artery recognizes all three isopeptides with equal affinity; thus, it has ET_B-like characteristics. It has also been recently reported that smooth muscle cells of the guinea pig intestine express an ET_B receptor that primarily mediates the contractile effect on smooth muscle cells.[108] Thus, the initial classification of ET_B as a nonvascular smooth muscle receptor may be somewhat misleading. Selective ET_B agonists such as ET[Ala1,3,11,15], ET-1[8–21, Ala11,15], and SRTX-6c are potent vasoconstrictors in rabbit pulmonary artery.[109–111] It is possible that the ET_B-like receptors in the brain and in endothelial cells are not identical. Recently it has been reported that contraction in the pig coronary artery is mediated by two receptors, one with an ET_A profile and the other recognizing SRTX-6c and ET-3 but not ET-1 and SRTX-6b, thus being different from the ET_B receptor subtype.[112] There has also been a report of endothelin receptors in the rat renal papilla with a high affinity for ET-3, in fact slightly higher than for ET-1 itself.[113] Recently, a third receptor subtype has been described from *Xenopus* dermal melanophores and is specific for ET-3.[114]

The distribution of ET receptors in various human and animal tissues has been reported[115–124] using autoradiographic techniques (Table 4) and there is some initial information regarding receptor subtype distribution using radiolabeled receptor-specific ligands.[125] Thus, ET_A receptors have been

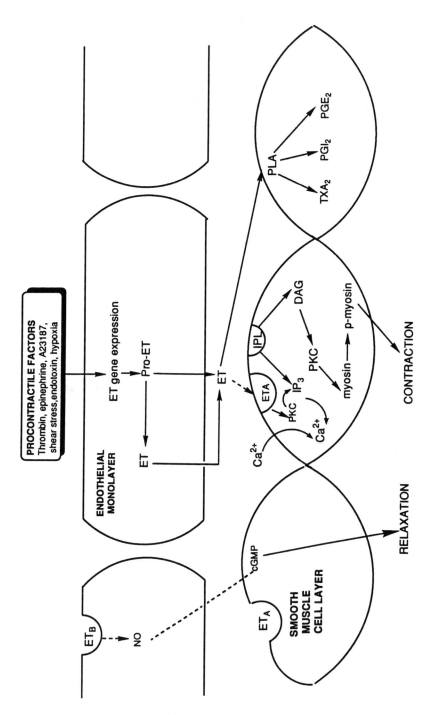

FIGURE 6. Intracellular signaling.

TABLE 5
Factors Affecting Endothelin Release

Factors Stimulating ET Release	Ref.
Thrombin	2,129–131
TGF-β	2,130,131
Epinephrine	2,131
Phorbol esters	2,129,130
A23187	2,131
AVP	132
Endotoxin	133
Oxyhemoglobin	134
Elevated glucose levels	135
Digitalis-like factor	136
Cytokines	130,137
Angiotensin II	132
Modified low-density lipoproteins	138
Shear stress	139
Hypoxia	140
Platelet aggregation	141
Insulin	142
Cyclosporine	143
Oxytocin and AVP agonists	144
Factors Inhibiting ET Release	
ANP	145, 146
BNP	145
Nitroprusside	146
8-Bromocyclic GMP	146
Endothelium-derived relaxing factor	147,148
Prostacyclin	146
ET-3	149
Angiotensin-converting enzyme inhibitors	150

reported to mediate the signaling and secretory actions in pituitary gonadotrophs,[126] and the mitogenic response in rat aortic vascular smooth muscle cells.[127]

VI. ENDOTHELIN RELEASE

There have been many studies attempting to understand the regulation of endothelin release (Table 5).[128–150] The induction of mRNA is relatively slow, suggesting that ET-1 may participate in long-term responses rather than as a circulating hormone.

VII. MECHANISM OF ACTION

There has been much research conducted to attempt to clarify the mechanism of action of the endothelins.[151–155] ET-1 stimulation of vascular smooth muscle (VSM) cells is associated with a phospholipase-mediated increase in the formation of diacylglycerol and inositol phosphates. Subsequent activation of protein kinase C (PKC) leads to an elevation in intracellular calcium concentration $[Ca^{2+}]_i$.[156] Interaction of Ca^{2+} with calmodulin and myosin light chain kinase, results in the phosphorylation of myosin and actin-myosin crossbridge formation, ultimately causing vasoconstriction (Figure 6). Although it was initially thought that ET-1 was a calcium channel agonist,[156] it was subsequently shown that the ET-1 receptors are distinct from the dihydropyridine-sensitive calcium channel.[157–159] Moreover, calcium channel antagonists only partially block the effects of ET.[157] The dose-dependent elevation in $[Ca^{2+}]_i$ has been reported to occur in two distinct phases.

Initially there is a rapid transient increase in $[Ca^{2+}]_i$, the mechanism of which is poorly understood, followed by a slower decrease to higher than basal values. Several other mechanisms have also been proposed for the action of ET-1 in VSM, including activation of phospholipase A_2, modulation of Na^+-K^+ ATPase, stimulation of Na^+/H^+ antiport and Na^+/K^+ exchange, and blockage of ATP-sensitive K^+ channels.[160,161] ET-1 stimulates PI hydrolysis with an EC_{50} of 0.3 to 0.8 nM in adult rat ventricular myocytes and is not inhibited by nifedipine or pertussis toxin. ET-1 also reduces cAMP in myocytes in response to isoproterenol and forskolin (EC_{50} = 1 nM). The cAMP lowering effect is sensitive to pertussis toxin and seems to be mediated via the inhibitory guanine nucleotide-binding protein G_i.[162]

Preliminary experiments suggest that a pertussis toxin-sensitive G-protein couples ET receptors to PLC and perhaps to PLD. PKC appears to inhibit ET-induced Ca^{2+} signaling, thereby serving as a negative feedback signal. There is some evidence that ET activates phospholipase A_2 (PLA_2) in cultured smooth muscle and mesangial cells, causing stimulation of the arachidonic acid cascade.[163] It is not known whether ET activates PLA_2 directly via a G-protein or indirectly by increasing intracellular Ca^{2+}.

A recent report, demonstrating the vasodilatory properties of ET-1, -2, and -3 *in vivo,* indicates that the pulmonary vasodilation observed depends, in part, on the potassium channel activation.[160] It has also been reported that nitric oxide (EDRF) plays a role in the regional vasodilator effects of endothelin.[164]

VIII. ENDOTHELIN ADMINISTRATION TO MAN

There have been a few isolated studies to evaluate the effects of administration of ET to humans.[165,166] When ET-1 was infused over 5-min periods into the nondominant brachial artery in two male subjects at doses from 5×10^{-14} to 5×10^{-9} M, there were no major changes in mean arterial pressure, heart rate, or right atrial pressure. An initial increase in forearm blood flow was observed followed by dose-dependent decreases.[165] At higher doses, given to one subject, edema formation in the forearm was observed, accompanied by sweating and emesis. The authors state that the adverse reactions observed in this subject are remarkably similar to those reported after a bite of the Israeli burrowing asp whose venom contains the closely related SRTX peptides![8]

In another study intravenous infusion (10^{-10} to 5×10^{-8} M) of human ET-1 to six healthy, sodium-loaded volunteers caused an increase in ET-1 plasma concentrations followed by an increase in mean arterial blood pressure, and serum potassium concentrations.[166] Plasma concentrations of renin, ANF, and aldosterone did not change during the infusion.

IX. CLINICAL RELEVANCE

The relevance of elevated plasma levels in different disease states is unclear at present. This is primarily due to the difficulty of accurately measuring ET levels, since the levels of extraction of ET from the plasma have been found to be low and variable. It is not known whether ET acts in a paracrine or autocrine manner and thus what local tissue concentrations are relevant. The concentrations of ET-1 in plasma from healthy control subjects has been measured at approximately 1.11 $\pm$ 0.2 pmol^{-1} and it has been suggested that careful extraction procedures are required for accurate measurement since ET-1 may bind to albumin.[167] Using monoclonal or polyclonal antibodies for ET-1, specific radioimmunoassays (RIAs) have been developed to measure plasma concentrations. However, the reported normal ranges have been found to vary, probably due to the specificity of the antibodies used in the assay and the difficulties with extraction from plasma. Elevated levels of endothelin have been reported in several disease states but the variation between groups makes it difficult to draw any real conclusions (Table 6).[168–195]

Lerman and coworkers recently demonstrated that exogenous infusion of ET (2.5 ng/kg/ml) to anesthetized dogs, producing a doubling of the circulating concentration, did have biological actions.[196] Thus, heart rate and cardiac output decreased in association with increased renal and

TABLE 6
Plasma Concentrations of ET-1 in Humans

Condition	Normal Control	ET Plasma Levels Reported (pg/ml)	Ref.
Atherosclerosis	1.4	3.2 (pmol/l)	168
Surgical operation	1.5	7.3	169
Buerger's disease	1.6	4.8	170
Takayasu's arthritis	1.6	5.3	170
Cardiogenic shock	0.3	3.7	171
Congestive heart failure (CHF)	9.7	20.4	162
Mild CHF	7.1	11.1	173
Severe CHF	7.1	13.8	173
Dilated cardiomyopathy	1.6	7.1	174
Preeclampsia	10.4 pmol/l	22.6 pmol/l	175
Pulmonary hypertension	1.45	3.5	176
Acute myocardial infarction	1.5	3.3	177
	6.0	11.0	178
	0.76	4.95	179
	0.50	3.8	180
Subarachnoid hemorrhage	0.4	2.2	181
Crohn's disease	0–24 fmol/mg	4–64 fmol/mg	182
Ulcerative colitis	0–24 fmol/mg	20–50 fmol/mg	182
Cold pressor test	1.2	8.4	183
Raynauld's phenomenon	1.7	5.3	184
Raynauld's/hand cooling	2.8	5.0	185
Hemodialysis	<7	10.9	186
	1.88	4.59	187
Chronic renal failure	1.88	10.1	187
Acute renal failure	1.5	10.4	188
Hypertensives	0.96	1.09	189
Uremia before hemodialysis	0.96	1.49	189
Uremia after hemodialysis	0.96	2.19	189
Essential hypertension	18.5	33.9	190
	0.2	1.1	191
Essential hypertension	1.6	1.7	192
Sepsis syndrome	6.1	19.9	193
Postoperative cardiac	6.1	11.9	193
Inflammatory arthritis	1.5	4.2	194
Malignant hemangioendothelioma	4.3 (after removal)	16.2	195

systemic vascular resistances and antinatriuresis. These studies support a role for endothelin in the regulation of cardiovascular, renal, and endocrine function.

In the anesthetized dog with congestive heart failure, a significant two- to threefold elevation of circulating ET levels has been reported,[197] and studies in humans have shown similar increases.[198] When ET was chronically infused into male rats, to determine whether a long-term increase in circulating ET levels would cause a sustained elevation in mean arterial blood pressure, significant, sustained, and dose-dependent increases in mean arterial BP were observed. Similar results were observed with ET-3, although larger doses were required.[199] There is some speculation that ET will be involved in acute or perhaps chronic renal failure due to its potent effects on the kidney. Cyclosporine-induced renal failure may be a useful *in vivo* model with which to study ET antagonists.

One of the problems in ET research at the current time is the lack of a relevant clinical model with which to study ET antagonists and ECE inhibitors. Most *in vivo* studies have reported the effects of ET antagonists after a big ET or ET-antibody infusion. Animal models that are relevant, where ET overproduction, either at the level of the vascular smooth muscle or in bloodstream and

TABLE 7
Biological Activities of Endothelin Analogs

| | Binding Assays (IC_{50}, nM) | | Functional Assays (EC_{50}, nM) | |
| | | Rabbit Pulmonary | | Rabbit Pulmonary |
Compound	Rat Aorta	Artery	Rat Aorta	Artery
ET-1	1.6 ± 0.1	0.5 ± 0.1	5.8 ± 0.2	5.5 ± 0.4
ET-2	2.6 ± 0.6	1.0 ± 0.1	9.0 ± 3.3	8.1 ± 0.6
ET-3	30.0 ± 3.0	7.1 ± 0.3	134.6 ± 2.7	4.9 ± 0.5
SRTX-6b	2.2 ± 0.1	0.5 ± 0.1	10.8 ± 1.7	4.5 ± 0.9
SRTX-6c	NT	0.6 ± 0.2	$>10,000$	2.1 ± 0.4
ET-1[12Pro]	3.6 ± 0.03	2.1 ± 0.7	99.4 ± 2.5	27.8 ± 5.1

Data from Reference 218.

in which ET is causal to the disease state, obviously need to be developed. Table 7 indicates some conditions in which ET levels have been measured and found to be elevated. It should be noted that ET-1 levels differ markedly between arterial and venous plasma, being significantly higher in the former.[200]

X. STUDIES WITH ENDOTHELIN ANTIBODIES

Several *in vivo* studies with ET antibodies have been reported in disease models. Left coronary artery ligation and reperfusion to induce myocardial infarction in the rat heart, caused a four- to sevenfold increase in endogenous endothelin levels.[201] Administration of ET antibody was reported to reduce the size of the infarction in a dose-dependent manner.[201] Thus, ET may be involved in the pathogenesis of congestive heart failure and myocardial ischemia.[13,201,202]

Studies by Kon and colleagues using anti-ET antibodies in an ischemic kidney model, to deactivate endogenous ET, indicated the peptide's involvement in acute renal ischemic injury.[203] In isolated kidneys, pre-exposed to specific anti-endothelin antibody and then challenged with cyclosporine, the renal perfusate flow and glomerular filtration rate increased, while renal resistance decreased as compared with isolated kidneys pre-exposed to a nonimmunized rabbit serum. The effectiveness and specificity of the anti-ET antibody were confirmed by its capacity to prevent renal deterioration caused by a single bolus dose (150 pmol) of synthetic ET, but not by infusion of angiotensin II, norepinephrine, or the thromboxane A_2 mimetic U-46619 in isolated kidneys.[204]

Others have reported inhibition of ET-1- or ET-2-induced vasoconstriction in rat isolated thoracic aorta using a monoclonal antibody to ET-1.[205]

Combined administration of ET-1 and ET-1 antibody to rabbits showed significant inhibition of the BP and renal blood flow responses. However, ET-1 antibody alone did not change either BP, renal BF, mesenteric BF, or carotid BF *in vivo*. Thus, the authors suggest that ET may not contribute to the control of BP and BF under normal conditions.[206]

In another study, anti-ET-1 antibody injected i.v. to SHR and WKY conscious rats showed no changes in blood pressure in both strains evaluated under acute and chronic conditions. These authors postulate that ET may not play a role in the maintenance of hypertension in SHR or in the regulation of blood pressure in WKY rats.[207,208] In contrast, other investigators have reported that infusion of ET-specific antibodies into SHR decreased mean arterial pressure (MAP), and increased glomerular filtration rate and renal blood flow. In the control study with WKY there were no significant changes in these parameters.[209]

Studies with monoclonal antibodies may be a powerful preliminary tool to investigate a pathophysiological significance for ET in different disease states. Antibody-directed drug design that utilizes the three-dimensional chemical information in antibodies to drug targets to design novel pharmacophores is an interesting approach to the discovery of receptor ligands. By using the idiotypic antibody, which is a mirror image of the target binding site, it may be possible to design new antagonists.[210]

XI. DESIGN OF ENDOTHELIN RECEPTOR ANTAGONISTS

There are a variety of approaches that medicinal chemists generally use when attempting to discover and design receptor agonists and antagonists for biologically active molecules. If one has some information about the binding site then it may be possible to design compounds that will optimize the key binding interactions required, particularly for agonists. Early in endothelin research the receptors had not been cloned and there was no structural information of the binding site. Thus, many groups studied the solution conformation of ET in order to elucidate important structural features that might be relevant in the bound state.

Thus, several structural studies of ET-1 by nuclear magnetic resonance (NMR), molecular dynamics simulation and energy minimization, fluorescence and circular dichroism (CD) have been reported in the last 4 years.[211–215] CD studies indicate that ET-1 is about 35% helical, the helicity existing between residues Lys9 and Cys15.[211–213]

There have been conflicting reports of the conformation of the biologically important C-terminal hexapeptide, but most studies are unable to define this apparently flexible region. The spatial proximity of residues Tyr[13] and Trp[21] is suggested from fluorescence energy transfer measurements and these authors thus suggest the possibility that the C-terminal region is located close to the bicyclic loop region.[214] The bioactive conformation may of course differ from the results indicated by these studies. A low-energy conformation of ET-1 derived from NMR and energy minimization is shown in Figure 7.

Endothelin-3 has considerable sequence homology with ET-1 (71%) and of the six amino acid changes, three are conservative in nature. However, the contractile potency of ET-3 is considerably less than ET-1 and thus it is of interest to know whether the conformation of the peptide differs. ET-3 has also been studied by NMR spectroscopy which revealed the same helical region between residues 9 and 15 as that reported previously for ET-1.[215] These authors also suggest the proximity of the C-terminal Trp to the bicyclic region.

Conformational studies of SRTX-6b in aqueous solution have also been reported. The structure derived from high-resolution NMR spectroscopy is very similar to that reported for ET-1. The C-terminal region of the peptide does not adopt a preferred conformation.[216]

Endothelin-1 in aqueous solution has been reported to aggregate through the formation of micelles with a critical micellar concentration of 2.2×10^{-5} M.[217] Thus, the solution conformation in water and organic solvents may not be relevant to the receptor-bound conformation.

Since the helical region appears to be well conserved among the ETs and SRTX-6b, disruption of this structural feature might be expected to cause a fall in biological activity. Thus, Val12 in the center of the helical region was replaced with Pro12 in ET-1.[218] The biological activities in a variety of tissues are shown in Table 7. Clearly disruption of the helical structure leads to a loss in binding affinity and functional activity at both ET_A and ET_B receptors, but surprisingly the analog is still a fairly potent agonist. Thus, the helical region between residues Lys9 and Cys15 of ET-1 is not critical for receptor binding and functional activity.

In an NMR study to determine the solution conformation of big ET, the tail region [Leu17-Ser38], reported to be important for cleavage by ECE, did not adopt a defined conformation.[219,220]

A. STRUCTURE-ACTIVITY RELATIONSHIPS

The first step towards the rational design of antagonists involves a knowledge of the residues important for agonist and antagonist activity. Thus, there have been many reports of the synthesis and pharmacological evaluation of ET analogs and fragments. Much of the early research on structure-activity relationships (SAR) is somewhat confusing because the compounds have been evaluated in a wide variety of both animal and human tissues of unknown receptor subtype populations.

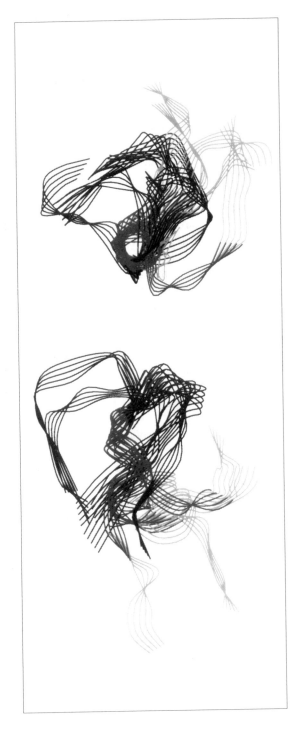

FIGURE 7.   Conformation of endothelin-1 derived from NMR data.

ET-3 is the weakest vasoconstrictor in the ET family, acting as a vasodilator in some vascular beds.[221] The sarafotoxins produce vasoconstriction in rat thoracic aorta, rat isolated perfused mesentery, and pithed rat. The potency order was ET-1 > SRTX6b > SRTX-6a > SRTX-6c at lower doses, but SRTX6b > ET-1 > SRTX-6a > SRTX-6c at higher doses.[222] The substitution of Ser2 with Thr2 in ET-3 is shared by two weak constrictor peptides in the closely related sarafotoxin (SRTX) family, SRTX-c and SRTX-d, suggesting its possible importance for receptor affinity.[223,224] However, structure-activity studies of various substituted SRTXs have indicated that the Lys9 to Glu9 substitution results in a much larger loss of biological activity than either the Ser2 to Thr2, Lys4 to Asn4, or Tyr13 to Asn13 substitutions.[225] Thus, the low lethality and vasoconstrictor activity of SRTX-d is somewhat unexpected in view of its structure.[10]

Structure-activity studies of ET-1-related peptides and fragments have indicated that the C-terminal Trp residue is important for the vasoconstrictor activity in porcine coronary artery strips (ET_A).[225] ET[1–20]ET[1–15,NH_2] has been shown to be inactive in binding in rabbit cardiac tissue,[226] while the hydrophobic C-terminal hexapeptide ET[16–21] has been reported as a partial agonist in guinea pig isolated bronchus, rat vas deferens, and rabbit pulmonary artery.[227–229] However, it has been reported that this fragment is devoid of activity in several tissue preparations, including rat aorta and rabbit pulmonary artery, guinea pig ileum, human urinary bladder, renal pelvis, and renal artery.[228,230,231] There have been several studies attempting to demonstrate the agonist activity originally reported for the ET[16–21] fragment, albeit with little success.[232–235] Maggi et al. have postulated that the agonist activity of ET[16–21] in certain tissues is due to its selective action at a single receptor subtype not expressed in cardiovascular tissues.[227] However, it is clear that the C-terminal hexapeptide, while being important for receptor recognition, possesses very weak binding affinity at both ET_A and ET_B receptors (IC_{50} = 50 to 70 μM).[230] In a binding assay to rabbit cardiac tissue, ET[16–21] was completely inactive at concentrations up to 10 μM.[236]

Monoamino acid substitutions in ET have revealed some information about the importance of individual amino acids to biological activity. Replacement of the Met7 of ET-1 with Leu7 caused no loss in agonist activity in isolated rabbit renal artery (ET_A) and rat pulmonary artery.[237] Acetylation of the N-terminus led to a complete loss in biological activity and binding affinity.[237,238] Replacement of the Trp21 with Ala21 led to a complete loss in binding affinity confirming the importance of the C-terminal residue Trp21. Other aromatic groups at the Trp21 position are only weakly tolerated. Clearly Asp8, Glu10, and Phe14 are key residues for biological activity in rat pulmonary artery (see Table 8).[237] It is interesting that in an energy minimized solution structure model of ET-1 these residues were all positioned on the same side of the molecule (Figure 7).

Systematic monoala (Table 9) and mono-D-amino acid (Table 10) replacements in ET-1 have revealed a number of interesting structure-activity relationships.[239,240] The peptides were tested for their binding affinity on human vascular smooth muscle cells and for agonist activity on the rabbit vena cava *in vitro*. The results were expressed as IC_{50} and ED_{50} values, respectively, and expressed as a percent potency relative to ET-1. Replacement of Ala tests the importance of each side chain functionality for receptor recognition. This study shows that residues essential for agonist activity include Asp8, Phe14, Leu17, and Trp21.[239] Substitution in the C-terminal region was generally much more sensitive than in the N-terminal residues. This is particularly interesting because most reports on the structure of ET-1 indicate that the C-terminus is completely flexible! While ET-1[Ala8] would appear to be an antagonist from this study, the authors state that this analog is not a competitive antagonist. Interestingly ET-1[Ala16] appears to be a more potent agonist than ET-1 itself.

In a mono-D-amino acid scan the orientation of the side chain and backbone conformation are probed.[240] From this study it was found that inversion of configuration is generally less well tolerated in the [16–21] tail of the molecule. The correct orientation of the charged residues Asp8 and Glu10 is also clearly important for agonist activity. The authors report that D-Phe14[ET] and D-Leu17[ET] exhibited partial agonism which is difficult to explain based on their very low binding affinities.

Replacement of the His16 residue with Phe16 led to a compound that was fivefold more potent as an agonist at both ET_A and rabbit pulmonary artery ET_B-like receptors.[218] This was clearly an

TABLE 8
Smooth Muscle Constrictor
Activity of ET-1 Analogs on
Rat Pulmonary Artery Ring
Preparations

Compound	Relative Activity
ET-1	100
ET(1–15)NH$_2$	0
ET[16–21]	0
Ac-ET	0.5
Des-Trp21-ET	0
ET-NH$_2$	5.6
Asn8-ET	0.8
Gln10-ET	0
Ala14-ET	0
Tyr21-ET	32.8
Phe21-ET	18.4
Met(O)7-ET	69.1
Big ET	1.8

Data from Reference 237.

TABLE 9
Monoala Scan of ET-1

Ala Replacement	% Binding	% Contraction
Ser2	100	70
Ser4	200	51.5
Ser5	142	24.8
Leu6	101	73
Met7	350	88.5
Asp8	103	0.8
Lys9	504	56
Glu10	162	121
Val12	16.2	29.4
Tyr13	0.8	87.3
Phe14	1.8	8.4
His16	339	382
Leu17	25	0.9
Asp18	50	14
Ile19	406	121
Ile20	22.0	22.5
Trp21	0.8	<0.1

Data from Reference 239.

important result for the later design of receptor antagonists derived from the C-terminal region of ET.

Receptor binding results in cultured rat smooth muscle cells (ET$_A$) revealed that ET[1–23], ET[1–26] and ET-1 were equipotent, although functional studies demonstrated that these C-terminal elongated peptides were weaker agonists.[241] Further research with this series could yield receptor antagonists.

TABLE 10
D-Amino Acid Scan of ET-1

D-Amino Acid	% Binding	% Contraction
Ser2	100	100
Ser4	10	0.1
Ser5	n.d.[a]	n.d.[a]
Leu6	126	58
Met7	126	20.4
Asp8	2.7	0.6
Lys9	210	418
Glu10	154	53.5
Val12	27	21.6
Tyr13	8	9.5
Phe14	2.2	1.5
His16	0.3	0.1
Leu17	12.6	0.1
Asp18	1.6	1.6
Ile19	0.05	<0.3
Ile20	3.2	5.0
Trp21	2.4	0.4

[a] n.d., Not determined.

Data from Reference 240.

B. THE IMPORTANCE OF THE DISULFIDES

The cyclic structure of the ETs would appear to be essential for binding and functional activity only in certain tissues,[226,242] for example, in the rat and porcine aorta (ET$_A$), and in rat pulmonary artery.[237] The outer disulfide bond Cys1-Cys15 would appear to be much more important than the inner Cys3-Cys11 bond for potent binding to the ET$_A$ receptor.[243] Reduction and carboxymethylation of the cysteine residues caused a complete loss of agonist activity in rat isolated perfused mesentery and also the tetra-alanyl analog, ET-1[Ala1,3,11,15], was functionally inactive in the rat mesenteric bed and rat isolated aorta.[244,245] However, the structural requirements for binding to rat cerebellum (ET$_B$ receptor) do not include the presence of the disulfide linkages, as illustrated by the equipotent binding of the tetra-alanyl-substituted analog ET-1[Ala1,3,11,15] and ET-1 (Table 11).[245–247] These results in various tissues clearly indicate a different distribution of receptor subtypes in the brain (ET$_B$) and in cardiac tissues (ET$_A$) (also see Table 3).[245–247]

The binding affinity of various monocyclic fragments of ET-1 in rabbit pulmonary artery (ET$_B$-like)[38] and aorta (ET$_A$) have indicated that the loop region 3–11 does not bind at concentrations up to 100 μM. A monocyclic analog without the 3–11 region (i.e., cyclo[Cys-Ser-Aoc-Val-Tyr-Phe-Cys]-His-Leu-Asp-Ile-Ile-Trp where Aoc = 8-aminooctanoic acid) binds with micromolar affinity but shows no functional activity in rabbit pulmonary artery up to 30 μM.[38] This series of compounds (Table 12) have, however, been found to antagonize ET-1-induced inositol phosphate accumulation in rat skin fibroblasts and to inhibit ET-induced arachidonic acid accumulation in rabbit renal artery smooth muscle cells indicative of ET$_A$ antagonist activity.[38,248]

Analogs of ET-1 (Ala^{3,11}, Nleu^7) containing alanine substitutions were prepared and assessed for receptor binding affinity in A10 vascular smooth muscle cells and vasoconstrictor activity in rabbit carotid rings.[249,250] All of the analogs bound with >75% inhibition at 10 μM, with the exception of the Ala21 compound, which was completely inactive. Comparison of the biological activities revealed that Ser2, Val12, His16, and Ile19 could all be replaced with alanine while still retaining reasonable agonist activity. Substitution of Asp8, Tyr13, Ile20, and Trp21 afforded analogs which

TABLE 11
Ki Binding Constants in
Rat Cerebellum

Peptide	Ki/nM
ET-1	0.83
ET-1[Ala1,3,11,15]	0.72
ET-1[Ala1,15]	11.5
ET-1[Ala3,11]	2.98

Data from Reference 247.

TABLE 12
Monocyclic Endothelin Analogs

Compound	Binding Affinity μM, IC_{50}		
	Rabbit Pulmonary Artery	Rabbit Aorta	Rat Ventricle
ET-1	0.0006	0.0013	0.0009
ET[16–21]	71	74	44
ET[3–11]-NH$_2$	>100	>100	>100
ET[3–11]-His-Leu-Asp-Ile-Ile-Trp	>100	>100	>100
Cys-Ser-Aoc-Val-Tyr-Phe-Cys-His-Leu-Asp-Ile-Ile-Trp	1.34	1.00	1.58

Data from Reference 38.

showed a greater than 1000-fold poorer affinity for the receptor than the parent monocycle, identifying these sites as important to receptor binding. Substitution of Glu10, Phe14, Leu17, and Asp18 produced analogs which showed appreciable affinity at the receptor at concentrations below those at which agonism was observed. Thus, these sites would appear useful for development of receptor antagonists.

C. ET RECEPTOR AGONISTS

Various groups have studied linear fragment analogs of endothelin in an attempt to discover agonists and antagonists. Many of the longer C-terminal-containing fragments are ET$_B$ agonists, the structural requirements for ET$_B$ agonism clearly being different from those required for ET$_A$ agonism. All known ET$_A$ agonists or nonselective ET$_A$/ET$_B$ agonists possess the bicyclic region of endothelin. ET$_B$ agonists appear to have been developed from systematic truncation of ET and the knowledge that Asp8 is particularly important in all structure-activity studies reported to date for binding and agonist activity.

Comparison of the receptor affinities of the ETs and SRTXs in rat aorta and atria (ET$_A$) or cerebellum and hippocampus (ET$_B$), indicate that SRTX-c is a selective ET$_B$ ligand (Tables 3 and 13).[251] A recent study showed that SRTX-c caused only vasodilation in the rat aortic ring, possibly through the release of EDRF from the endothelium.[109] Clearly the structural requirements for binding to the ET$_B$ receptor do not require the presence of the disulfide linkages (Table 13).[252,253] Thus, reported selective ET$_B$ agonists, for example, the linear analog ET-1[Ala 1,3,11,15] and truncated analogs ET-1[6–21, Ala 1,3,11,15], ET-1[8–21, Ala 11, 15], and N-acetyl-ET-1[10–21, Ala 1,3,11,15] caused vasorelaxation in isolated, endothelium-intact porcine pulmonary arteries.[253] However, some of these analogs ET-1[Ala 1,3,11,15] and ET-1[8–21, Ala 11,15] are potent vasoconstrictors in the rabbit pulmonary artery, a tissue that appears to possess an ET$_B$, nonselective type of receptor. ET-1[6–21, Ala-1,3,11,15] and ET-1[11–21, Ala 1,3,11,15] were considerably less

TABLE 13

**Receptor Subtype Binding in a Variety of Rat (a)
or Porcine (b) Tissues**

Ligand	Aorta	Atrium	Cerebellum	Hippocampus
(a) Binding in Rat Tissues (Ki/nM)				
ET-1	0.11	0.034	0.015	0.010
ET-3	2.50	1.60	0.021	0.036
SRTX-6b	0.24	0.087	0.013	0.012
SRTX-6c	>5000	4200	0.016	0.023
(b) Binding in Porcine tissues (IC$_{50}$/nM)				
ET-1	0.16		0.11	
ET-2	0.22		0.11	
ET-3	5.7		0.07	
SRTX-6a	9.4		0.41	
SRTX-6b	0.95		0.13	
SRTX-6c	>200		3.00	
ET[1,3,11,15-Ala]	570		0.33	
N-AcET[1,3,11,15-Ala]	600		0.34	
4AlaET(6–21)	330		0.32	
ET[8–21,11,15-Ala]	1400		1.2	
N-acetyl-ET[10–21]	9900		12.0	
IRL 1620	1900[a] (Ki)		0.016 (Ki)	

[a] porcine lung

Data from Reference 253.

potent at the ET$_B$ receptor (IC$_{50}$ = 22 and 1.2 μM, respectively). BQ-3020 is a linear ET-1 analog [Ac-Leu-Met-Asp-Lys-Glu-Ala-Val-Tyr-Phe-Ala-His-Leu-Asp-Ile-Ile-Trp] recently reported as a selective ET$_B$ agonist.[254] IRL 1620 [Suc-Asp-Glu-Glu-Ala-Val-Tyr-Phe-Ala-His-Leu-Asp-Ile-Ile-Trp] is also a selective ET$_B$ agonist causing relaxation of norepinephrine-stimulated aorta. This effect was abolished by removal of the endothelium or by adding an inhibitor of NO synthesis, N-monomethyl-L-arginine.[255] IRL 1620 administered to rats (0.1 to 3 nmol/kg i.v.) produced an immediate dose-related fall in arterial pressure with a maximum reduction of −26 ± 3 mmHg in WKY and 60 ± 3 mmHg in SHR with a 1-min duration. A secondary more sustained response then occurred. In contrast, when IRL 1620 was given i.c.v. (0.12 to 120 pmol) an immediate and transient pressor response was observed (14 mmHg). The pressor response to i.v. IRL 1620 was not blocked by the ET$_A$ selective antagonist BQ-123, demonstrating that selective ET$_B$ activation elicits a biphasic BP response that is not mediated via the ET$_A$ receptor.[256] Interestingly, the ET$_B$ agonists ET-1[Ala 1,3,11,15] and IRL 1620, however, have been shown to reduce pulmonary vascular resistance without systemic hemodynamic effects (at all doses) in a new-born lamb model of pulmonary hypertension.[257]

D. ET RECEPTOR ANTAGONISTS

1. Bicyclic Analogs

Initial reports of ET receptor antagonists have begun to appear. A full-length ET analog, ET-1[Dpr1-Asp15] (where Dpr = diaminopropionic acid), has been reported as a specific ET$_A$ antagonist by Spinella and coworkers (Figure 8).[258,259] Inhaled ET-1 and, to a lesser extent, ET-3 cause dose-related bronchoconstriction in conscious sheep. The response to ET-1 can be blocked by the ET-1 receptor antagonist analog [diaminopropionic acid1-Asp15]ET-1.[260] [Thr18, Leu19]-ET-1 and [Thr18,

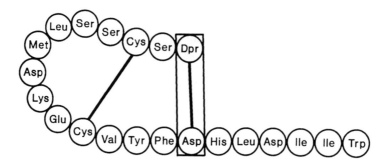

FIGURE 8. Structure of the ET_A antagonist ET-1[Dpr1-Asp15].

γ-MeLeu19]-ET-1 have been reported as ET_A/ET_B nonselective antagonists and the analog [Asn8, Gln10]ET-1 has also been shown to possess antagonistic activity at the ET_A receptor.[261,262]

2. Rational Design of Hexapeptide Antagonists

Based upon the importance of the C-terminal region for binding affinity and agonist activity several groups have focused on this region for the development of receptor antagonists. Recent structure-activity studies on the C-terminal hexapeptide of endothelin in the guinea pig bronchus have indicated that the Trp21, Asp18, His16, and Leu17 residues are important for biological activity in this tissue.

A mono D-amino acid scan has revealed that substitution at the His16, Asp18, and Ile20 with the D-amino acids led to compounds that were more potent than ET[16–21] itself (Table 14).[263] Indeed the hexapeptide Ac-D-His-Leu-Asp-Ile-Ile-Trp, exhibited a 13-fold enhancement in binding affinity compared with the natural ET[16–21] (see Table 14).[264] This compound is also a functional antagonist of endothelin-induced inositol phosphate accumulation in rat skin fibroblasts and inhibits ET-induced arachidonic acid release in vascular smooth muscle cells.[264,265] Unfortunately ET[16–21, D-His16] did not inhibit endothelin-induced vasoconstrictor activity in a bioassay with either known ET_A or ET_B tissues. It was previously reported that ET-1[16-Phe] was a fivefold more potent agonist than ET-1 itself in the rat aorta (ET_A) and in rabbit pulmonary artery (ET_B).[218] Substitution of the D-His residue with D-Phe (Table 15) showed a further enhancement in binding affinities and functional antagonism. Comparison of the binding affinities in vascular smooth muscle cells (ET_A) and in rat cerebellum (ET_B) indicates that these hexapeptides are nonselective ET_A/ET_B ligands. D-stereochemistry at position 16 affords a large increase in binding affinity and antagonist activity at both ET_A and ET_B receptors. This was found to be a fairly general phenomenon throughout the series. Extensive SAR around this series has been carried out and Table 15 lists a selection of compounds with differing selectivity for ET_A vs. ET_B receptors.[235] Generally the leucine can be replaced with a variety of residues without a loss of activity at either ET_A or ET_B receptors. Indeed others have used this observation to carry out an extensive SAR of Ac-D-Phe-Orn-Asp-Ile-Ile-Trp using multipin peptide synthetic technology.[266,267] Replacement of the Leu17 and Asp18 positions with aromatic residues has led to increased ET_B selectivity. Few changes are well tolerated in the Ile19-Ile20-Trp21 tripeptide without marginal losses in antagonist activity at the ET_A receptor. However, some modifications in the C-terminal tripeptide have maintained ET_B binding. Clearly, shortening the hexapeptide generally leads to some loss in activity at both receptor subtypes.

SAR studies around the D-Phe16 position with different aromatic substitutions have led to the potent functional ET_A/ET_B antagonist PD 142893 (Table 16).[268–270] PD 142893 displays moderate binding selectivity (10-fold) for the ET_A receptor over the ET_B subtype. Moreover, PD 142893 is a functional antagonist of endothelin-induced arachidonic acid release in vascular smooth muscle cells (ET_A) and antagonizes endothelin-stimulated constriction in the rabbit femoral, renal (ET_A), and pulmonary arteries (ET_B). We have further explored the structure-activity relationships of the 16-position with the substitution of tricyclic amino acids in this position. In particular, PD 145065,

TABLE 14
Analogs of ET[16–21]

Compound	Binding[a] ($IC_{50}/\mu M$)	IP[b]
Ac-His-Leu-Asp-Ile-Ile-Trp	58	50
Ac-His-Leu-Asp-Ile-Ile-D-Trp	>50	c
Ac-His-Leu-Asp-Ile-D-Ile-Trp	16.9[d]	>100
Ac-D-His-Leu-Asp-D-Ile-Ile-Trp	169[d]	c
Ac-His-Leu-D-Asp-Ile-Ile-Trp	37	>50
Ac-His-D-Leu-Asp-Ile-Ile-Trp	39	c
Ac-D-His-Leu-Asp-Ile-Ile-Trp	4.1[d]	1.4

[a] Rat heart ventricle.
[b] Inhibition of ET-1-stimulated inositol phosphate accumulation in rat skin fibroblasts.
[c] Not tested.
[d] n = 2 IC_{50} determinations. All other values represent one IC_{50} determination.

Data from Reference 263.

TABLE 15
Analogs of the ET[16–21,16-D-Phe] Series

Compound	Binding ET-A[a]	$IC_{50}/\mu M$ ET-B[b]	Functional AAR[d]
Ac-D-Phe-Leu-Asp-Ile-Ile-Trp	2.8	3.3	3.1
Ac-Phe-Leu-Asp-Ile-Ile-Trp	8.4	18.5	c
Ac-D-Phe-Glu-Asp-Ile-Ile-Trp	0.65	1.3	0.60
Ac-D-Phe-Orn-Asp-Ile-Ile-Trp	0.75	4.0	2.0
Ac-D-Phe-Phe-Asp-Ile-Ile-Trp	0.20	0.05	2.6
Ac-D-Phe-D-Phe-Asp-Ile-Ile-Trp	0.93	0.03	1.6
Ac-D-Phe-Leu-Phe-Ile-Ile-Trp	1.18	0.035	4.5
Ac-D-Phe-Asp-Ile-Ile-Trp	9.1	9.3	c
Ac-D-Phe-Leu-Asp-Ile-Trp	>10	>10	c
Ac-D-Phe-Ile-Ile-Trp	>10	>10	c
Ac-D-Phe-Ile-Trp	8.0	10	c

[a] Rabbit renal artery vascular smooth muscle cells.
[b] Rat cerebellum.
[c] Not tested.
[d] Functional assay measures inhibition of ET-1-induced arachidonic acid release in rabbit renal vascular smooth muscle cells.

Data from References 263 and 266.

with D-Bhg (10,11-dihydro-5H-dibenzo[*a,d*]cycloheptene glycine) in position 16 (Table 16) possessed low nanomolar binding affinity at each receptor and was a functional antagonist of vasoconstriction in tissues specifically expressing each receptor subtype (pA_2 = 6.8 to 7.0).[271,272] In anesthetized or conscious rats pretreatment with PD 142893 or PD 145065 antagonized both the ET-1-induced pressor and depressor responses dose-dependently, indicative of nonselective antagonists in this species.[273] Duration of action studies with various antagonists (10 μmol/kg i.v. bolus) in conscious chronically prepared normotensive rats over 5 d indicate that PD 142893 and PD 145065 both showed some blockade of the depressor component of the ET challenge (1.0 nmol/kg) at 2 h post-

TABLE 16
Hexapeptide Endothelin Antagonists

Compound	ET-A[a]	ET-B[b] (IC$_{50}$/μM)	AAR[c]
Ac-D-Phe-Leu-Asp-Ile-Ile-Trp	1.8[e]	2.0[e]	3.1
Ac-D-Tyr-Leu-Asp-Ile-Ile-Trp	0.40	7.0	0.25
Ac-D-Trp-Leu-Asp-Ile-Ile-Trp	0.13	1.8	0.45
Ac-(β-Phe)D-Phe-Leu-Asp-Ile-Ile-Trp·2Na PD142893	0.04[f]	0.06[f]	0.07[f]
Ac-D-Bhg-Leu-Asp-Ile-Ile-Trp·2Na PD 145065	0.0035[e]	0.015[e]	0.049[e]
Ac-D-2-Nal-Leu-Asp-Ile-Ile-Trp	1.0	1.0	1.9
Ac-D-Bip-Leu-Asp-Ile-Ile-Trp	4.4	3.5	d

a   Rabbit renal artery vascular smooth muscle cells.
b   Rat cerebellum.
c   Functional assay measures inhibition of ET-1-induced arachidonic acid release in rabbit renal vascular smooth muscle cells.
d   Not tested.
e   n = 2 IC$_{50}$ determinations. All other values represent one IC$_{50}$ determination.
f   n = 5 IC$_{50}$ determinations. All other values represent one IC$_{50}$ determination.

Bip, 4,4′-biphenylalanine; 2-Nal, 2-naphthylalanine.
Data from Reference 268.

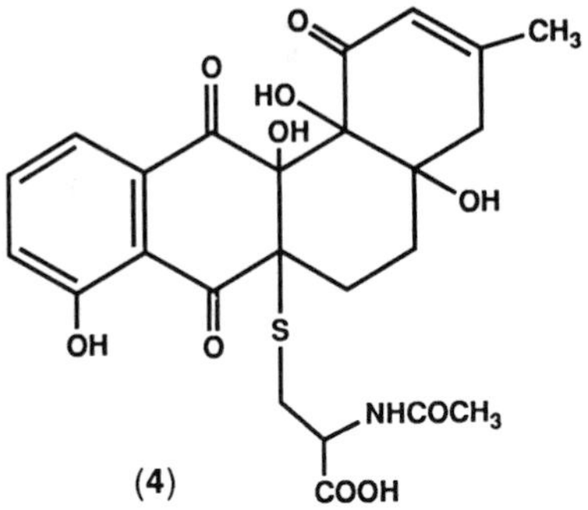

FIGURE 9.   Anthraquinone ET receptor antagonist from *Streptomyces* sp. no. 89009.

dose.[272] In the anesthetized guinea pig, PD 145065 inhibited ET-1-induced bronchoconstriction dose-dependently.[274]

3. ET Antagonists Discovered by Random Screening

A nonpeptide series of ET receptor antagonists of unknown specificity, discovered by random screening from a *Streptomyces* strain, have been disclosed.[275] The most potent compound (FR901367) (binding affinity in porcine aorta of IC$_{50}$ = 0.67 μM: presumably ET$_A$) inhibited the contractile response of ET-1 in rabbit thoracic aorta very weakly (75% at 10^{-4} M) (Figure 9).[275]

A report of a cyclic pentapeptide ET$_A$ receptor antagonist discovered by random screening of fermentation products from *Streptomyces misakiensis* has recently appeared.[265–267] Structure-activity studies around this peptide BE-18257B cyclo[D-Trp-D-Glu-Ala-D-allo-Ile-Leu]) have led to more potent analogs BQ-123 and BQ-153 (Table 17).[276–278] Both BQ-123 and BQ-153 antagonize ET-1-induced vasoconstriction of isolated porcine coronary arteries.[277–279] There was a small amount of ET-1-induced vasoconstriction resistant to these antagonists, presumably mediated by the ET$_B$ receptor subtype. In conscious rats, pretreatment with these antagonists antagonized ET-1-induced

TABLE 17
Selective ET$_A$ Antagonists: Cyclo[D-Trp-X-Y-Z-L-Leu]

Compound	X	Binding Affinity Y	Z	Functional (IC$_{50}$/nM) ET$_A$	ET$_B$	pA$_2$(ET$_A$)
BE-18257A	D-Glu	L-Ala	D-Va	1,400	>10^5	
BE-18257B	D-Glu	D-allo-Ile	D-allo-Ile	470	>10^5	
BQ-123	D-Asp	L-Pro	D-Val	7.3	18,000	7.4
BQ-153	D-Glu	L-Pro	D-Val	8.6	54,000	7.4
BQ-162	D-sulfoalanine	L-Pro	D-Val	230	>10^5	

ET$_A$, porcine aortic smooth muscle (SM) cell; ET$_B$, rat cerebellum; pA$_2$, values measured in porcine left anterior descending coronary arteries.
Data from Reference 279.

sustained pressor responses dose-dependently without affecting the transient depressor action of ET-1.[279,280] It has previously been reported that this vasodilator component is mediated by the ET$_B$ receptor, as indicated by the reported action of selective ET$_B$ agonists in certain tissues.[253]

BQ-123 antagonizes ET-1-induced mitogenic responses in rat aortic VSMC which may be an important aspect of vascular hypertrophy and proliferation associated with development of hypertension and atherosclerosis.[281] BQ-123 has also been shown to completely antagonize both ET-1 and angiotensin II-induced contractions in rabbit aorta.[282]

Intravenous infusion of BQ-123 (1.2 to 30 mg/kg/h) produced a significant dose-related decrease in blood pressure (BP) in 20- to 29-week-old stroke-prone SHR but did not alter BP in 13- to 16-week-old WKY or in 18- to 19-week-old and 40-week-old SHR. It has been suggested that ET-1 may thus be involved in the pathogenesis of malignant hypertension.[283]

Pretreatment with BQ-123 (10 mg/kg) inhibited the pressor response evoked by administration of big ET (1 nmol/kg) while it had no effect on the ET-3-mediated pressor response. In WKY, BQ-123 reversed the hypertension produced by an infusion of ET-1 (0.01 nmol/kg/min). Administration of BQ-123 produced a mild antihypertensive effect in normal- to low-renin models of hypertension, but no blood pressure lowering was observed in high-renin models of hypertension.[284]

It has been shown, however, that BQ-123 infused at 50 mg/kg/h for 5 h significantly lowered MAP (by 25 mmHg) in SHRs while it had no effect at 10 and 30 mg/kg/h.[285]

BQ-123 failed to reduce MABP in normotensive rats but BQ-123 decreased MABP in conscious, spontaneously hypertensive rats.[281] In another study BQ-123 (30 mg/kg/h, infusion pump for 6 weeks i.v.) significantly reduced BP and the incidence of stroke in SHRSP but had no effect in SHR or WKY under the same experimental conditions.[286] BQ-153, an ET$_A$-selective receptor antagonist analog of BQ-153, attenuated the pressor actions of ET-1 in pig kidney *in vivo*. A residual component of the pressor action and of the complete depressor action were unaffected by BQ-153, consistent with ET$_B$ receptors also mediating vasoconstriction and vasodilation.[287]

BQ-123 has been studied in a rat model of acute cyclosporine A (CsA) toxicity and is reported to be protective. While selective infusion of BQ-123 (0.76 mg/h continued throughout the experimental period) into the left renal artery following systemic CsA treatment had no effect on the renal hemodynamics, intrarenal infusion of BQ-123 prior to systemic administration of CsA protected against glomerular dysfunction.[288] However, in another report ET blockade by either BQ-123 or a mixture of BQ-123 and PD 142893 failed to affect the acute hemodynamic response to CsA in anesthetized rats.[289]

Several groups have studied the solution structure of BQ-123 by NMR spectroscopy and molecular modeling.[290–292] In both dimethyl sulfoxide (DMSO) and 20% acetonitrile in water, there are close contacts between the DTrp benzene ring protons and the Leu methyl groups. The structure contains a β-turn with Pro as the i + 1 residue and DVal as the i + 2 residue. A γ-turn was centered

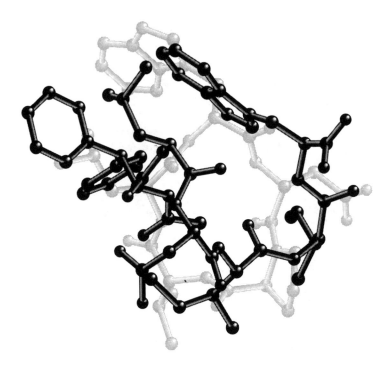

FIGURE 10.   Possible overlap between low-energy structures for BQ-123 and PD 142893.

on the DTrp residue. The DVal amide proton is internal and is involved in a transannular hydrogen bond to the carbonyl oxygens of the DAsp residue. Interestingly a very similar series of cyclic pentapeptide antagonists to those covered in the Banyu patent was very recently reported by the Fujisawa group.[293-295]

It is of interest to understand how PD 142893 and BQ-123 differ in their binding interactions. Clearly the linear peptide can accomodate ET_B structural requirements, while the cyclic pentapeptide BQ-123 does not. NMR studies of the constrained peptide BQ-123 allow one to speculate as to the features important for ET_A receptor binding. Clearly from the SAR reported for the BQ-123 series,[277] the Trp, Asp, and Leu residues appear important for receptor binding. The Pro and Val residues may serve to orient the important functional groups by inducing a defined β-turn. There are several possible low-energy conformations where PD 142893 can be overlaid on BQ-123. Three possibilities are shown in Figures 10, 11, and 12.

The bombesin antagonist [D-Arg[1], D-Phe, D-Trp[7,9], Leu[11]] substance P (SP), previously reported to inhibit [^{125}I]ET-1 to rat cardiac membranes, appears not to be characterized as competitive inhibition, due to the large increase in nonspecific binding observed at higher concentrations.[33,296]

Cyclodepsipeptides, known as the cochinmicins (I–III) produced by *Microbispora* sp., were discovered by random screening techniques[297] (Figure 13).

Linear tripeptidic compounds are reported as ET_A antagonists (Table 18).[298] FR 139317 (Table 18) is a functional antagonist inhibiting the specific binding of [^{125}I]ET-1 to porcine and human aortic microsomes in a concentration-dependent, monophasic fashion with IC_{50}s of 0.53 and 2.5 n*M*, respectively.[299] In isolated rabbit aorta, FR 139317 inhibited the ET-1-induced vasoconstriction with a pA_2 value of 7.2.[299] FR 139317 inhibited ET_A-mediated phosphatidylinositol hydrolysis and arachidonic acid release and produced a parallel shift in the dose-response curve for ET-1, with respective pA_2 values of 8.2 and 7.7.[300] An i.v. administration (bolus) of FR 139317 (1 mg/kg) completely inhibited the pressor response to ET-1, while it had no effect on the initial depressor response, similar to the effects seen with BQ-123 and BQ-153. FR 139317 also suppressed ET-1-induced arrythmia in rats. In a subarachnoid hemorrhage canine model, intracisternal administration of FR 139317 (0.1 mg) significantly inhibited the vasoconstriction of the basilar artery after 7 d.[299]

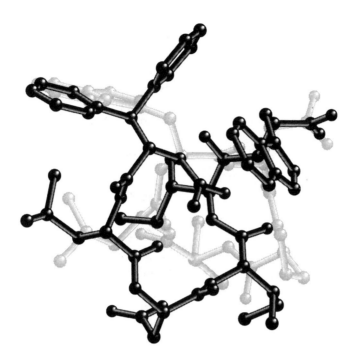

FIGURE 11. Possible overlap between low-energy structures for BQ-123 and PD 142893.

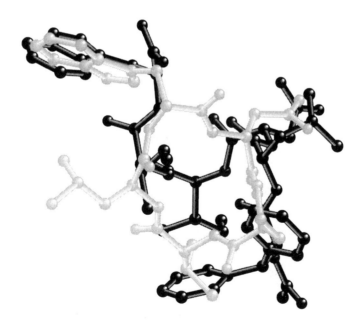

FIGURE 12. Possible overlap between low-energy structures for BQ-123 and PD 142893.

N-Pentanoyl tryptophan has very recently been reported as a functional antagonist inhibiting ET-1-induced vasoconstriction in rat pulmonary artery.[301] Since no binding data were noted it is not known whether the effects reported are due to ET antagonism. Very recently, the first nonpeptide ET$_A$ receptor antagonist (myriceron caffeoyl ester) was reported, isolated from bayberry, *Myrica cerifera*. This compound (Figure 14) selectively antagonized specific binding of [^{125}I]ET-1, but not of [^{125}I]ET-3 to rat cardiac membranes. It also antagonized ET-1-induced increase in intracellular

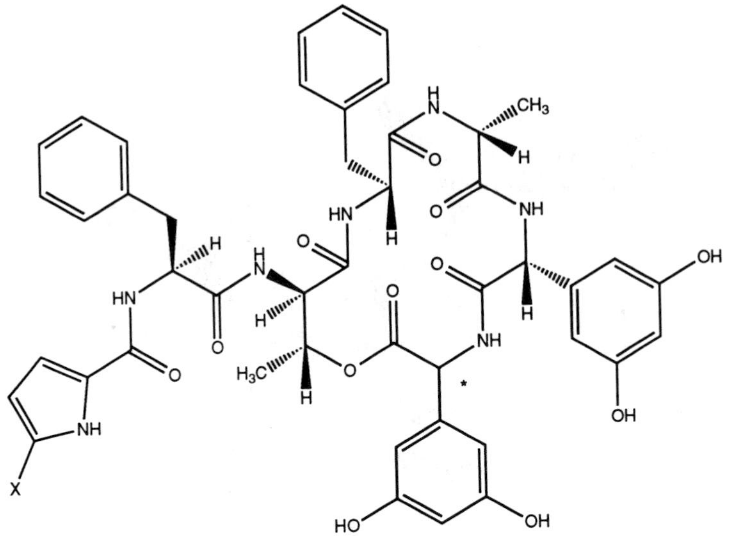

Cochinmicin	X	*
I	H	*R*
II	Cl	*S*
III	Cl	*R*

FIGURE 13. Structure of the cyclic hexapeptide L-290,936.

TABLE 18
Tripeptidic Endothelin Antagonists

Structure	Binding[a] (IC$_{50}$/nM)
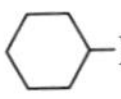 ◯–HNCO–Leu–D–Trp(Me)–D–Pya–OH.HCl	2.3
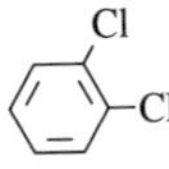 Cl-substituted phenyl–CH$_2$CO–Leu–D–Trp(Me)–D–Leu–OH	32
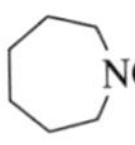 ◯NCO–Leu–D–Trp(Me)–D–Pya–Sar–ONa	7.6
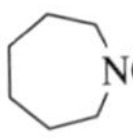 ◯NCO–Leu–D–Trp(Me)–D–Pya–Phe–ONa	21
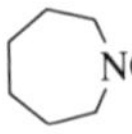 ◯NCOCH$_2$CH(CH$_2$CH(Me)$_2$)CO–D–Trp(Me)–D–Pya–NHMe (S)	7.6
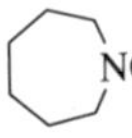 ◯NCO–Leu–D–Trp(Me)–D–Pya–OH (FR-139317)	0.53

[a] Binding affinity in porcine aortic membranes.

D-Pya, D-(2-pyridyl)alanine.
Data from Reference 299.

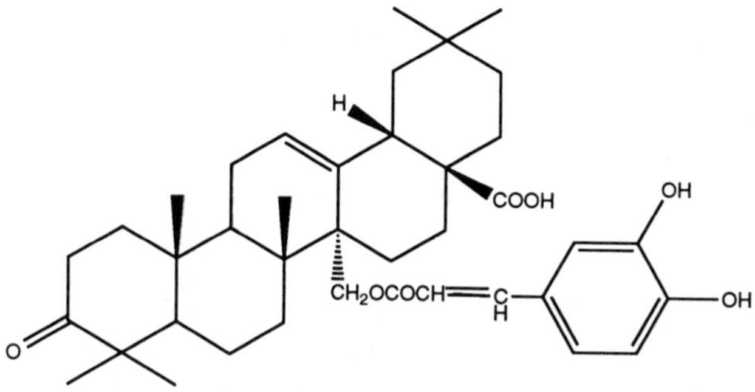

FIGURE 14. Steroid ET antagonist isolated from *Myrica cerifera*.

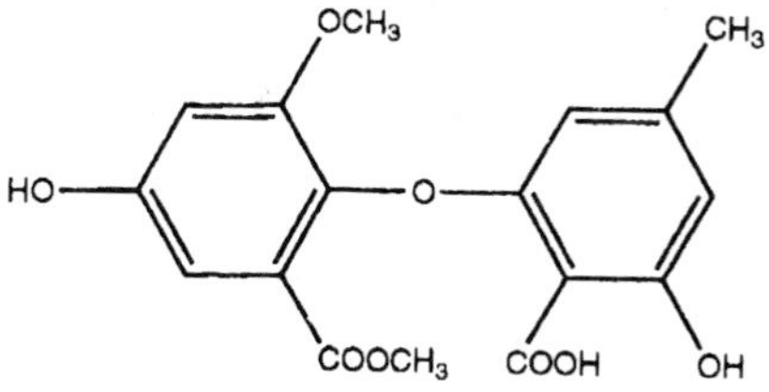

FIGURE 15. Nonpeptide antagonist Ro 46–2005.

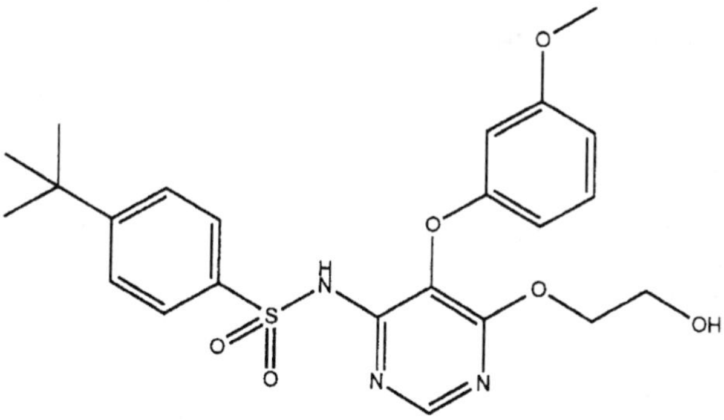

FIGURE 16. Nonpeptide ET antagonists.

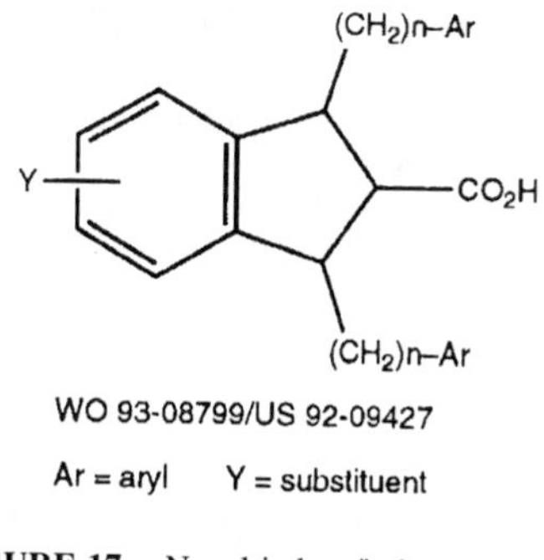

WO 93-08799/US 92-09427

Ar = aryl   Y = substituent

FIGURE 17. Novel indane/indene structures.

free calcium concentration in Swiss 3T3 fibroblasts, and ET-1-induced contraction of rat aortic strips. A series of substituted diphenyl ethers, including asterric acid (Figure 15), discovered by random screening of fungal broths, were reported to possess ET antagonist activity in the micro-molar range.[302] Other nonpeptide antagonists that have been reported are N-(4-pyrimidinyl)phenyl sulfonamides,[303,304] including the nonselective ET_A/ET_B antagonist Ro 46–2005[304] (Figure 16), and novel indane and indene structures (Figure 17).[305] Ro 46–2005 is reported as a nonselective ET_A/ET_B antagonist inhibiting binding of [^{125}I]ET-1 on human smooth muscle cells (ET_A) and rat aortic endothelial cells (ET_B) with IC_{50}s of 180 and 1100 nm, respectively. In normotensive rats, Ro 46–2005 has no effect on MABP, but inhibited the pressor effect of big ET-1 and the depressor and pressor effects of ET-1. Ro 46–2005 (3 mg/kg i.v.) is reported to prevent renal vasoconstriction after 45 min of renal ischemia and also the cerebral vasoconstriction which follows subarachnoid hemorrhage in rats.[304]

The availability of these small peptidic antagonists will aid in the elucidation of the physiological and/or pathological roles for endothelin and its isopeptides.

ACKNOWLEDGMENTS

I would like to thank Dr. J. Dunbar for providing Figures 7, 10, 11, and 12 and Mr. N. Mendoza for his invaluable technical assistance in the preparation of this chapter.

REFERENCES

1. **Hickey, K.A., Rubanyi, G., Paul, R.J., and Highsmith, R.F.,** Characterization of a coronary vasoconstrictor produced by endothelial cells in culture, *Am. J. Physiol.,* 248, C550, 1985.
2. **Yanagisawa, M., Kurihara, H., Kimura, S., Tomobe, Y., Kobayashi, M., Mitsui, Y., Yazaki, Y., Goto, K., and Masaki, T.,** A novel potent vasoconstrictor peptide produced by vascular endothelial cells, *Nature (London),* 332, 411, 1988.
3. **Inoue, A., Yanagisawa, M., Kimura, S., Kasuya, Y., Miyauchi, T., Goto, K., and Masaki, T.,** The human endothelin family: three structurally and pharmacologically distinct isopeptides predicted by three separate genes, *Proc. Natl. Acad. Sci. U.S.A.,* 86, 2863, 1989.
4. **Saida, K., Mitsui, Y., and Ishida, N.,** A novel isopeptide, vasoactive intestinal contractor, of a new (endothelin) peptide family. Molecular cloning, expression and biological activity, *J. Biol. Chem.,* 264, 14613, 1989.
5. **Itoh, Y., Yanagisawa, M., Ohkubo, S., Kimura, C., Kosaka, T., Inoue, A., Ishida, N., Mitsui, Y., Onda, H., Fujino, M., and Masaki, T.,** Cloning and sequence analysis of cDNA encoding the precursor of a human endothe-lium-derived vasoconstrictor peptide, endothelin: identity of human and porcine endothelin, *FEBS Lett.,* 231, 440, 1988.
6. **Doherty, A.M. and Weishaar, R.E.,** Endogenous vasoactive peptides, *Annu. Rep. Med. Chem.,* 25, 89, 1990.
7. **Doherty, A.M.,** Endogenous vasoactive peptides, *Annu. Rep. Med. Chem.,* 26, 83, 1991.
8. **Kochva, E., Viljoen, C.C., and Botes, D.P.,** A new type of toxin in the venom of snakes of the genus *Atractaspis* (Atractaspidinae), *Toxicon,* 20, 581, 1982.
9. **Kloog, Y., Amber, I., Sokolovsky, M., Kochva, E., Wollberg, Z., and Bdolah, A.,** Sarafotoxin, a novel vasocon-strictor peptide: phosphoinositide hydrolysis in rat heart and brain, *Science,* 242, 268, 1988.
10. **Bdolah, A., Wollberg, Z., Fleminger, G., and Kochva, E.,** SRTX-d, a new native peptide of the endothelin/sarafotoxin family, *FEBS Lett.,* 256, 1, 1989.
11. **Masaki, T. and Yanagisawa, M.,** Cardiovascular effects of the endothelins, *Cardiovasc. Drug Rev.,* 8 (4), 373, 1990.
12. **Lerman, A., Hildebrand, F.L., Margulies, K.B., O'Murchu, B., Perrella, M.A., Heublein, D.M., Schwab, T.R., and Burnett, J.C.,** Endothelin: a new cardiovascular regulatory peptide, *Mayo Clin. Proc.,* 65, 1441, 1990.
13. **Randall, M.D.,** Vascular activities of the endothelins, *Pharm. Ther.,* 50, 73, 1991.
14. **Nayler, W.G. and Gu, X.H.,** Endothelin in the heart, in *Endothelin,* Rubuanyi, G.M., Ed., Oxford Press, Oxford, 1992, 144.
15. **Schini, V.B. and Vanhoutte, P.M.,** Endothelin-1: a potent vasoactive peptide, *Pharmacol. Toxicol.,* 69, 303, 1991.
16. **Clarke, J.G., Benjamin, N., Larkin, S.W., Webb, D.J., Davies, G.J., and Maseri, A.,** Endothelin is a potent long-lasting vasoconstrictor in man, *Am. J. Physiol.,* 257, H2033, 1989.
17. **Fukuda, K., Hori, S., Kusuhara, M., Satoh, T., Kyotani, S., Handa, S., Nakamura, Y., Oono, H., and Yamaguchi, K.,** Effect of endothelin as a coronary vasoconstrictor in the Langendorff perfused rat heart, *Eur. J. Pharmacol.,* 165, 301, 1989.
18. **Kramer, B.K., Nishida, M., Kelly, R.A., and Smith, T.W.,** Endothelins. Myocardial actions of a new class of cytokines, *Circulation,* 85, 351, 1992.

19. **Nakamoto, H., Suzuki, H., Murakami, M., Kageyama, Y., Naitoh, M., Sakamaki, Y., Ohishi, A., and Saruta, T.,** Different effects of low and high doses of endothelin on hemodynamics and hormones in the normotensive conscious dog, *J. Hypertension,* 9, 337, 1991.

20. **Wright, C.E. and Fozard, J.R.,** Regional vasodilation is a prominent feature of the haemodynamic response to endothelin in anaesthetized, spontaneously hypertensive rats, *Eur. J. Pharmacol.,* 155, 201, 1988.

21. **Hoffman, A., Grossman, E., Ohman, K.P., Marks, E., and Keiser, H.R.,** The initial vasodilation and the later vasoconstriction of endothelin-1 are selective to specific vascular beds, *Am. J. Hypertension,* 3, 789, 1990.

22. **Fukuda, N., Soma, M., Izumi, Y., Minato, M., Watanabe, M., and Hatano, M.,** Low doses of endothelin-3 elicit endothelium dependent vasodilation which is accompanied by elevation of cyclic GMP, *Jpn. Circ. J.,* 55, 618, 1991.

23. **Moravec, C.S., Reynolds, E.E., Stewart, R.W., and Bond, M.,** Endothelin is a positive inotropic agent in human and rat heart in vitro, *Biochem. Biophys. Res. Commun.,* 159, 14, 1989.

24. **Ishikawa, T., Yanagisawa, M., Kimura, S., Goto, K., and Masaki, T.,** Positive chronotropic effects of endothelin, a novel endothelium-derived vasoconstrictor peptide, *Pflügers Arch.,* 413, 108, 1988.

25. **Takanashi, M. and Endoh, M.,** Characterization of the positive inotropic effect of endothelin on mammalian ventricular myocardium, *Am. J. Physiol.,* 261, H611, 1991.

26. **Le Monnier de Gouville, A.C., Mondot, S., Lippton, H., Hyman, A., and Cavero, I.,** Hemodynamic and pharmacological evaluation of the vasodilator and vasoconstrictor effects of endothelin-1 in rats, *J. Pharmacol. Exp. Ther.,* 252, 300, 1990.

27. **Yorikane, R., Shiga, H., Miyake, S., and Koike, H.,** Evidence for the direct arrhythmogenic action of endothelin, *Biochem. Biophys. Res. Commun.,* 173, 457, 1990.

28. **Hirata, Y., Takagi, Y., Fukuda, Y., and Marumo, F.,** Endothelin is a potent mitogen for rat vascular smooth muscle cells, *Atherosclerosis,* 78, 225, 1989.

29. **Weissburg, P.L., Witchell, C., Davenport, A.P., Hesketh, T.R., and Metcalfe, J.C.,** The endothelin peptides ET-1, ET-2, ET-3 and sarafotoxin S6b are co-mitogenic with platelet-derived growth factor for vascular smooth muscle cells, *Atherosclerosis,* 85, 257, 1990.

30. **Bobik, A., Grooms, A., Millar, J.A., Mitchell, A., and Grinpukel, S.,** Growth factor activity of endothelin on vascular smooth muscle, *Am. J. Physiol.,* 258, C408, 1990.

31. **MacCumber, M.W., Ross, C.A., and Snyder, S.H.,** Endothelin in brain: receptors, mitogenesis, and biosynthesis in glial cells, *Proc. Natl. Acad. Sci. U.S.A.,* 87, 2359, 1990.

32. **Neuser, D., Knorr, A., Stasch, J.P., and Kazda, S.,** Mitogenic activity of endothelin-1 and -3 on vascular smooth muscle cells is inhibited by atrial natriuretic peptides, *Artery,* 17, 311, 1990.

33. **Fabregat, I. and Rozengurt, E.,** [D-Arg^1,D-Phe^5,D-Trp^{7,9},Leu^{11}]substance P, a neuropeptide antagonist, blocks binding, Ca^{2+}-mobilizing, and mitogenic effects of endothelin and vasoactive intestinal contractor in mouse 3T3 cells, *J. Cell Physiol.,* 145, 88, 1990.

34. **Ito, H., Hirata, Y., Hiroe, M., Tsujino, M., Adachi, S., Takamoto, T., Nitta, M., Taniguchi, K., and Marumo, F.,** Endothelin-1 induces hypertrophy with enhanced expression of muscle specific genes in cultured neonatal rat cardiomyocytes, *Circ. Res.,* 69, 209, 1991.

35. **Edlund, A. and Wennmalm, A.,** Endothelin does not affect aggregation of human platelets, *Clin. Physiol.,* 10, 585, 1990.

36. **Thiemermann, C., Lidbury, P.S., Thomas, G.R., and Vane, J.R.,** Endothelin-1 inhibits ex vivo platelet aggregation in the rabbit, *Eur. J. Pharmacol.,* 158, 182, 1988.

37. **Wright, C.D., Cody, W.L., Dunbar, J.B., Doherty, A.M., and Rapundalo, S.T.,** Characterization of the chemotactic activity of endothelins and peptide fragments for human neutrophils, *FASEB J.,* 5(4), A637, 1991.

38. **Cody, W.L., Doherty, A.M., He, X., Rapundalo, S.T., Hingorani, G.P., Panek, R.B., and Major, T.C.,** Monocyclic endothelins: examination of the individual disulfide rings, *J. Cardiovasc. Pharmacol.,* 17 (Suppl. 7), S62, 1991.

39. **Lagente, V., Touvay, C., Mencia-Huerta, J., Chabrier, P.E., and Braquet, P.,** Bronchopulmonary effects of endothelin, *Clin. Exp. Allergy,* 20, 343, 1990.

40. **Battistini, B., Sirios, P., Braquet, P., and Filep, J.G.,** Endothelin-induced constriction of guinea-pig airways: role of platelet-activating factor, *Eur. J. Pharmacol.,* 186, 307, 1990.

41. **Filep, J.G., Battistini, B., and Sirois, P.,** Endothelin induces thromboxane release and contraction of isolated guinea-pig airways, *Life Sci.,* 47, 1845, 1990.

42. **Lagente, V., Touvay, C., Mencia-Huerta, J.-M., Chabrier, P.-E., and Braquet, P.,** Bronchopulmonary effects of endothelin, *Clin Exp. Allergy,* 20, 343, 1990.

43. **Rakugi, H., Nakamaru, M., Saito, H., Higaki, J., and Ogihara, T.,** Endothelin inhibits renin release from isolated rat glomeruli, *Biochem. Biophys. Res. Commun.,* 155, 1244, 1988.

44. **Miura, K., Yukimura, T., Yamashita, Y., Shichino, K., Shimmen, T., Saito, M., Okumura, M., Imanishi, M., Yamanaka, S., and Yamamoto, K.,** Effects of endothelin on renal hemodynamics and renal function in anesthetized dogs, *Am. J. Hypertension,* 3, 632, 1990.

45. **Lopez-Farre, A., Gomez-Garre, D., Bernabeu, F., and Lopez-Novoa, J.M.,** A role for endothelin in the maintenance of postischemic renal failure in the rat, *J. Physiol.,* 444, 513, 1991.

46. **Miller, W.L., Redfield, M.M., and Burnett, J.C.,** Integrated cardiac, renal and endocrine actions of endothelin, *J. Clin. Invest.,* 83, 317, 1989.

47. **Makino, S., Hashimoto, K., Hiraswa, R., Hattori, T., and Ota, Z.,** Central interaction between endothelin and brain natriuretic peptide on vasopressin secretion, *J. Hypertension,* 10, 25, 1992.

48. **Calvo, J.J., Gonzalez, R., De Carvalho, L., Takahashi, K., Kanse, S.M., Hart, G.R., Ghatei, M.A., and Bloom, S.R.,** Release of substance P from rat hypothalamus and pituitary by endothelin, *Endocrinology,* 126, 2288, 1990.

49. **Cozza, E.N., Gomez, S.C., Foecking, M.F., and Chiou, S.,** Endothelin binding to cultured calf adrenal zona glomerulosa cells and stimulation of aldosterone secretion, *J. Clin. Invest.,* 84, 1032, 1989.

50. **Boarder, M.R. and Marriott, D.B.,** Characterization of endothelin-1 stimulation of catecholamine release from adrenal chromaffin cells, *J. Cardiovasc. Pharmacol.,* 13 (Suppl. 5), S223, 1989.

51. **Reid, J.J., Wong, D.H., and Rand, M.J.,** The effect of endothelin on noradrenergic transmission in rat and guinea-pig atria, *Eur. J. Pharmacol.,* 168, 93, 1989.

52. **Ito, S., Juncos, L.A., Nushiro, N., Johnson, C.S., and Carretero, O.A.,** Endothelium-derived relaxing factor modulates endothelin action in afferent arterioles, *Hypertension,* 17, 1052, 1991.

53. **Hu, J.R., Berninger, U.G., and Lang, R.E.,** Endothelin stimulates atrial natriuretic peptide (ANP) release from rat atria, *Eur. J. Pharmacol.,* 158, 177, 1988.

54. **Winquist, R.J., Scott, A.L., and Vlasuk, G.P.,** Enhanced release of atrial natriuretic factor by endothelin in atria from hypertensive rats, *Hypertension,* 14, 111, 1989.

55. **Rakugi, H., Nakamaru, M., Tabuchi, Y., Nagano, M., Mikami, H., and Ogihara, T.,** Endothelin stimulates the release of prostacyclin from rat mesenteric arteries, *Biochem. Biophys. Res. Commun.,* 160, 924, 1989.

56. **Miura, S., Kurose, I., Fukumura, D., Suematsu, M., Sekizuka, E., Tashiro, H., Serizawa, H., Asako, H., and Tsuchiya, M.,** Ischemic bowel necrosis induced by endothelin-1: an experimental model in rats, *Digestion,* 48, 163, 1991.

57. **Wallace, J.L., Cirino, G., De Nucci, G., McKnight, W., and MacNaughton, W.K.,** Endothelin has potent ulcerogenic and vasoconstrictor actions in the stomach, *Am. J. Physiol.,* 256, G661, 1989.

58. **Kanyicska, B., Burris, T.P., and Freeman, M.E.,** Endothelin-3 inhibits prolactin and stimulates LH, FSH and TSH secretion from pituitary cell culture, *Biochem. Biophys. Res. Commun.,* 174, 338, 1991.

59. **Stojilkovic, S.S., Merelli, F., Iida, T., Krsmanovic, L.Z., and Catt, K.J.,** Endothelin stimulation of cytosolic calcium and gonadotropin secretion in anterior pituitary cells, *Science,* 248, 1663, 1990.

60. **Stojilkovic, S.S., Balla, T., Fukuda, S., Cesnjaj, M., Merelli, F., Krsmanovic, L.Z., and Catt, K.J.,** Endothelin ET_A receptors mediate the signalling and secretory actions of endothelins in pituitary gonadotrophs, *Endocrinology,* 130, 465, 1992.

61. **Serradeil-Le Gal, C., Jouneaux, C., Sanchez-Bueno, A., Raufaste, D., Roche, B., Preaux, A.M., Maffrand, J.P., Cobbold, P.H., Hanourne, J., and Loterstajn, S.,** Endothelin action in rat liver, *J. Clin. Invest.,* 87, 133, 1991.

62. **Gandhi, C.R., Stephenson, K., and Olson, M.S.,** Endothelin, a potent peptide agonist in the liver, *J. Biol. Chem.,* 265, 17432, 1990.

63. **Jaffer, F.E., Knauss, T.C., Poptic, E., and Abboud, H.E.,** Endothelin stimulates PDGF secretion in cultured human mesangial cells, *Kidney Int.,* 38, 1193, 1990.

64. **Takayanagi, R., Kitazumi, K., Takasaki, C., Ohnaka, K., Aimoto, S., Tasaka, K., Ohashi, M., and Nawata, H.,** Presence of non-selective type of endothelin receptor on vascular endothelium and its linkage to vasodilation, *FEBS Lett.,* 282, 103, 1991.

65. **Tabuchi, Y., Nakamaru, M., Rakugi, H., Nagano, M., and Ogihara, T.,** Endothelin enhances adrenergic vasoconstriction in perfused rat mesenteric arteries, *Biochem. Biophys. Res. Commun.,* 159, 1304, 1989.

66. **Yang, Z., Richard, V., von Segesser, L., Bauer, E., Stulz, P., Turina, M., and Luscher, T.F.,** Threshold concentrations of endothelin-1 potentiate the effects of norepinephrine and serotonin in human arteries: a new mechanism of vasospasm?, *Circulation,* 82, 188, 1990.

67. **De Nucci, G., Thomas, G.R., D'Orleans-Juste, P., Antunes, E., Walder, C., Warner, T.D., and Vane, J.R.,** Pressor effects of circulating endothelin are limited by its removal in the pulmonary circulation and by the release of prostacyclin and endothelium-derived relaxing factor, *Proc. Natl. Acad. Sci. U.S.A.,* 85, 9797, 1988.

68. **Kimura, S., Kasuya, Y., Sawamura, T., Shinmi, O., Sugita, Y., Yanagisawa, M., Goto, K., and Masaki, T.,**

73. **Bird, J.E., Waldron, T.L., Little, D.K., Asaad, M.M., Dorso, C.R., DiDonato, G., and Norman, J.A.,** The effects of novel cathepsin E inhibitors on big endothelin pressor responses in conscious rats, *Biochem. Biophys. Res. Commun.,* 182, 224, 1992.

74. **Shields, P.P., Gonzales, T.A., Charles, D., Gilligan, J.P., and Stern, W.,** Accumulation of pepstatin in cultured endothelial cells and its effect on endothelin processing, *Biochem. Biophys. Res. Commun.,* 177, 1006, 1991.

75. **Matsumura, Y., Ikegawa, R., Tsukahara, Y., Takaoka, M., and Morimoto, S.,** Conversion of big endothelin-1 to endothelin-1 by two types of metalloproteinases derived from porcine aortic endothelial cells, *FEBS Lett.,* 272, 166, 1990.

76. **Sawamura, T., Kasuya, Y., Matsushita, Y., Suzuki, N., Shinmi, O., Kishi, N., Sugita, Y., Yanagisawa, M., Goto, K., Masaki, T., and Kimura, S.,** Phosphoramidon inhibits the intracellular conversion of big endothelin-1 to endothelin-1 in cultured endothelial cells, *Biochem. Biophys. Res. Commun.,* 174, 779, 1991.

77. **Takada, J., Okada, K., Ikenaga, T., Matsuyama, K., and Yano, M.,** Phosphoramidon-sensitive endothelin-converting enzyme in the cytosol of cultured bovine endothelial cells, *Biochem. Biophys. Res. Commun.,* 176, 860, 1991.

78. **Ahn, K., Beningo, K., Olds, G., and Hupe, D.,** Endothelin-converting enzyme from bovine and human endothelial cells. 2nd International Symposium on endothelium-derived vasoactive factors, *J. Vasc. Res.,* 29, 76, 1992.

79. **Fukuroda, T., Noguchi, K., Tsuchida, S., Nishikibe, M., Ikemoto, F., Okada, K., and Yano, M.,** Inhibition of biological actions of big endothelin-1 by phosphoramidon, *Biochem. Biophys. Res. Commun.,* 172, 390, 1990.

80. **Matsumura, Y., Hisaki, K., Takaoka, M., and Morimoto, S.,** Phosphoramidon, a metalloproteinase inhibitor, suppresses the hypertensive effect of big endothelin-1, *Eur. J. Pharmacol.,* 185, 103, 1990.

81. **McMahon, E.G., Palomo, M.A., Moore, W.M., McDonald, J.F., and Stern, M.K.,** Phosphoramidon blocks the pressor activity of porcine big endothelin-1-(1–39) in vivo and conversion of big endothelin-1–1(1–39) to endothelin-1(1–21) in vitro, *Proc. Natl. Acad. Sci. U.S.A.,* 88, 703, 1991.

82. **Biochot, E., Pons, F., Lagente, V., Touvay, C., Mencia-Huerta, J., and Braquet, P.,** Phosphoramidon potentiates the endothelin-1-induced bronchopulmonary response in guinea-pigs, *Neurochem. Int.,* 18, 477, 1991.

83. **Modin, A., Pernow, J., and Lundberg, J.M.,** Phosphoramidon inhibits the vasoconstrictor effects evoked by big endothelin-1 but not the elevation of plasma endothelin-1 in vivo, *Life Sci.,* 49, 1619, 1991.

84. **Okada, K., Takada, J., Arai, Y., Matsuyama, K., and Yano, M.,** Importance of the C-terminal region of big endothelin-1 for specific conversion by phosphoramidon-sensitive endothelin converting enzyme, *Biochem. Biophys. Res. Commun.,* 180, 1019, 1991.

85. **Giaid, A., Polak, J.M., Gaitonde, V., Hamid, Q.A., Moscoso, G., Legon, S., Uwanogho, D., Roncalli, M., Shinmi, O., and Sawamura, T.,** Distribution of endothelin-like immunoreactivity and mRNA in the developing and adult human lung, *Am. J. Respir. Cell. Mol. Biol.,* 4, 50, 1991.

86. **Ohkubo, S., Ogi, K., Hosoya, M., Matsumoto, H., Suzuki, N., Kimura, C., Ondo, H., and Fujino, M.,** Specific expression of human endothelin-2 (ET-2) gene in a renal adenocarcinoma cell line. Molecular cloning of cDNA encoding the precursor of ET-2 and its characterization, *FEBS Lett.,* 274, 136, 1990.

87. **Shichiri, M., Hirata, Y., Emori, T., Ohta, K., Nakajima, T., Sato, K., Sato, A., and Marumo, F.,** Secretion of endothelin and related peptides from renal epithelial cell lines, *FEBS Lett.,* 253, 203, 1989.

88. **Martin-Nizard, F., Houssaini, H.S., Lestavel-Delattre, S., Duriez, P., and Fruchart, J.-C.,** Modified low density lipoproteins activate human macrophages to secrete immunoreactive endothelin, *FEBS Lett.,* 293, 127, 1991.

89. **Kusuhara, M., Yamaguchi, K., Nagasaki, K., Hayashi, C., Suzuki, A., Hori, S., Handa, S., Nakamura, Y., and Abe, K.,** Production of endothelin in human cancer cell lines, *Cancer Res.,* 50, 3257, 1990.

90. **Nunez, D.J.R., Brown, M.J., Davenport, A.P., Neylon, C.B., Schofield, J.P., and Wyse, R.K.,** Endothelin-1 mRNA is widely expressed in porcine and human tissues, *J. Clin. Invest.,* 85, 1537, 1990.

91. **Gaiad, A., Gibson, S.J., Herrero, M.T., Gentleman, S., Legon, S., Yanagisawa, M.T.M., Ibrahim, N.B.N., Roberts, G.W., Rossi, M.L., and Polak, M.,** Topographical localisation of endothelin mRNA and peptide immunoreactivity in neurones of the human brain, *Histochemistry,* 95, 303, 1991.

92. **Togashi, T., Ando, K., Kameya, T., and Kawakami, M.,** Regional distribution of immunoreactive endothelin-1 in the human central nervous system, *Biomed. Res.,* 12, 161, 1991.

93. **Yamaji, T., Johshita, H., Ishibashi, M., Takaku, F., Ohno, H., Suzuki, N., Matsumoto, H., and Fujino, M.,** Endothelin family in human plasma and cerebrospinal fluid, *J. Clin. Endocrinol.,* 71, 1611, 1990.

94. **Davenport, A.P., Cameron, I.T., Smith, S.K., and Brown, M.J.,** Binding sites for iodinated endothelin-1, endothelin-2 and endothelin-3 demonstrated on human uterine glandular epithelial cells by quantitative high resolution autoradiography, *J. Endocrinol.,* 129, 149, 1991.

95. **Onda, H., Ohkubo, S., Ogi, K., Kosaka, T., Kimura, C., Matsumoto, H., Suzuki, N., and Fujino, M.,** One of the endothelin gene family, endothelin 3 gene, is expressed in the placenta, *FEBS Lett.,* 261, 327, 1990.

96. **Van Papendorf, C.L., Cameron, I.T., Davenport, A.P., King, A., Barker, P.J., Huskisson, N.S., Gilmour, R.S., Brown, M.J., and Smith, S.K.,** Localization and endogenous concentration of endothelin-like immunoreactivity in human placenta, *J. Endocrinol.,* 131, 507, 1991.

97. **Arai, H., Hori, S., Aramori, I., Ohkubo, H., and Nakanishi, S.,** Cloning and expression of a cDNA encoding an endothelin receptor, *Nature (London),* 348, 730, 1990.

98. **Sakurai, T., Yanagisawa, M., Takuwa, Y., Miyazaki, H., Kimura, S., Goto, K., and Masaki, T.,** Cloning of a cDNA encoding a non-isopeptide-selective subtype of the endothelin receptor, *Nature (London),* 348, 732, 1990.

99. **Williams, D.L., Jones, K.L., Colton, C.D., and Nutt. R.F.,** Identification of high affinity endothelin-1 receptor subtypes in human tissues, *Biochem. Biophys. Res. Commun.,* 180, 475, 1991.

100. **Nakamuta, M., Takayanagi, R., Sakai, Y., Sakamoto, S., Hagiwara, H., Mizuno, T., Saito, Y., Hirose, S., Yamamoto, M., and Nawata, H.,** Cloning and sequence analysis of a cDNA encoding human non-selective type of endothelin receptor, *Biochem. Biophys. Res. Commun.,* 177, 34, 1991.

101. **Cyr, C., Huebner, K., Druck, T., and Kris, R.,** Cloning and chromosomal localization of a human ETA receptor, *Biochem. Biophys. Res. Commun.,* 181, 184, 1991.

102. **Adachi, M., Yang, Y.-Y., Furuichi, Y., and Miyamoto, C.,** Cloning and characterization of cDNA encoding human A-type endothelin receptor, *Biochem. Biophys. Res. Commun.,* 180, 1265, 1991.

103. **Hosoda, K., Nakao, K., Arai, H., Suga, S., Ogawa, Y., Mukoyama, M., Shirakami, G., Saito, Y., Nakanishi, S., and Imura, H.,** Cloning and expression of human endothelin-1 receptor cDNA. H, *FEBS Lett.,* 287, 23, 1991.

104. **Ogawa, Y., Nakao, K., Arai, H., Nakagawa, O., Hosoda, K., Suga, S., Nakanishi, S., and Imura, H.,** Molecular cloning of a non-isopeptide-selective human endothelin receptor, *Biochem. Biophys. Res. Commun.,* 178, 248, 1991.

105. **Sakamoto, A., Yanagisawa, M., Sakurai, T., Takuwa, Y., Yanagisawa, H., and Masaki, T.,** Cloning and functional expression of human cDNA for the ET_B endothelin receptor, *Biochem. Biophys. Res. Commun.,* 178, 656, 1991.

106. **Lin, H.Y., Kaji, E.H., Winkel, G.K., Ives, H.E., and Lodish, H.F.,** Cloning and functional expression of a vascular smooth muscle endothelin 1 receptor, *Proc. Natl. Acad. Sci. U.S.A.,* 88, 3185, 1991.

107. **Elshourbagy, N.A., Lee, J.A., Korman, D.R., Nuthalaganti, P., Sylvester, D.R., Dilella, A.G., Sutiphong, J.A., and Kumar, C.S.,** Molecular cloning and characterization of the major endothelin receptor subtype in porcine cerebellum, *Mol. Pharmacol.,* 41, 465, 1992.

108. **Yoshinaga, M., Chijiiwa, Y., Misawa, T., Harada, N., and Nawata, H.,** Endothelin B receptor on guinea pig small intestinal smooth muscle cells, *Am. J. Physiol.,* 262, G308, 1992.

109. **Williams, D.L., Jones, K.L., Pettibone, D.J., Lis, E.V., and Clineschmidt, B.V.,** Sarafotoxin S6c: an agonist which distinguishes between endothelin receptor subtypes, *Biochem. Biophys. Res. Commun.,* 175, 556, 1991.

110. **Saeki, T., Ihara, M., Fukuroda, T., Yamagiwa, M., and Yano, M.,** [Ala1,3,11,15]Endothelin-1 analogs with ET_B agonistic activity, *Biochem. Biophys. Res. Commun.,* 179, 286, 1991.

111. **Panek, R.L., Major, T.C., Hingorani, G.P., Doherty, A.M., Taylor, D.G., and Rapundalo, S.T.,** Endothelin and structurally related analogs distinguish between endothelin receptor subtypes, *Biochem. Biophys. Res. Commun.,* 183(2), 566, 1992.

112. **Harrison, V.J., Randriantsoa, A., and Schoeffter, P.,** Heterogeneity of endothelin-sarafotoxin receptors mediating contraction of pig coronary artery, *Br. J. Pharmacol.,* 105, 511, 1992.

113. **Woodcock, E.A. and Land, S.,** Endothelin receptors in rat renal papilla with a high affinity for endothelin-3, *Eur. J. Pharmacol.,* 208, 255, 1991.

114. **Karne, S., Jayawickreme, C.K., and Lerner, M.R.,** Cloning and characterization of ET-3 specific receptor from *Xenopus* melanophores, *Soc. Neurosci. Abstr.,* 118, Abstract 193.19, 1992.

115. **Takahashi, K., Jones, P.M., Kanse, S.M., Lam, H.C., Spokes, R.A., Ghatei, M.A., and Bloom, S.R.,** Endothelin in the gastrointestinal tract. Presence of endothelin like immunoreactivity, endothelin-1 messenger RNA, endothelin receptors, and pharmacological effect, *Gastroenterology,* 99, 1660, 1990.

116. **Takahashi, K., Ghatei, M.A., Jones, P.M., Murphy, J.K., Lam, H.C., O'Halloran, D.J., and Bloom, S.R.,** Endothelin in human brain and pituitary gland: presence of immunoreactive endothelin, endothelin messenger ribonucleic acid, and endothelin receptors, *J. Clin. Endocrinol. Metab.,* 72, 693, 1991.

117. **Ehrenreich, H., Kehrl, J.H., Anderson, R.W., Reickmann, P., Vitkovic, L., Coligan, J.E., and Fauci, A.S.,** A vasoactive peptide, endothelin-3, is produced by and specifically binds to primary astrocytes, *Brain Res.,* 538, 54, 1991.

118. **Hemsen, A., Franco-Cereceda, A., Maaatrin, R., Rudehill, A., and Lundberg, J.M.,** Occurrence, specific binding sites and functional effects of endothelin in human cardiopulmonary tissue, *Eur. J. Pharmacol.,* 191, 319, 1990.

119. **Dashwood, M.R., Sykes, R.M., Collins, M.J., Prehar, S., Theodoropoulos, S., and Yacoub, M.H.,** Identification of [125I]endothelin binding sites in human coronary tissue, *Neurochem. Int.,* 18, 439, 1991.

120. **Waeber, C., Hoyer, D., and Palacios, J.-M.,** Similar distribution of [125I]sarafotoxin-6b and [125I]endothelin-1,2,-3 binding sites in the human kidney, *Eur. J. Pharmacol.,* 176, 233, 1990.

121. **Imai, T., Hirata, Y., Eguchi, S., Kanno, K., Ohta, K., Emori, T., Sakamoto, A., Yanagisawa, M., Masaki, T., and Marumo, F.,** Concomitant expression of receptor subtype and isopeptide of endothelin by the human adrenal gland, *Biochem. Biophys. Res. Commun.,* 182, 1115, 1992.

122. **Henry, P.J., Rigby, P.J., Self, G.J., Preuss, J.M., and Goldie, R.G.,** Relationship between endothelin-1 binding site densities and constrictor activities in human and airway smooth muscle, *Br. J. Pharmacol.,* 100, 786, 1990.

123. **Robault, C., Mondon, F., Bandet, J., Ferre, F., and Cavero, I.,** Regional distribution and pharmacological characterization of [125I]endothelin-1 binding sites in human fetal placental vessels, *Placenta,* 12, 55, 1991.

124. **Gu, J., Pinheiro, J.M.B., Yu, C.-Z., D'Andrea, M., Muralidharan, S., and Malik, A.,** Detection of endothelin-like immunoreactivity in epithelium and fibroblasts of the human umbilical cord, *Tissue Cell,* 23, 437, 1991.

125. **Nakamichi, K., Ihara, M., Kobayahi, M., Saeki, T., Ishikawa, K., and Yano, M.,** Different distribution of endothelin receptor subtypes in pulmonary tissues revealed by the novel selective ligands BQ-123 and [Ala1,3,11,15] ET-1, *Biochem. Biophys. Res. Commun.,* 182, 144, 1992.

126. **Stojilkovic, S.S., Balla, T., Fukuda, S., Acesnjaj, M., Merelli, F., Krsmanovic, L.Z., and Catt, K.J.,** Endothelin ET-A receptors mediate the signalling and secretory actions of endothelins in pituitary gonadotrophs, *Endocrinology,* 130, 465, 1992.

127. **Ohlstein, E.H., Arleth, A., Elliott, J.D., and Sung, C.P.,** Endothelin-mediated mitogenesis is mediated by selective stimulation of the ET-A receptor, *FASEB J.,* 6, A973, 1992.

128. **Philips, P.P., Cade, C., Parker Botelho, L.H., and Rubanyi, G.M.,** Molecular biology of the endothelins, in *Endothelin,* Rubanyi, G.M., Ed., Oxford University Press, Oxford, 1992, 31.

129. **Fukunaga, M., Fujiwara, Y., Ochi, S., Yokoyama, K., Fujibayashi, M., Orita, Y., Fukuhara, Y., Ueda, N., and Kamada, T.,** Stimulatory effect of thrombin on endothelin-1 production in isolated glomeruli and cultured mesangial cells of rats, *J. Cardiovasc. Pharmacol.,* 17 (Suppl. 7), S411, 1991.

130. **Ohta, K., Hirata, Y., Imai, T., Kanno, K., Emori, T., Shichiri, M., and Marumo, F.,** Cytokine-induced release of endothelin-1 from porcine renal epithelial cell line, *Biochem. Biophys. Res. Commun.,* 169, 578, 1990.

131. **Yanagisawa, M. and Masaki, T.,** Molecular biology and biochemistry of the endothelins, *Trends Pharmacol. Sci.,* 10, 374, 1989.

132. **Emori, T., Hirata, Y., Ohta, K., Kanno, K., Eguchi, S., Imai, T., Schichiri, M., and Marum, O.F.,** Cellular mechanism of endothelin-1 release by angiotensin and vasopressin, *Hypertension,* 18, 165, 1991.

133. **Ninoyama, H., Uchida, Y., Ishii, Y., Nomura, A., Kameyasma, M., Saotome, M., Endo, T., and Hasegawa, S.,** Endotoxin stimulates endothelin release from cultured epithelial cells of guinea pig trachea, *Eur. J. Pharmacol.,* 203, 299, 1991.

134. **Cocks, T.M., Malta, E., Woods, R.L., King, S.J., and Angus, J.A.,** Oxyhaemoglobin increases the production of endothelin-1 by endothelial cells in culture, *Eur. J. Pharmacol.,* 196, 177, 1991.

135. **Yamauchi, T., Ohnaka, K., Takayanagi, R., Umeda, F., and Nawata, H.,** Enhanced secretion of endothelin-1 by elevated glucose levels from cultured bovine aortic endothelial cells, *FEBS Lett.,* 267, 16, 1990.

136. **Yamada, K., Goto, A., Hui, C., and Sugimoto, T.,** Endogenous digitalis-like factor as a stimulator of endothelin secretion from endothelial cells, *Biochem. Biophys. Res. Commun.,* 172, 178, 1990.

137. **Kanse, S.M., Takahashi, K., Warren, J.B., Ghatei, M., and Bloom, S.R.,** Glucocorticoids induce endothelin release from vascular smooth muscle cells but not endothelial cells, *Eur. J. Pharmacol.,* 199, 99, 1991.

138. **Martin-Nizard, F., Houssaini, H.S., Lestavel-Delattre, S., Duriez, P., and Fruchart, J.C.,** Modified low density lipoproteins activate human macrophages to secrete immunoreactive endothelin, *FEBS Lett.,* 293, 127, 1991.

139. **Milner, P., Bodin, P., Loesch, A., and Burnstock, G.,** Rapid release of endothelin and ATP from isolated aortic endothelial cells exposed to increased flow, *Biochem. Biophys. Res. Commun.,* 170, 649, 1990.

140. **Kourembanas, S., Marsden, P., McQullan, L.P., and Faller, D.V.,** Hypoxia induces endothelin gene expression and secretion in cultured human endothelium, *J. Clin. Invest.,* 88, 1054, 1991.

141. **Ohlstein, E.H., Storer, B.L., Butcher, J.A., Debouck, C., and Feuerstein, G.,** Platelets stimulate expression of endothelin mRNA and endothelin biosynthesis in cultured endothelial cells, *Circ. Res.,* 69, 832, 1991.

142. **Oliver, F.J., de la Rubia, G., Feener, E.P., Lee, M.-E., Loeken, M., Shiba, T., Quertermous, T., and King, G.L.,** Stimulation of endothelin-1 expression by insulin in endothelial cells, *J. Biol. Chem.,* 266, 23251, 1991.

143. **Brunchman, T.E. and Brookshire, C.A.,** Cyclosporine-induced synthesis of endothelin by cultured human endothelial cells, *J. Clin. Invest.,* 88, 310, 1991.

144. **Orlando, C., Brandi, M.L., Peri, A., Giannini, S., Fantoni, G., Calabresi, E., Serio, M., and Maggi, M.,** Neurohypophyseal hormone regulation of endothelin secretion from rabbit endometrial cells in primary culture, *Endocrinology,* 1990, 126, 1780.

145. **De Feo, M.L., Bartolini, O., Orlando, C., Maggi, M., Serio, M., Pines, M., Hurwitz, S., Fujii, Y., Sakaguchi, K., Aurbach, G.D., and Brandi, M.L.,** Natriuretic peptide receptors regulate endothelin synthesis and release from parathyroid cells, *Proc. Natl. Acad. Sci. U.S.A.,* 88, 6496, 1991.

146. **Saijonmaa, O., Ristimaki, A., and Fyhrquist, F.,** Atrial natriuretic peptide, nitroglycerine, and nitroprusside reduce basal and stimulated endothelin production from cultured endothelial cells, *Biochem. Biophys. Res. Commun.,* 173, 514, 1990.

147. **Boulanger, C. and Luscher, T.F.,** Release of endothelin from the porcine aorta. Inhibition by endothelium-derived nitric oxide, *J. Clin. Invest.,* 85, 587, 1990.

148. **Saijonmaa, O., Nyman, T., Hohenthal, U., and Fyhrquist, F.,** Endothelin-1 is expressed and released by a human endothelial hybrid cell line, *Biochem. Biophys. Res. Commun.,* 181, 529, 1991.

149. **Yokokawa, K., Kohno, M., Yasunari, K., Murakawa, K., and Takeda, T.,** Endothelin-3 regulates endothelin-1 production in cultured human endothelial cells, *Hypertension,* 18, 304, 1991.

150. **Yoshida, H. and Nakamura, M.,** Inhibition by angiotensin converting enzyme inhibitors of endothelin secretion from cultured human endothelial cells, *Life Sci.,* 50, 195, 1992.

151. **Highsmith, R.F., Blackburn, K., and Schmidt, D.J.,** Endothelin and calcium dynamics in vascular smooth muscle, *Annu. Rev. Physiol.,* 54, 257, 1992.

152. **Simonson, M.S. and Dunn, M.J.,** Endothelin: pathways of transmembrane signalling, *Hypertension,* 15 (Suppl. I), I-5, 1990.

153. **Takuwa, Y., Yanagisawa, M., Takuwa, N., and Masaki, T.,** Endothelin, its diverse biological activities and mechanisms of action, *Prog. Growth Factors,* 1, 195, 1989.

154. **Simonson, M.S., Jones, J.M., and Dunn, M.J.,** Cytosolic and nuclear signaling by endothelin peptides: mesangial response to glomerular injury, *Kidney Int.,* 41, 542, 1992.

155. **Danthuluri, N.R. and Brock, T.A.,** Endothelin receptor-coupling mechanisms in vascular smooth muscle: a role for protein kinase C, *J. Pharmacol. Exp. Ther.,* 254, 393, 1990.

156. **Goto, K., Kasuya, Y., Matsuki, N., Takuwa, Y., Kurihara, H., Ishikawa, T., Kimura, S., Yanagisawa, M., and Masaki, T.,** Endothelin activates the dihydropyridine-sensitive voltage dependent Ca^{2+} channel in vascular smooth muscle, *Proc. Natl. Acad. Sci. U.S.A.,* 86, 3915, 1989.

157. **Criscione, L., Thomann, H., Rodriguez, C., Egleme, C., and Chiesi, M.,** Blockade of endothelin-induced contractions by dichlorobenzamil: mechanism of action, *Biochem. Biophys. Res. Commun.,* 163, 247, 1989.

158. **Ohlstein, E.H., Horohonich, S., and Hay, D.W.,** Cellular mechanisms of endothelin in rabbit aorta, *J. Pharmacol. Exp. Ther.,* 250, 548, 1989.

159. **Topouzis, S., Pelton, J.T., and Miller, R.C.,** Effects of calcium entry blockers on contractions evoked by endothelin-1. [Ala^{3,11}]endothelin-1 and [Ala^{1,15}] endothelin-1 in rat isolated aorta, *Br. J. Pharmacol.,* 98, 669, 1989.

160. **Lippton, H.L., Cohen, G.A., McMurtry, I.F., and Hyman, A.L.,** Pulmonary vasodilation to endothelin isopeptides in vivo is mediated by potassium channel activation, *J. Appl. Physiol.,* 70(2), 947, 1991.

161. **Miyoshi, Y., Nakaya, Y., Wakatsuki, T., Nakaya, S., Fujino, K., Saito, K., and Inoue, I.,** Endothelin blocks ATP-sensitive K+ channels and depolarizes smooth muscle cells of porcine coronary artery, *Circ. Res.,* 70, 612, 1992.

162. **Hilal-Dandan, R.T., Urasawa, K., and Brunton, L.L.,** Endothelin inhibits adenylate cyclase and stimulates phosphoinositide hydrolysis in adult cardiac myocytes, *J. Biol. Chem.,* 267, 10620, 1992.

163. **Reynolds, E. and Mok, L.,** Endothelin-stimulated arachidonic acid release in vascular smooth muscle cells, *FASEB J.,* 5, A1066, 1991.

164. **Fozard, J.R. and Part, M.,** The role of nitric oxide in the regional vasodilator effects of endothelin-1 in the rat, *Br. J. Pharmacol.,* 105, 744, 1992.

165. **Dahlof, B., Gustafsson, D., Hedner, T., Jern, S., and Hansson, L.,** Regional haemodynamic effects of endothelin-1 in rat and man: unexpected adverse reactions, *J. Hypertension,* 8, 811, 1990.

166. **Vierhapper, H., Wagner, O., Nowotny, P., and Waldhausl, W.,** Effect of endothelin-1 in man, *Circulation,* 81, 1415, 1990.

167. **Sorenson, S.S.,** Radio-immunoassay of endothelin in human plasma, *Scand. J. Clin. Lab. Invest.,* 51, 615, 1991.

168. **Lerman, A., Edwards, B.S., Hallett, J.W., Heublein, D.M., Sandberg, S.M., and Burnett, J.C.,** Circulating and tissue immunoreactivity in advanced atherosclerosis, *N. Engl. J. Med.,* 325, 997, 1991.

169. **Hirata, Y., Itoh, K., Ando, K., Endo, M., and Marumo, F.,** Plasma endothelin levels during surgery, *N. Engl. J. Med.,* 321, 1686, 1989.

170. **Kanno, K., Hirata, Y., Numano, F., Emori, T., Ohta, K., Schichiri, M., and Marumo, F.,** Endothelin-1 and vasculitis, *JAMA,* 264, 2868, 1990.

171. **Cernacek, P. and Stewart, D.J.,** Immunoreactive endothelin in human plasma: marked elevation in patients with cardiogenic shock, *Biochem. Biophys. Res. Commun.,* 161, 562, 1989.

172. **Marguiles, K.B., Hildebrand, F.L., Lerman, A., Perrella, M.A., and Burnett, J.C.,** Increased endothelin in heart failure, *Circulation,* 82, 2226, 1990.

173. **Rodeheffer, R.J., Lerman, A., Heublein, D.M., and Burnett, J.C.,** Circulating plasma endothelin correlates with the severity of congestive heart failure in humans, *Am. J. Hypertension,* 4, 9A, 1991.

174. **Hiroe, M., Hirata, Y., Fujita, N., Umezawa, S., Ito, H., Tsujino, M., Koike, A., Nogami, A., Takamoto, T., and Marumo, F.,** Plasma endothelin-1 levels in idiopathic dilated cardiomyopathy, *Am. J. Cardiol.,* 68, 1114, 1991.

175. **Clark, B.A., Halvorson, L., Sachs, B., and Epstein, F.H.,** Plasma endothelin levels in preeclampsia: elevation and correlation with uric acid levels and renal impairment, *Am. J. Obstet. Gynecol.,* 166, 962, 1992.

176. **Stewart, D.J., Levy, R.D., Cernacek, P., and Langleben, D.,** Increased plasma endothelin-1 in pulmonary hypertension: marker or mediator of disease?, *Ann. Int. Med.,* 114, 464, 1991.

177. **Miyauchi, T., Yanagisawa, M., Tomizawa, T., Sugushita, Y., Susuki, N., Fujino, M., Ajisaka, R., Goto, K., and Masaki, T.,** Increased plasma concentration of endothelin-1 and big endothelin-1 in acute myocardial infarction, *Lancet,* 2, 53, 1989.

178. **Salminen, K., Tikkanen, I., Saijonmaa, O., Nieminen, M., Fyhrquist, F., and Frick, M.H.,** Modulation of coronary tone in acute myocardial infarction by endothelin, *Lancet,* 2, 747, 1989.

179. **Stewart, D.J., Kubac, G., Costello, K.B., and Cernacek, P.,** Increased plasma endothelin-1 in the early hours of acute myocardial infarction, *J. Am. Coll. Cardiol.,* 18, 38, 1991.

180. **Yasuda, M., Kohno, M., Tahara, A., Itagane, H., Toda, I., Akioka, K., Teragaki, M., Oku, H., Takeuchi, K., and Takeda, T.,** Circulating immunoreactive endothelin in ischemic disease, *Am. Heart J.,* 119, 801, 1990.

181. **Suzuki, H., Sato, S., Suzuki, Y., Takekoshi, K., Ishihara, N., and Shimoda, S.,** Increased endothelin concentration in CSF from patients with subarachnoid hemorrhage, *Acta Neurol. Scand.,* 81, 553, 1990.

182. **Murch, S.H., Braegger, C.P., Sessa, W.C., and Macdonald, T.T.,** High endothelin-1 immunoreactivity in Crohn's disease and ulcerative colitis, *Lancet,* 339, 381, 1992.

183. **Fyhrquist, F., Saijonmaa, O., Metsarinne, K., Tikkanen, I., Rosenlof, K., and Tikkanen, T.,** Raised plasma endothelin-1 concentration following cold pressor test, *Biochem. Biophys. Res. Commun.,* 169, 217, 1990.

184. **Zamura, M.R., O'Brien, R.F., Rutherford, R.B., and Weil, J.V.,** Serum endothelin-1 concentrations and cold provocation in primary Raynaud's phenomenon, *Lancet,* 336, 1144, 1990.

185. **Kanno, K., Hirata, Y., Emori, T., Ohta, K., Schichiri, M., Shinohara, S., Chida, Y., Tomura, S., and Marumo, F.,** Endothelin and Raynaud's phenomenon, *Am. J. Med.,* 90, 130, 1991.

186. **Totsune, K., Mouri, T., Takahashi, K., Ohneda, M., Sone, M., Saito, T., and Yoshinaga, K.,** Detection of immunoreactive endothelin in plasma of hemodialysis patients, *FEBS Lett.,* 249, 239, 1989.

187. **Stockenhuber, F., Gottsauner-Wolf, M., Marosi, L., Liebisch, B., Kurz, R.W., and Balcke, P.,** Plasma levels of endothelin in chronic renal failure and after renal transplantation impact on hypertension and cyclosporine A associated nephrotoxicity, *Clin. Sci.,* 82, 255, 1992.

188. **Tomita, K., Ujiie, K., Nakanishi, T., Tomura, S., Matsuda, O., Ando, K., Shichiri, M., Hirata, Y., and Marumo, F.,** Plasma endothelin levels in patients with acute renal failure, *N. Engl. J. Med.,* 321, 1127, 1989.

189. **Shichiri, M., Hirata, Y., Ando, K., Emori, T., Ohta, K., Kimoto, S., Ogura, M., Inoue, A., and Marumo, F.,** Plasma endothelin levels in hypertension and chronic renal failure, *Hypertension,* 15, 493, 1990.

190. **Saito, Y., Nakao, K., Mukoyama, M., and Imura, H.,** Increased plasma endothelin level in patients with essential hypertension, *N. Engl. J. Med.,* 322, 205, 1990.

191. **Kohno, M., Yasunari, K., Murakama, K., Yokoyama, K., Horio, T., Fukui, T., and Takeda, T.,** Plasma immunoreactive endothelin in essential hypertension, *Am. J. Med.,* 88, 614, 1990.

192. **Miyauchi, T., Yanagisawa, M., Suzuki, N., Iida, K., Sigishita, Y., Fujino, M., Goto, K., and Masaki, T.,** Venous plasma concentrations of endothelin in normal and hypertensive subjects, *Circulation,* 80, II-573, 1989.

193. **Pittet, J.F., Morel, D.R., Hemsen, A., Gunning, K., Lacroix, J.S., Suter, P.M., and Lundberg, J.M.,** Elevated plasma endothelin-1 concentrations are associated with the severity of illness in patients with sepsis, *Ann. Surg.,* 213, 261, 1991.

194. **Miyasaka, N., Hirata, Y., Ando, K., Sato, K., Morita, H., Shichiri, M., Kanno, K., Tomita, K., and Marumo, F.,** Increased production of endothelin-1 in patients with inflammatory arthritides, *Arthritis Rheum.,* 35, 397, 1992.

195. **Nakagawa, K., Nishimura, T., Shindo, K., Kobayashi, H., Hamada, T., and Yokokawa, K.,** Measurement of immunoreactive endothelin-1 in plasma of a patient with malignant hemangioendothelioma, *Nippon Hifuka Gakkai Zasshi,* 100, 1453, 1990.

196. **Lerman, A., Hildebrand, F.L., Aarhus, L.L., and Burnett, J.C.,** Endothelin has biological actions at pathophysiological concentrations, *Circulation,* 83, 1808, 1991.

197. **Cavero, P.G., Miller, W.L., Heublein, D.M., Margulies, K.B., and Burnett, J.C.,** Endothelin in experimental congestive heart failure in the anesthetized dog, *Am. J. Physiol.,* 259, F312, 1990.

198. **Rodeheffer, R.J., Lerman, A., Heublein, D.M., and Burnett, J.C.,** Circulating plasma endothelin correlates with the severity of congestive heart failure in humans, *Am. J. Hypertension,* 4, 9A, 1991.

199. **Mortenson, L.H., Pawloski, C.M., Kanagy, N.L., and Fink, G.D.,** Chronic hypertension produced by infusion of endothelin in rats, *Hypertension,* 15, 729, 1990.

200. **Moldovan, F., Boudaoud, S., Jacob, L., Carron, S.A., Boudou, P., Villette, J.M., Eurin, B., and Fiet, J.,** Significant differences between arterial and venous endothelin-like immunoreactivity in human plasma, *Clin. Chem.,* 37, 2012, 1991.

201. **Watanabe, T., Suzuki, N., Shimamoto, N., Fujino, M., and Imada, A.,** Endothelin in myocardial infarction, *Nature (London),* 344, 114, 1990.

202. **Margulies, K.B., Hildebrand, F.L., Lerman, A., Perrella, M.A., and Burnett, J.J.C.,** Increased endothelin in experimental heart failure, *Circulation,* 82, 2226, 1990.

203. **Kon, V., Yoshioka, T., Fogo, A., and Ichikawa, I.,** Glomerular actions of endothelin in vivo, *J. Clin. Invest.,* 83, 1762, 1989.

204. **Perico, N., Dadon, J., and Remuzzi, G.,** Endothelin mediates the renal vasoconstriction induced by cyclosporine in the rat, *J. Am. Soc. Nephrol.,* 1, 76, 1990.

205. **Koshi, T., Torii, T., Arai, K., Edano, T., Hirata, M., Ohkuchi, M., and Okabe, T.,** Inhibition of endothelin (ET)-1 and ET-2-induced vasoconstriction by anti-ET-1 monoclonal antibody, *Chem. Pharm. Bull.,* 39, 1295, 1991.

206. **Miyamori, I., Itoh, Y., Matsubara, T., Koshida, H., and Takeda, R.,** Systemic and regional effects of endothelin in rabbits: effects of endothelin antibody, *Clin. Exp. Pharmacol. Physiol.,* 17, 691, 1990.

207. **Kinoshita, O., Kawano, Y., Yoshimi, H., Ashida, T., Yoshida, K., Akabane, S., Kuramochi, M., and Omae, T.,** Acute and chronic effects of anti-endothelin-1 antibody on blood pressure in spontaneously hypertensive rats, *J. Cardiovasc. Pharmacol.,* 17 (Suppl. 7), S511, 1991.

208. **Saito, Y., Nakao, K., Mukoyama, M., Shirakami, G., Itoh, H., Yamada, T., Arai, H., Hosoda, K., Suga, S., Jougasaki, M., Ogawa, Y., Nakajima, S., Ueda, M., and Imura, H.,** Application of monoclonal antibodies for endothelin to hypertensive research, *Hypertension,* 15, 734, 1990.

209. **Ohno, A.,** Effects of endothelin-specific antibodies and endothelin in spontaneously hypertensive rats, *J. Tokyo Womens' Med. Coll.,* 61, 951, 1991.

210. **Wolff, M.E. and Maggio, E.T.,** Antibody directed drug design and screening, *Med. Chem. Res.,* 1, 101, 1991.

211. **Reily, M.D. and Dunbar, J.B.,** The conformation of endothelin-1 in aqueous solution. NMR-derived constraints combined with distance geometry and molecular mechanics calculations, *Biochem. Biophys. Res. Commun.,* 178, 570, 1991.

212. **Endo, S., Inooka, H., Ishibashi, Y., Kitada, C., Mizuta, E., and Fujino, M.,** Solution conformation of endothelin determined by nuclear magnetic resonance and distance geometry, *FEBS Lett.,* 257, 149, 1989.

213. **Saudek, V., Hoflack, J., and Pelton, J.T.,** ^{1}H-NMR study of endothelin, sequence-specific assignment of the spectrum and a solution structure, *FEBS Lett.,* 257, 145, 1989.

214. **Pelton, J.T.,** Fluorescence studies of endothelin-1, *Neurochem. Int.,* 18, 485, 1991.

215. **Bortman, P., Hoflack, J., Pelton, J.T., and Saudek, V.,** Solution conformation of ET-3 by ^{1}H NMR and distance geometry calculations, *Neurochem. Int.,* 18, 491, 1991.

216. **Mills, R.G., Atkins, A.R., Harvey, T., Junius, F.K., Smith, R., and King, G.F.,** Conformation of sarafotoxin-6b in aqueous solution determined by NMR spectroscopy and distance geometry, *FEBS Lett.,* 282, 247, 1991.

217. **Bennes, R., Calas, B., Chabrier, P.-E., Demaille, J., and Heitz, F.,** Evidence for aggregation of endothelin-1 in water, *FEBS Lett.,* 276, 21, 1990.

218. **Panek, R.L., Major, T.C., Taylor, D.G., Hingorani, G.P., Dunbar, J.B., Doherty, A.M., and Rapundalo, S.T.,** Importance of secondary structure for endothelin binding and functional activity, *Biochem. Biophys. Res. Commun.,* 183, 572, 1992.

219. **Okada, K., Takada, J., Arai, Y., Matsuyama, K., and Yano, M.,** Importance of the C-terminal region of big endothelin-1 for specific conversion by phosphoramidon-sensitive endothelin converting enzyme, *Biochem. Biophys. Res. Commun.,* 180, 1019, 1991.

220. **Inooka, H., Endo S., Kikuchi, T., Wakimasu, M., and Fujino, M.,** in Solution Confirmation of human big endothelin-1 (BIG ET-1), *Peptide Chemistry 1990,* Shimonishi, Y., Ed., Protein Research Foundation, Osaka, 1991.

221. **Crawley, D.E., Liu, S.F., Barnes, P.J., and Evans, T.W.,** ET-3 is a potent pulmonary vasodilator in the rat, *J. Appl. Physiol.,* 72, 1425, 1992.

222. **Kitazumi, K., Shiba, T., Nishiki, K., Furukawa, Y., Takasaki, C., and Tasaka, K.,** Structure-activity relationship in vasoconstrictor effects of sarafotoxins and endothelin-1, *FEBS Lett.,* 260, 269, 1990.

223. **Takasaki, C., Aimoto, S., Kitazumi, K., Tasaka, K., Shiba, T., Nishiki, K., Furukawa, Y., Takayanagi, R., Ohnaka, K., and Nawata, H.,** Structure-activity relationships of sarafotoxins: chemical syntheses of chimera peptides of sarafotoxins S6b and S6c, *Eur. J. Pharmacol.,* 198, 165, 1991.

224. **Wollberg, A., Bdolah, A., and Kochva, E.,** Vasoconstrictor effects of sarafotoxins in rabbit aorta: structure-function relationships, *Biochem. Biophys. Res. Commun.,* 162, 371, 1989.

225. **Kimura, S., Kasuya, Y., Sawamura, T., Shinmi, O., Sugita, Y., Yanagisawa, M., Goto, K., and Masaki, T.,** Structure-activity relationships of endothelin; importance of the C-terminal moiety, *Biochem. Biophys. Res. Commun.,* 156, 1182, 1988.

226. **Takayanagi, R., Hashiguchi, T., Ohashi, M., and Nawata, H.,** Regional distribution of endothelin receptor in porcine cardiovascular tissues, *Regul. Pept.,* 27, 247, 1990.

227. **Maggi, C.A., Giuliani, S., Patacchini, R., Santicioli, P., Rovero, P., Giachetti, A., and Meli, A.,** The C-terminal hexapeptide, endothelin-(16–21), discriminates between different receptors, *Eur. J. Pharmacol.,* 166, 121, 1989.

228. **Maggi, C.A., Giuliani, S., Patacchini, R., Rovero, P., Giachetti, A., and Meli, A.,** The activity of peptides of the endothelin family in various mammalian smooth muscle preparations, *Eur. J. Pharmacol.,* 174, 23, 1989.

229. **Rovero, P., Patacchini, R., and Maggi, C.A.,** Structure-activity studies on endothelin (16–21), the C-terminal hexapeptide of the endothelins, in the guinea pig bronchus, *Br. J. Pharmacol.,* 101, 232, 1990.

230. **Doherty, A.M., Cody, W.L., Leitz, N.L., DePue, P.L., Taylor, M.D., Rapundalo, S.T., Hingorani, G.P., Major, T.C., Panek, R.L., and Taylor, D.G.,** Structure-activity studies of the C-terminal region of the endothelins and the sarafotoxins, *J. Cardiovasc. Pharmacol.,* 17 (Suppl. 7), S59, 1991.

231. **Randall, M.D., Douglas, S.A., and Hiley, C.R.,** Vascular activities of endothelin-1 and some alanyl substituted analogues in resistance beds of the rat, *Br. J. Pharmacol.,* 98, 685, 1989.

232. **Lecci, A., Maggi, C.A., Rovero, P., Giachetti, A., and Meli, A.,** Effect of endothelin-1, endothelin-3 and C-terminal hexapeptide, endothelin (16–21) on motor activity in rats, *Neuropeptides,* 16, 21, 1990.

233. **Tschirhart, E.J., Pelton, J.T., and Jones, C.R.,** Binding characteristics of ^{125}I endothelin-1 in guinea-pig trachea and its displacement by endothelin-1, endothelin-3 and endothelin (16–21), in *Mediators in Airway Hyperreactivity, Agents in Actions Supplements,* Vol. 31, Nijkamp, F.P., Engles, F., Henricks, P.A.J., and Oosterhout, A.J.M., Eds., Birkhauser Verlag, Basel, Switzerland, 1990, 233.

234. **Douglas, S.A. and Hiley, C.R.,** Endothelium-dependent mesenteric vasorelaxant effects and systemic actions of endothelin (16–21) and other endothelin related peptides in the rat, *Br. J. Pharmacol.,* 104, 311, 1991.

235. **Doherty, A.M., Cody, W.L., He, J.X., DePue, P.L., Leonard, D.M., Dunbar, J.B., Hill, K.E., Flynn, M.A., and Reynolds, E.E.,** Design of C-terminal peptide antagonists of endothelin: structure-activity relationships of ET-1[16–21, D-His^{16}], *Bioorg. Med. Chem. Lett.,* 4, 497, 1993.

236. **Johansen, N.L., Lundt, B.F., Madsen, K., Olson, V.V., Suzdak, P., Thogerson, H., and Weiss, J.U.,** Structure-activity relationships of endothelin analogs, in *Peptides 1991, Proceedings 21st European Peptide Symposium,* Giralt, E. and Anderson, D., Eds., ESCOM, Leiden, The Netherlands, 1991, 680.

237. **Nakajima, K., Kubo, S., Kumagaye, S., Nishio, H., Tsunemi, M., Inui, T., Kuroda, H., Chino, N., Watanabe, T.X., Kimura, T., and Sakakibara, S.,** Structure-activity relationship of endothelin: importance of charged groups, *Biochem. Biophys. Res. Commun.,* 163, 424, 1989.

238. **Magazine, H.I., Malik, A.B., Bruner, C.A., and Anderson, T.T.,** Acetylated endothelin-1 is a constrictor in guinea pig lung vasculature but not in isolated vascular strips, *J. Pharmacol. Exp. Ther.,* 260, 632, 1992.

239. **De Castiglione, R., Tam, J.P., Liu, W., Zhang, J.-W., Galantino, M., Bertolero, F., and Vaghi, F.,** Alanine scan of endothelin, *Peptides: Chemistry and Biology; Proceedings of the 12th American Peptide Symposium,* Smith, J.A., and Rivier, J.E., Eds., ESCOM, Leiden, The Netherlands, 1992, 402.

240. **Galantino, M., De Castiglione, R., Tam, J.P., Liu, W., Zhang, J.-W., Cristiani, C., and Vaghi, F.,** D-Amino acid scan of endothelin, in *Peptides: Chemistry and Biology; Proceedings of the 12th American Peptide Symposium,* Smith, J.A. and Rivier, J.E., Eds., ESCOM, Leiden, The Netherlands, 1992, 404.

241. **Watanabe, T.X., Itahara, Y., Nakajima, K., Kumagaye, S.I., Kimura, T., and Sakakibara, S.,** Receptor binding affinity and biological activity of varied length of peptides elongated from C-terminal of endothelin, *Jpn. J. Pharmacol.,* 52, P-O40, 1990.

242. **Hirata, Y., Yoshimi, H., Marumo, F., Watanabe, T.X., Kumagaye, S., Nakajima, K., Kimura, T., and Sakakibara, S.,** Interaction of synthetic sarafotoxin with rat vascular endothelin receptors, *Biochem. Biophys. Res. Commun.,* 162, 441, 1989.

243. **Takasaki, C., Aimoto, S., Takayanagi, R., Ohashi, M., and Nawata, H.,** Structure-receptor binding relationships of sarafotoxin and endothelin in porcine cardiovascular tissues, *Biochem. Int.,* 21, 1059, 1990.

244. **Kitazumi, K., Shiba, T., Nishiki, K., Furukawa, Y., Takasaki, C., and Tasaka, K.,** Structure-activity relationship in vasoconstrictor effects of sarafotoxins and endothelin-1, *FEBS Lett.,* 260, 269, 1990.

245. **Randall, M.D., Douglas, S.A., and Hiley, C.R.,** Vascular activities of endothelin-1 and some alanyl substituted analogues in resistance beds of the rat, *Br. J. Pharmacol.,* 98, 685, 1989.

246. **Pelton, J.T. and Miller, R.C.,** The role of the disulfide bonds in endothelin-1, *J. Pharm. Pharmacol.,* 43, 43, 1990.

247. **Hiley, C.R., Jones, C.R., Pelton, J.T., and Miller, R.C.,** Binding of [^{125}I]-endothelin-1 to rat cerebellar homogenates and its interactions with some analogues, *Br. J. Pharmacol.,* 101, 319, 1990.

248. **Cody, W.L., He, J.X., DePue, P.L., Rapundalo, S.T., Hingorani, G.P., Dudley, D.T., Hill, K.E., Reynolds, E.E., and Doherty, A.M.,** Structure-activity relationships in a series of monocyclic endothelin analogs, *Bioorg. Med. Chem. Lett.,* 4, 567, 1994.

249. **Hunt, J.T., Lee, V.G., Stein, P.D., Hedberg, A., Liu, E.C., McMullen, D., and Moreland, S.,** Structure-activity relationships of monocyclic endothelin analogues, *Bioorg. Med. Chem. Lett.,* 1, 33, 1991.

250. **Hunt, J.T.,** SAR of endothelin deduced from monocyclic analogs, *Drug News Perspectives,* 5(2), 78, 1992.

251. **Williams, D.L., Jones, K.L., Pettibone, D.J., Lis, E.V., and Clineschmidt, V.,** Sarafotoxin S6c: an agonist which distinguishes between endothelin receptor subtypes, *Biochem. Biophys. Res. Commun.,* 175, 556, 1991.

252. **Moreland, S., McMullen, D., Delaney, C.L., Lee, V.G., and Hunt, J.T.,** Vascular muscle contains vasoconstrictor ET$_B$ receptors, *J. Vasc. Biol.,* 184 (Abstr. 279), 171, 1992.

253. **Saeki, T., Ihara, M., Fukuroda, T., Yamagiwa, M., and Yano, M.,** [Ala1,3,11,15]Endothelin-1 analogs with ET$_B$ agonistic activity, *Biochem. Biophys. Res. Commun.,* 179, 286, 1991.

254. **Davenport, A.P., Kuc, R.E., Molenaar, P., Plumpton, C., Hiley, C.R., and Brown, M.J.,** Distribution of ET$_A$ and ET$_B$ receptor subtypes in cardiovascular, renal and central nervous system revealed using BQ-123, BQ3020 and [Ala1,3,11,15]ET-1: comparison with ^{125}I endothelin-1 binding, *J. Vasc. Biol.,* 25 (Abstr. 73), 73, 1992.

255. **Karada, H., Sudjarwo, S.A., Hori, M., Urade, Y., Takai, M., and Okada, T.,** Effects of IRL 1620, a selective agonist of endothelin B receptors, on vascular endothelium and smooth muscle, *J. Vasc. Res.,* 2nd International symposium on endothelium-derived vasoactive factors. 1992, P. 147, Abstract 206.

256. **Webb, R.L., Okada, R.T., Barclay, B.W., and Lappe, R.W.,** Central and peripheral actions of IRL 1620, an ET$_B$ selective endothelin (ET) receptor agonist in conscious rats (Abstract), *Hypertension,* 20(3), 425, 1992.

257. **Wong, J., Winters, J.W., Vanderford, P.A., Fineman, J.R., and Soifer, S.J.,** ET$_B$ receptor agonists produce potent pulmonary vasodilation in the intact newborn lamb, *Ped. Res.,* 4, 71A, 1993.

258. **Spinella, M.J., Malik, A.B., Everitt, J., and Anderson, T.T.,** Design and synthesis of a specific endothelin 1 antagonist: effects on pulmonary vasoconstriction, *Proc. Natl. Acad. Sci. U.S.A.,* 88, 7443, 1991.

259. **Werber, A.W., Spinella, M.J., and Andersen, T.T.,** DPR-1 ASP-15 endothelin-1 does not antagonize endothelin-1 in the superior cerebellar artery of rats, *FASEB J.,* 6, A1005, 1992.

260. **Abraham, W.H., Ahmed, A., Cortes, A., Spinella, M.J., Malik, A.B., and Andersen, T.T.,** A specific endothelin antagonist blocks inhaled endothelin-1-induced bronchoconstriction in sheep, *J. Appl. Physiol.,* 74(5), 2537, 1993.

261. **Wakimasu, M., Kikuchi, T., Asami, T., Ohtaki, T., and Fujino, M.,** Studies on endothelin antagonists, in *Perspectives in Medicinal Chemistry,* Barnard, T., Ed., Verlag Helvetica Chim. Acta, Basel, Switzerland, 1993, 165.

262. **Akiyama, H., Inagaki, Y., Kashiwabara, T., Ohta, H., Fushima, H., and Nishikori, K.,** *Jpn. J. Pharmacol.,* 58 (Suppl. 1), 336P, 1992.

263. **Doherty, A.M., Cody, W.L., DePue, P.L., He, J.X., Waite, L.A., Leonard D.M., Leitz, N.L., Dudley, D.T., Rapundalo, S.T., Hingorani, G.P., Haleen, S.J., LaDouceur, D.M., Hill, K.E., Flynn, M.A., Reynolds, E.E.,** Structure-activity relationships of C-terminal endothelin hexapeptide antagonists, *J. Med. Chem.,* 36, 2585, 1993.

264. **Doherty, A.M.,** Endothelin: a new challenge, *J. Med. Chem.,* 35, 1493, 1992.

265. **Muldoon, L.L., Rodland, K.D., Forsythe, M.L., and Magun, B.E.,** Stimulation of phosphatidylinositol hydrolysis, diacylglycerol release and gene expression in response to endothelin, a potent new agonist for fibroblasts and smooth muscle cells, *J. Biol. Chem.,* 264, 8529, 1989.

266. **Spellmeyer, D.C., Brown, S., Stauber, G.B., Geysen, H.M., and Valerio, R.,** Endothelin receptor ligands — multiple D-amino acid replacement net approach, *Bioorg. Med. Chem. Lett.,* 6, 1253, 1993.

267. **Spellmeyer, D.C., Brown, S., Stauber, G.B., Geysen H.M., and Valerio, R.,** Endothelin receptor ligands — replacement net approach to SAR determination of potent hexapeptides, *Bioorg. Med. Chem. Lett.,* 4, 519, 1993.

268. **Cody, W.L., Doherty, A.M., He, J.X., DePue, P.L., Rapundalo, S.T., Hingorani, G.A., Major, T.C., Panek, R.L., Haleen, S., LaDouceur, D., Reynolds, E.E., Hill, K.E., and Flynn, M.A.,** Design of a functional antagonist of endothelin, *J. Med. Chem.,* 35, 3301, 1992.

269. **Hingorani, G., Major, T., Panek, R., Reynolds, E., He, X., Cody, W., Doherty, A., and Rapundalo, S.,** In vitro pharmacology of a non-selective ET_A/ET_B endothelin receptor antagonist, PD 142893 (Ac-(β-phenyl)D-Phe-L-Leu-L-Asp-L-Ile-L-Ile-L-Trp trifluoroacetate), *FASEB J.,* 6 (Part 1, No. 4), 392, A1003, 1992.

270. **LaDouceur, D.M., Davis, L.S., Keiser, J.A., Doherty, A.M., Cody, W.L., He, F.X., and Haleen, S.J.,** Effects of the endothelin receptor antagonist PD 142893 (Ac-(β-phenyl)D-Phe-L-Leu-L-Asp-L-Ile-L-Ile-L-Trp trifluoroacetate) on endothelin-1 (ET-1) induced vasodilation and vasoconstrictor in regional arterial beds of the anesthetized rat, *FASEB J.,* 6 (Part 1, No. 4), 390, A1003, 1992.

271. **Cody, W.L., Doherty, A.M., He, J.X., DePue, P.L., Waite, L.A., Topliss, J.G., Haleen, S.J., LaDouceur, D., Flynn, M.A., Hill, K.E., and Reynolds, E.E.,** The rational design of a highly potent combined ET_A and ET_B receptor antagonist (PD 145065) and related analogues, *Med Chem. Res.,* 3, 154, 1993.

272. **Cody, W.L., Doherty, A.M., He, J.X., Waite, L.A., Topliss, J.G., Haleen, S.J., LaDouceur, D., Flynn, M.A., Hill, K.E., and Reynolds, E.E.,** Structure-activity relationships in the C-terminus of endothelin-1 (ET-1): the discovery of potent antagonists, in *Proceedings of the 22nd European Peptide Symposium,* Schneider, C.H. and Eberle, A.N., Eds., ESCOM Science, Leiden, The Netherlands, 1993, 687.

273. **Doherty, A.M., Cody, W.L., He, J.X., DePue, P.L., Welch, K., Flynn, M.A., Reynolds, E.E., LaDouceur, D.M., Davis, L.S., and Haleen, S.J.,** In vitro and in vivo studies with a series of hexapeptide endothelin antagonists, *J. Cardiovasc. Pharmacol.,* 22(Suppl. 8), 598, 1993.

274. **Warner, T.D., Sorrentino, R., Allcock, G.H., Battistini, B., and Vane, J.R.,** Inhibition by a non-selective endothelin receptor antagonist of bronchoconstrictions induced by endothelin-1 or sarafotoxin 6c in the anaesthetized guinea-pig, *Pharmacol. Commun.,* submitted.

275. **Oohata, N., Nishikawa, M., Kiyoto, S., Takase, S., Hemmi, K., Murai, H., and Okuhara, M.,** Anthraquinone derivatives and preparation thereof. European patent application 90112076.6, filed June 26, 1990.

276. **Ihara, M., Fukuroda, T., Saeki, T., Nishikibe, M., Kojiri, K., Suda, H., and Yano, M.,** An endothelin receptor (ET_A) antagonist isolated from Streptomyces misakiensis, *Biochem. Biophys. Res. Commun.,* 178, 132, 1991.

277. **Ishikawa, K., Fukami, T., Nagase, T., Fujita, K., Hayama, T., Niyama, K., Mase, T., Ihara, M., and Yano, M.,** Cyclic pentapeptide endothelin antagonists with high ET_A selectivity. Potency and solubility enhancing modifications, *J. Med. Chem.,* 35, 2139, 1992.

278. **Kiyofumi, I., Takehiro, F., Takashi, H., Kenji, N., Toshio, N., Toshiaki, Mase., Kagari, F., Masaru, N., Masaki, I., and Yano, M.,** Endothelin antagonistic cyclic pentapeptides, EPA 0 436 189 A1; filed December 20, 1990.

279. **Ihara, M., Noguchi, K., Saeki, T., Fukuroda, T., Tsuchida, S., Kimura, S., Fukami, T., Ishikawa, K., Nishikibe, M., and Yano, M.,** Biological profiles of highly potent novel endothelin antagonists selective for the ET_A receptor, *Life Sci.,* 50, 247, 1992.

280. **Fukuroda, T., Nishikibe, M., Ohta, Y., Ihara, N., Yano, M., Ishikawa, K., Fukami, T., and Ikemoto, F.,** Analysis of responses to endothelins in isolated porcine blood vessels by using a novel endothelin antagonist, BQ-153, *Life Sci.,* 50, 107, 1992.

281. **Ohlstein, E.H., Arleth, A., Bryan, H., Elliott, J.D., and Po Sung, C.,** The selective endothelin ET_A receptor antagonist BQ-123 antagonizes endothelin-1-mediated mitogenesis, *Eur. J. Pharmacol.,* 225, 347, 1992.

282. **Webb, M.L., Dickenson, K.E.J., Delaney, C.L., Liu, E. C.-K., Serafino, R., Cohen, R.B., Monshizadegan, H., and Moreland, S.,** The endothelin receptor antagonist, BQ-123, inhibits angiotensin II-induced contractions in rabbit aorta, *Biochem. Biophys. Res. Commun.,* 185(3), 887, 1992.

283. **Nishikibe, M., Tsuchida, S., Okada, M., Fikuroda, T., Shimamoto, K., Yano, M., Ishikawa, K., and Ikemoto, F.,** Antihypertensive effect of a newly synthesized endothelin antagonist, BQ-123, in a genetic hypertensive model, *Life Sci.,* 52, 717, 1993.

284. **Bazil, M.K., Lappe, R.W., and Webb, R.L.,** Pharmacologic characterization of an endothelin$_A$ (ET_A) receptor antagonist in conscious rats, *J. Cardiovasc. Pharmacol.,* 20, 940, 1992.

285. **McMahon, E.G., Palomo, M.A., Brown, M., and Carter, J.S.,** Phosphoramidon (ECE inhibitor) and BQ-123 $(ET_A$ receptor antagonist) lower blood pressure in conscious SHRs. 46th Annu. Conf. Scientific Sessions for the Council of High Blood Pressure, Poster no. 47, *Hypertension* (Dallas) 2093, 431, 1992.

286. **Nishikibe, M., Okada, M., Tsuchida, S., Fukuroda, T., Shimamoto, K., Kobayashi, M., Yano, M., and Ikamoto, F.,** *J. Hypertension,* 10 (Suppl. 4), S50, Abstract No. P53, 1992.

287. **Cirino, M., Motz, C., Maw, J., Ford-Hutchinson, A.W., and Yano, M.,** BQ-153, a novel endothelin (ET)$_A$ antagonist, attenuates the renal vascular effects of endothelin-1, *J. Pharm. Pharmacol.,* 44, 782, 1992.

288. **Fogo, A., Hellings, S.E., Inagami, T., and Kon, V.,** Endothelin receptor antagonism is protective in *in vivo* acute cyclosporine toxicity, *Kidney Int.,* 42, 770, 1992.

289. **Davis, L.S., Haleen, S.J., Doherty, A.M., Cody, W.L., and Keiser, J.A.,** Effects of selective endothelin antagonists on the hemodynamic response to cyclosporin, *Am. J. Nephrol.,* 4(7), 1448, 1993.

290. **Reily, M.D., Thanabal, V., Omecinsky, D.O., Dunbar, J.B., Doherty, A.M., and DePue, P.L.,** The solution structure of a cyclic endothelin antagonist, BQ-123, based on the 1H-1H and 13C-1H three bond coupling constants, *FEBS Lett.,* 300, 136, 1992.

291. **Krystek, S.R., Bassolino, D.A., Bruccoleri, R. E., Hunt, J.T., Porubcan, M.A., Wandler, C.F., and Anderson, N.H.,** Solution conformation of a cyclic pentapeptide endothelin antagonist, *FEBS Lett.,* 299, 255, 1992.

292. **Atkinson, R.A. and Pelton, J.T.,** Conformational study of cyclo[D-Trp-D-Asp-Pro-D-Val-Leu], and endothelin-A receptor selective antagonist, *FEBS Lett.,* 296, 1, 1992.

293. **Miyata, S., Hashimoto, M., Masui, Y., Ezaki, M., Takase, S., Nishikawa, M., Kiyoto, S., Okuhara, M., and Kohsaka, M.,** WS-7338, new endothelin receptor antagonists isolated from Streptomyces sp. No. 7338, *J. Antibiot.,* 45, 74, 1992.

294. **Miyata, S., Hashimoto, M., Fujie, K., Nishikawa, M., Kiyoto, M., Okuhara, M., and Kohsaka, M.,** WS-7338, new endothelin receptor antagonists isolated from Streptomyces sp. No. 7338, *J. Antibiot.,* 45, 83, 1992.

295. **Hashimoto, M., Nishikawa, M., Esaki, M., Kiyoto, S., Okuhara, M., Takase, S., Henmi, K., Neya, M., Fukami, N., and Hashimoto, M.,** New peptide prepared by culturing *streptomyces* sp. used as endothelin antagonist, JO 3130–299-A, filed August 8, 1990.

296. **Gu, X.H., Casley, D.J., and Naylor, W.G.,** The inhibitory effect of [D-Arg, D-Phe, D-Trp^{7,9}, Leu^{11}] substance P on endothelin-1 binding sites in rat cardiac membranes, *Biochem. Biophys. Res. Commun.,* 179, 130, 1991.

297. **Lam, Y.K., Williams, D.L., Sigmund, J.M., Sanchez, M., Genilloud, O., Kong, Y.L., Stevens-Miles, S., Huang, L., and Garrity, G.M.,** *J. Antibiot.,* 45, 1709, 1992.

298. **Keiji, H., Masahiro, N., Naoki, F., Masashi, H., Tanaka, H., and Kayakiri, N.,** Peptides having endothelin antagonist activity, a process for the preparation thereof and pharmaceutical compositions comprising the same, EPA 0457 195 A2, May 9, 1991.

299. **Sogabe, K., Nirei, H., Shoubo, M., Hamada, K., Nomoto, A., Henmi, K., Notsu, Y., and Ono, T.,** A novel endothelial receptor antagonist: studies with FR 139317, *J. Vasc. Res.,* 2nd International symposium on endothelium-derived vasoactive factors. 29(2), 201, Abstract No. 367, 1992.

300. **Aramori, I., Nirei, H., Shoubo, M., Sogabe, K., Nakamura, K., Kojo, H., Notsu, Y., Ono, T., and Nakanishi, S.,** Subtype selectivity of a novel endothelin antagonist, FR 139317, for the two endothelin receptors in transfected Chinese hamster ovary cells, *Mol. Pharmacol.,* 43, 127, 1993.

301. **Mazaki, M., Morfuji, N., and Yamamoto, M.,** N-Pentanoyltryptophan, Japanese Patent Application No. 2[1990]-45,305, filed November 11, 1991.

302. **Ohashi, H., Akiyama, H., Nishikori, K., and Mochizuki, J.-I.,** Asterric acid, a new endothelin binding inhibitor, *J. Antibiot.,* 45(10), 1684, 1992.

303. **Burri, K., Clozel, M., Fischli, W., Hirth, G., Loffler, B.M., and Ramuz, H.,** Application of sulfonamides as therapeutics and new sulfonamides, EP 0510526 A1, filed April 25, 1991.

304. **Clozel, M., Breu, V., Burri, K., Cassal, J.-M., Fischli, W., Gray, G.A., Hirth, G., Loffler, B.-M., Muller, M., Neidhart, W., and Ramuz, H.,** Pathophysiological role of endothelin revealed by the first orally active endothelin receptor antagonist, *Nature,* 365, 759, 1993.

305. **Cousins, R., Elliott, J.D., Lago, M.A., Leber, J.D., and Peishoff, C.E.,** Endothelin receptor antagonists, WO 93/08799, November 1991.

Chapter 5

MOLECULAR TARGETING CHEMISTRY IN RATIONAL DRUG DESIGN

Alan R. Fritzberg, Linda M. Gustavson, Mark D. Hylarides, and John M. Reno

CONTENTS

I. RATIONALE OF MOLECULAR TARGETING

Drugs can act in a direct manner in which a molecule results in a desired cytocidal, cytostatic, or stimulatory effect on target cells. Alternatively, a molecule can be developed for indirect therapy using a targeting vehicle which has high affinity and selectivity for a desired target cell. Attachment of radionuclides, drugs, or toxins to targeting molecules selectively delivers the effector to the target cells. In addition, if *in vivo* identification of a target cell provides useful information for patient management, attachment of gamma-emitting radiotracers allows *in vivo* measurement of the target cells by external gamma camera imaging. Beyond imaging, is a process referred to as radioimmunoguided surgery (RIGS) in which further delineation of tumor tissue by probe detection during surgery potentially provides added ability to detect and eliminate pathologic tissue.[1]

A. PROTEIN TARGETING AGENTS

The use of proteins as targeting agents has been of interest for many years. Many proteins bind to target cells with high affinity, although with varying degrees of selectivity. Targeting proteins include antibodies, growth factors, and growth inhibitors, as well as proteins involved in processes such as clot formation and dissolution. Because of the nature of antibodies in terms of specificity for binding and potential for development of antibodies reactive toward selected antigens, much effort has gone into their development as targeting agents.[2,3] While they may not meet the original expectations of the imagined "magic bullet", much has been learned from antibody-based studies that can be applied to molecular targeting with other molecules.

B. PEPTIDE TARGETING AGENTS

A number of peptides are known with high affinity for binding to receptors that present opportunities for targeting.[4–6] Compared to proteins, peptides exhibit rapid targeting and blood disappearance. The higher target to background ratios also results in improved images. Therapeutic applications of peptides, however, may present a more complicated challenge. Typically, receptor numbers range in the hundreds on normal cells to a few tens of thousands on upregulated cells.[7] Thus, too few receptors may be present to provide the basis for delivery of sufficient numbers of effector molecules to exert the desired pharmacologic effects. In contrast, antigen numbers for many antibodies can be in the 10^5 to 10^6 per cell range.[8] Other concerns include normal tissue or cell cross-reactivity and stability. Solutions to these concerns can be found, as exemplified by efforts with somatostatin in which *in vivo* stability has been improved and selectivity and affinity for tumor cells is being increased by structural modifications.[4]

C. NONPEPTIDE TARGETING AGENTS

Efforts to use nonpeptide or protein-based defined delivery agents are at an earlier stage. High affinities and selectivity are obtained from antisense polynucleotides and thus they in principle could target a payload. However, issues of serum stability, pharmacokinetics, and efficiency of these molecules to reach target nucleic acids which are intracellular need to be addressed.[9]

II. PHARMACOKINETIC AND DELIVERY CAPACITY CONSIDERATIONS OF CONJUGATES

The effector moieties under investigation include a broad range: radiation, drugs, enzymatic toxins, and immune-active peptides and proteins. The numbers of molecules to be delivered per cell and pharmacologic impact during circulation in the blood and bone marrow during the targeting process also vary widely. Basic considerations of each general type will be described.

A. RADIATION

Imaging agents utilizing gamma ray-emitting radionuclides, such as ^{99m}Tc, ^{111}In, ^{123}I, or ^{131}I primarily require a severalfold target-to-background ratio to be successful. Selection of radionuclides must be matched with physical half-life. For example, the 6-h half-life of ^{99m}Tc requires targeting and clearance of radioactivity from blood and normal tissues within 24 h. Thus, peptides and small proteins such as Fab fragments of antibodies are best suited with ^{99m}Tc. For longer clearing biologicals, ^{111}In with its 2.7-d half-life or ^{123}I with an intermediate 13-h half-life may be more suitable.

Particulate emission provides a potent form of cell killing. Thus, a cytocidal dose ranges from only a few tens of cell traversals for alpha particles (helium +2 atoms: e.g., ^{212}Bi, ^{211}At) to a few thousand or greater for beta particles, depending on closeness of cells and crossfire buildup. These factors are well discussed by Humm.[10] The pharmacokinetics of target cell buildup and retention and blood (marrow cell exposure) disappearance need to be analyzed for optimal half-life matching.[11] In addition, the type of particle (alpha, beta) and energy of emission have been considered from microdosimetric and radiation biology points of view and matching of tumor or lesion size and best radionuclide energy match suggested.[12]

Radiotherapeutic radionuclides present a complicated situation with respect to selection. Particulate emission results in toxicity to radiation-sensitive bone marrow cells during the time of systemic circulation. Thus, the objective is to get enough radiation delivered to the target cells to realize the desired response without undue bone marrow toxicity.

B. DRUGS

Drug conjugates for targeting proteins such as antibodies have been appealing due to the great amount of supporting data stemming from the clinical use of the drugs themselves. The addition of targeting should improve the therapeutic index and such improvement would be expected to be relatively readily approved for clinical use. Unfortunately, until recently, significant results showing antibody- or targeting agent-dependent cytotoxicity and xenograft regressions have been lacking. This is due to several hurdles inherent in drug conjugates. The numbers of drug molecules needed for cytotoxicity are large for standard chemotherapy agents and either exceed the carrying capacity of antibodies (unless an additional carrier is used), impractically high doses of drug conjugates are used, or, alternatively, very high-potency drugs are used.[8] Drugs must be delivered intracellularly to be effective and either require release at the surface, compromising specificity, or need an internalizing targeting molecule to deliver the drugs to the actual site of action. Through the process of selection of high-potency drugs and internalizing antibodies, success at the preclinical level is now being realized.

C. PROTEIN TOXINS

The protein toxins of interest as targeted conjugates have been based on enzymatic activity.[13] Thus, one active toxin molecule should be adequate to kill a cell. Since several steps of cell entry, translocation, and protein modification are necessary prior to engaging the cellular substrate, the number required for targeting is certainly more than one. Nevertheless, very high potency has typically been seen.

Because of the high potencies of toxins, there is no need for more than a 1:1 ratio of targeting molecule to toxin. However, these proteins typically have normal tissue binding sites. Until recently these have resulted in toxicity and limited the circulation time and hence the limited ability for efficient solid tumor targeting of toxin conjugates *in vivo*.[14]

III. DESIGN OF PEPTIDE AND PROTEIN CONJUGATES

A. DESIGN OF RADIOIMAGING AGENTS

Radiolabeled peptides and proteins have proven clinically useful for the detection of malignant tumors, sites of inflammation caused by infection and thrombi.[15–18] In these applications, radiolabel is attached to the targeting protein which binds to the site of pathology and clears from normal tissue. A major problem using radiolabeled proteins is that only a very small fraction of the radioactivity administered to patients actually localizes in the lesion, with the remainder found largely in blood and normal tissues such as liver, spleen, and kidneys.[19] One approach to reduce accumulation of radioactivity in normal tissue is to improve the methods for attachment of the radiolabel to the protein. The type of the protein carrier and the use of direct vs. indirect labeling methods determine the extent of normal tissue accumulation. Since antibodies have been studied extensively as targeting vehicles, they will be used to illustrate chemistry affecting their ability to selectively target the radioactivity.

Two general approaches have been utilized for radiolabeling antibodies: direct labeling of antibody functional groups or amino acid moieties activated toward label incorporation, or indirect labeling of a carrier moiety which stably carries the radionuclide and contains linkage chemistry for peptide or protein attachment. While the former approach offers the advantage of simplicity, the ability to modify the carrier moiety provides a means to control such factors as site of attachment, stability, hydrophilicity/lipophilicity, and cleavability and thus control target cell uptake as well as the secondary distribution of catabolites of the labeled conjugate.

1. Radioiodines (^{123}I, ^{131}I)

Iodine-131, with its 364-keV gamma emission and 8.05-d half-life, has a long history of use for imaging. Drawbacks of this isotope are the high energy of its gamma rays and the beta emission that adds to the radiation dose of the patient. In contrast to ^{131}I, ^{123}I (159 keV gamma and 13-h half-life) decays by electron capture and is preferable for imaging studies. However, ^{123}I is only available by cyclotron production and is therefore expensive.

The most commonly employed and long established methods for attachment of radioiodine to proteins involve direct labeling procedures using oxidants such as chloramine-T or iodogen which oxidize iodine to cationic species for electrophilic replacement of a hydrogen atom on phenolic tyrosine residues. There are several potential disadvantages to direct radioiodination. If underivatized tyrosine residues are present in the binding region of the peptide or protein, a loss in binding reactivity may result upon such direct labeling. Some proteins are sensitive to the oxidants required for iodine activation and may thus be damaged. An additional, major limitation of the direct methods which involve iodination of tyrosine moieties is that extensive *in vivo* deiodination may occur. The deiodination is thought to be enzyme mediated since iodinated tyrosines are structurally similar to units of thyroxine which is rapidly deiodinated by enzymes found in the liver, kidney, and thyroid.[20]

The Bolton-Hunter reagent, *N*-succinimidyl 3-(4'-hydroxphenyl)-propionate was developed as an alternative radiodiodination method which avoids exposing proteins to oxidants by conjugating a radioiodinated compound to amino groups on proteins. Antibodies labeled by this method surprisingly showed only 7% of the thyroid uptake observed with iodogen-labeled antibody despite the presence of the activating hydroxyl group.[21] One possible explanation for the reduced thyroid uptake is that catabolism of the Bolton-Hunter modified lysine residues releases an unnatural substrate which is not recognized by deiodinase enzymes.

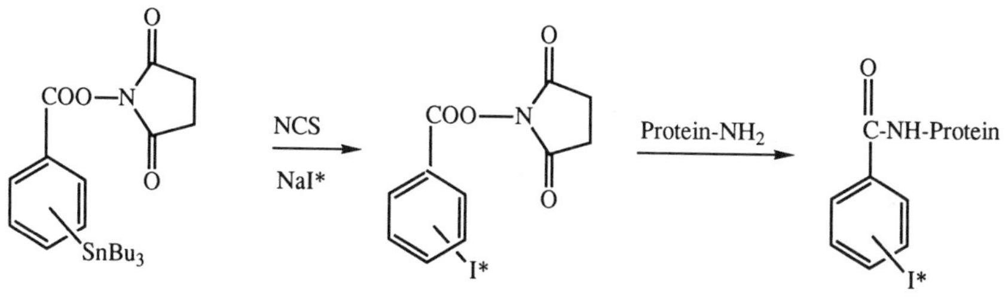

SCHEME 1. Indirect radioiodination PIB and MIB reagents.

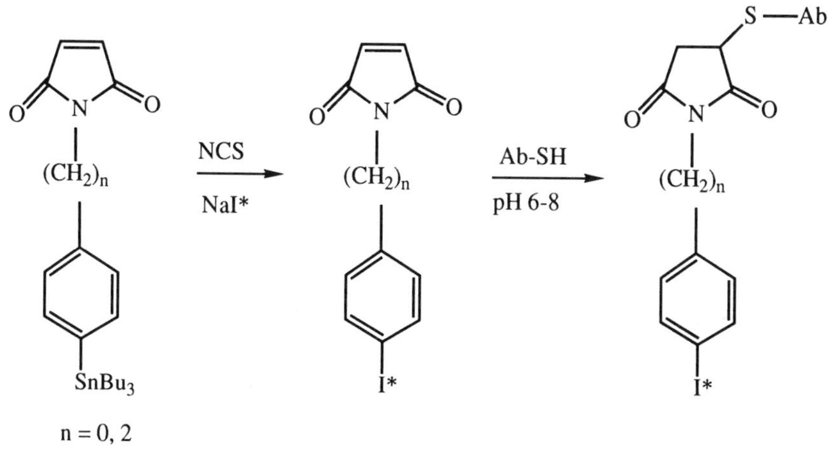

n = 0, 2

SCHEME 2. Indirect radioiodination with iodphenylmaleimides.

The use of phenyl rings as nonactivated carriers of radioiodine has been developed using organometallics as a way to more completely avoid deiodination. The corresponding *para* (PIB)[22,23] or *meta* (ATE)[24] iodobenzoate reagents are conveniently prepared by radioiododestannylation of the corresponding *para* or *meta* N-succinimidyl-(tri-*n*-butylstannyl)benzoate, as shown in Scheme 1.

In a separate step, antibody lysine amino groups are acylated by the succinimidyl ester to provide the radioiodinated conjugate. Both PIB and ATE conjugates have proven to be stable towards *in vivo* dehalogenation in mice, as demonstrated by markedly decreased uptake by stomach and thyroid compared to chloramine-T labeled antibody.[22] Comparison of the *p*- and *m*-iodobenzoyl conjugates showed similar normal tissue and tumor biodistributions. Thus, attachment of iodine to aromatic rings lacking an adjacent hydroxy group is a successful approach to minimize deiodinase activity in animals and patients.

Recently, radioiodinated maleimide derivatives were prepared to selectively react with protein sulfhydryls as an alternative to acylation of protein lysine residues.[25–27] The method involves radioiododestannylation followed by conjugation to sulfhydryl containing antibody such as antibody-reduced dithiothreitol (DTT) to give the thioether conjugate, as shown in Scheme 2.

Biodistribution studies in mice of the thioether conjugate also showed stability toward *in vivo* dehalogenation and tumor uptake similar to the chloramine-T or *p*-iodobenzoyl-labeled conjugates.[27]

2. Indium (^{111}In)

Indium (^{111}In) has been widely used in radioimmunodiagnosis because of its availability and favorable nuclear properties ($t_{1/2} = 67$ h). The use of bifunctional chelates has been necessary since

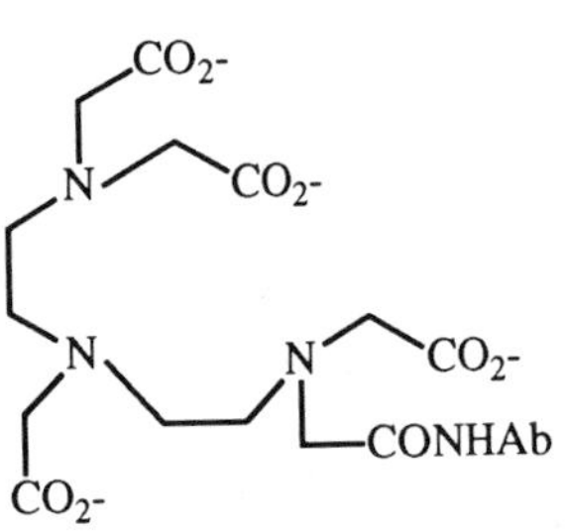

1 Ab-DTPA
(amide, direct carboxylate attachment)

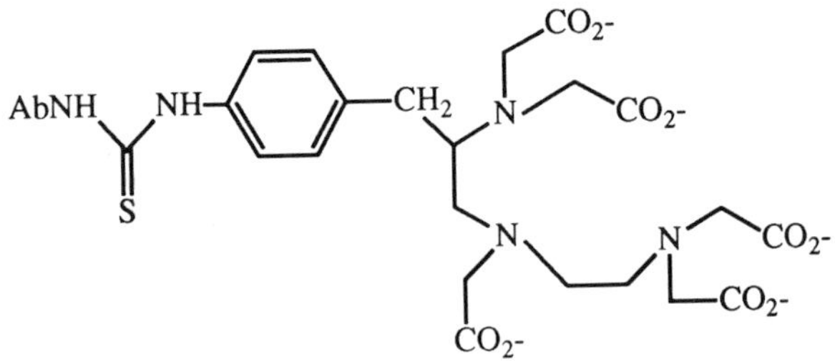

2 Ab-benzyl DTPA
(thiourea, backbone attachment)

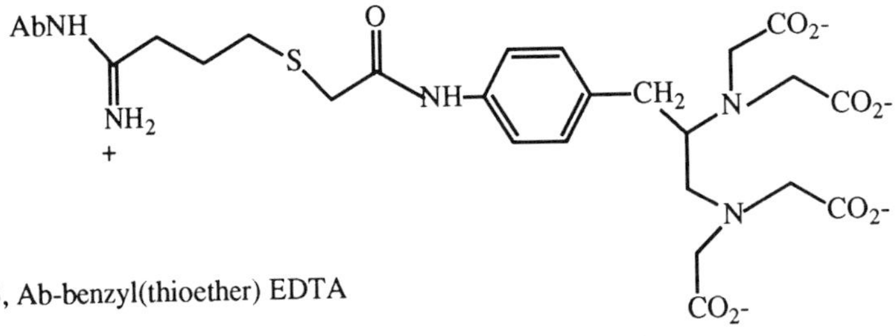

3, Ab-benzyl(thioether) EDTA

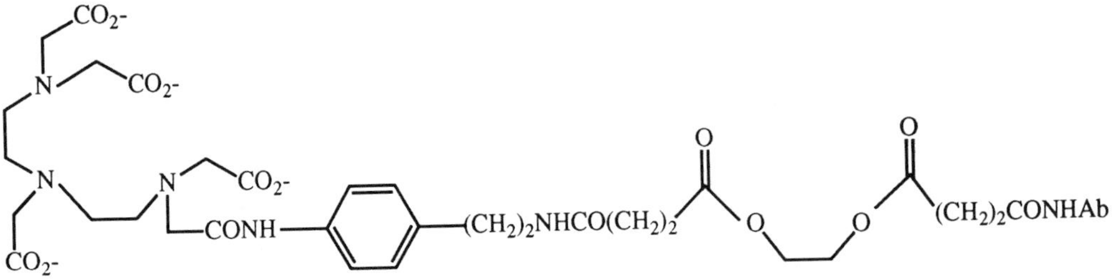

4, Ab(amide) diesterlinker to DTPA benzylamide.

FIGURE 1.   Various indium-chelate conjugates.

indium does not bind to native non-iron binding proteins. Substituted polyaminocarboxylate ligands, such as ethylene diamine tetraacetate (EDTA) and DTPA have typically been used to chelate indium. The method of attachment of these ligands to antibodies influences both the stability of the metal in the chelate and the susceptibility of the linkage between antibody and the chelate to metabolic cleavage. Variations in the linkages between the indium chelate and antibody are illustrated in Figure 1.

Direct attachment of DTPA (**1**, Figure 1) and EDTA to proteins through formation of an amide bond between a carboxylate group of the chelate and amino groups on the antibody has been widely studied.[28–31] Attachment of DTPA by direct bonding of one of the carboxylates reduces the number of groups available for coordination to indium. The reduced stability of the chelate is thought to be one of the reasons antibodies labeled with indium by direct attachment of DTPA show high levels of accumulation in normal organs such as liver and kidney.[32] Transchelation in serum of indium from DTPA-coupled antibodies to circulating transferrin results in increased liver localization.[33,34]

Stability studies *in vitro* and in serum have shown that DTPA attachment of DTPA or EDTA through a linker bonded to the ethylene backbone (structures **2** and **3**, Figure 1) produces chelates which bind indium more tightly than directly bound DTPA conjugates such as **1** (Figure 1).[34,35]

Substituted DTPA conjugate **2** is prepared from *p*-isothiocyanatobenzyl DTPA and DTPA conjugate **3** is prepared from *p*-bromoacetamidobenzyl DTPA derivatives. Biodistribution studies in mice showed that the antibody conjugate derived from SCN-benzyl-DTPA (**2**, Figure 1) had lower liver accumulation than antibody conjugates obtained by direct coupling to DTPA (**1**, Figure 1).[35,36] Since ^{111}In-Bz-EDTA was detected in the urine of mice that received the antibody conjugate derived from SCN-Bz-EDTA, retention of indium in the chelate and instability of the thiourea bond is suggested.[37]

The location of chelate attachment on antibody molecules has been considered. The carbohydrate groups are known not to be typically in the area of the complementary determining regions. Oxidation of the carbohydrates with periodate followed by reaction with amines using hydrazine Schiff base chemistry has been used to attach a DTPA chelate.[38] Retention of immunoreactivity to a high degree and *in vivo* targeting has been shown.

Reduction of normal tissue accumulation was also investigated using bifunctional chelates containing enzyme metabolizable linkages.[39,40] Introduction of a metabolizable linkage between the metal chelate and antibody was expected to increase tumor to nontarget organ ratios on the basis that cleavage at the tumor occurred more slowly than cleavage in nontarget organs. The linkage was designed to produce cleavage products with the radionuclide on small, polar fragments which are rapidly excreted by the kidneys, thus reducing background activity and whole-body radiation. The diester **4** (Figure 1) prepared from DTPA-*p*-(aminoethyl) anilide, antibody and ethylene glycolbis (succinimidyl succinate) (EGS) as cross-linking agent,[39] showed faster blood clearance and lower normal tissue accumulations compared to the amide-linked conjugate. Tumor retention of ester-linked conjugate was greater than or similar to the amide conjugate over the same time period. Thus, a greater than twofold increase of tumor to blood, liver, and kidney ratios was obtained using the cleavable linkage. Decrease in liver uptake was also achieved using the thioether conjugate **3** (Figure 1) compared to the corresponding peptide conjugate.[39] A systematic study on the effect of EDTA-antibody conjugates containing disulfide, thioether, thiourea, and peptide linkages gave corresponding liver values of 2.2, 13.4, 7.6, and 20% 72 h after injection.[40] The disulfide was found to be rapidly cleaved in serum. The other three conjugates exhibited similar blood disappearance rates. Thus, the use of cleavable linkers shows promise for improved specificity for radioactivity targeting.

A novel application of antibodies radiolabeled with indium was achieved in the successful imaging of patients with acute myocardial infarction. Indium-labeled antimyosin Fab specifically identifies necrotic regions which have been irreversibly damaged by acute infarction.[41] The radio-labeled antibody does not accumulate in old infarcts or in normal, noninfarcted myocardium. The indium label was attached by transchelation from citrate to DTPA-amide Fab obtained by mixed anhydride conjugation. The transchelation method facilitates labeling of the antibody at a pH that minimizes damage to the protein.[42] A limitation of this conjugate is the persistent blood pooling observed in patients 24 h postinjection, which resulted in 13 to 21% nondiagnostic scans.[42]

3. Technetium (^{99m}Tc)

The favorable nuclear properties, low cost, and wide availability make technetium the radioisotope of choice for diagnostic imaging. The gamma energy of 140 keV is ideal for efficient external detection and low scatter. As mentioned earlier, the half-life of 6.02 h, however, requires completion of studies within 24 h. As a result, preferred targeting forms for ^{99m}Tc are peptides and proteins that rapidly target and are cleared from the circulation.

Due to the simplicity of direct attachment of ^{99m}Tc to proteins, much effort has gone into the development of direct labeling methods. Because proteins contain donor atoms/groups that bind to reduced Tc such as sulfhydryl, carboxylate, amide, amino, and hydroxyl, a proclivity of Tc binding to these groups in proteins exists. The sulfhydryl groups exhibit the strongest binding interaction while the other donor atoms interact similarly and at a weaker level. Thus, all direct methods involve at some level, production of sulfhydryl groups as the key point of metal protein interaction. As additional functional group/donor atoms are available at more or less favorable positions with respect to the Tc metal center stereochemistry, variable levels of Tc binding stability result. Much of what has been done then, is to optimize binding of ^{99m}Tc to proteins with favorable donor atom environments for direct binding. The use of labile technetium complexes facilitates exchange of the metal to protein donor atom sites of higher stability.

An early development with respect to direct antibody labeling was the pretinning formulation of Rhodes et al.[43] Protein was incubated for an extended time with stannous ion, which resulted in slow disulfide sulfhydryl reduction as well as an availability of excess stannous ion for reduction of ^{99m}Tc as pertechnetate. This approach has been used with some success for imaging melanoma,[44] as well as clots with an anti-fibrin antibody.[45] More recently, standard disulfide reducing agents such as dithiothreitol (DTT),[46] 2-mercaptoethanol,[47–49] as well as high concentrations of ascorbic acid[50] have been used to rapidly produce sulfhydryl groups. Various weak chelating agents, including gluconate,[46] glucarate,[51] and methylene diphosphonate,[47] have been used to optimize exchange of ^{99m}Tc to higher affinity sites. Treatment of an anti-CEA antibody F (ab′)$_2$ fragment to cleave the bridging disulfide into Fab′ has resulted in a ^{99m}Tc Fab′ that has completed Phase III clinical trials.[52] Thus, a combination of suitable antibody fragment and ^{99m}Tc labeling formulation has had apparent clinical utility demonstrated.

More recently, studies have appeared that have analyzed the impact of disulfide reducing agents on proteins in more detail.[48,49] Reduction of protein by 2-mercaptoethanol resulted in protein readily labeled by ^{99m}Tc via diphosphonate exchange.[48] The number of sulfhydryls obtained after antibody reduction far exceeds the theoretical maximum, even including the interchain sulfhydryls. Although HPLC showed only one 150-kDa radiolabeled protein, SDS PAGE showed the presence of a range of fragments, indicating that interchain disulfides are reduced. The presence of residual mercaptoethanol, which affects accurate sulfhydryl determination, is required, since rigorous removal of this reagent by HPLC or dialysis renders the product unreactive toward Tc. Apparently mercaptoethanol prevents reoxidation of the reduced disulfide bonds.

While direct methods of Tc labeling are appealing for their simplicity, they are not suitable for antibodies and fragments which would lose binding affinity during the reduction process required to generate thiols. Many peptides possess cyclic disulfide groups which impart the necessary conformational restrictions for binding. For example, somatostatins exhibit loss of biological activity if the disulfide is disrupted. With decreasing size of the protein fragment or peptide, the likelihood that functional groups are part of the binding domain increases. For these reasons, the development of bifunctional chelates (BFC) has been an area of intense investigation. Indirect labeling with BFC offers the advantage that the *in vivo* distribution can be directed by appropriate drug design including the addition of hydrophilic groups and cleavable linkers.

There are two ways that BFCs have been used to label proteins: (1) the preformed approach, in which a preformed Tc complex is allowed to react with antibody; and (2) the post-formed approach, in which the BFC is conjugated to the protein first and subsequently labeled with Tc. The preformed approach offers the advantage that the Tc complex, a small molecule, can be fully characterized and nonspecific binding as described above for direct labeling can be avoided. Maximum control over the Tc labeling process and purity is achieved by this sequence. A drawback of this approach is that multiple steps and postlabeling purification are required to prepare the radiolabeled conjugate. A potential problem with the post-formed antibody chelator conjugate is the nonspecific binding of Tc to protein donor groups instead of the antibody-chelator. A brief survey of some of the BFCs developed for labeling proteins with Tc is given below.

Efforts to extend the DTPA labeling method for indium to technetium revealed the Tc-DTPA chelate to be far less stable *in vivo* since DTPA fails to block nonspecific binding of Tc to proteins.[53,54]

The dithiosemicarbazone 5 (CE-DTS, Figure 2), an example of a post-formed BFC, was conjugated to IgG by the phosphorylazide method.[55] It was necessary to conduct subsequent labeling with ^{99m}Tc at pH 4.5 to 6.2, since pH outside this range impaired the immunoreactivity of the IgG. The labeling proceeded slowly, with an optimal yield achieved only after 3 h. The labeled conjugate showed good stability upon incubation with mice sera and similar biodistribution to indium-labeled IgG. The notable differences in %ID/G at 1 h postinjection were higher stomach (2.57 vs. 0.87%) and liver (13.07 vs. 10.58%) accumulations of Tc compared to In, suggestive of some *in vivo*

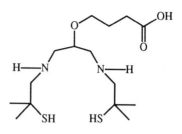

5, CE-DTS
p-carboxyethylphenylglyoxal-di-
(N-methylthiosemicarbazone)

6, Diaminodithiol

7, Hydrazino-Nicotinamide Derivative

8, Isothiocyanatoaryl phosphinimine

FIGURE 2. Various Tc bifunctional chelates (BFCs).

instability of the Tc label. Increasing the labeling efficiency by increasing the loading of DTS groups per antibody resulted in decreased targeting. The diaminodithiol BFC **6** (Figure 2) was reported to complex Tc under mild conditions and the resultant labeled product was stable to human serum.[56] However, results of ^{99m}Tc labeling of the antibody conjugate have not been reported.

Post-formed labeling via Tc-oxo group replacement by hydrazine has been developed as a novel approach to Tc labeling.[57,58] The NHS ester **7** (Figure 2) was conjugated to IgG (pH 7.8) and the conjugate was labeled with Tc by exchange with Tc-gluocoheptonate for 1 h at room temperature. The binding of ^{99m}Tc was highly selective since a control experiment in which the nicotinic free acid derivative was substituted for the succinimidyl ester showed only 6% labeling of protein. The conjugate was stable upon incubation with mouse serum. Comparison of the biodistribution of ^{99m}Tc- and ^{111}In-labeled IgG for targeting abscess were similar and showed the ^{99m}Tc images of infection to be equivalent to ^{111}In images. The status of structural characterization of the Tc complex is limited.

The phosphinimine BFC (**8**, Figure 2) was shown to readily complex Tc and the preformed Tc-BFC conjugated to IgG in 70% yield.[59] While the Tc-labeled IgG showed no release of ^{99m}Tc after incubation of the conjugate at physiological pH for 24 h, stability in the presence of more stringent and relevant challenges such as serum and endogenous strong Tc chelators including cysteine and glutathione were not tested.[59]

The novel thiolactone **9** (Scheme 3) reacts with amino groups on proteins to simultaneously give an amide bond and a diaminodithiol ligand for post-formed labeling with Tc.[60,61] The thiolactone requires no prior activation and reacts with proteins at pH 9 at room temperature. The coupled protein was labeled directly with stannous ion and pertechnetate or by exchange with ^{99m}Tc-glucoheptonate. The exchange labeling technique indicated little nonspecific incorporation of Tc onto protein since control protein showed no bound ^{99m}Tc radioactivity. Limited *in vivo* studies of the radiolabeled conjugate showed stability *in vitro* and *in vivo* as evidenced by the similar blood clearance and normal organ uptake of ^{99m}Tc-3DADT:HSA and ^{125}I-HSA.[62]

The application of ester-linked conjugates as a cleavable linkage to improve target to blood ratios was extended to ^{99m}Tc in the N$_3$S-maleimide BFC shown in Scheme 4.[63] The BFC maleimide **10** undergoes Michael addition with sulfhydryls of the C-terminus of antibody Fab' fragments. The

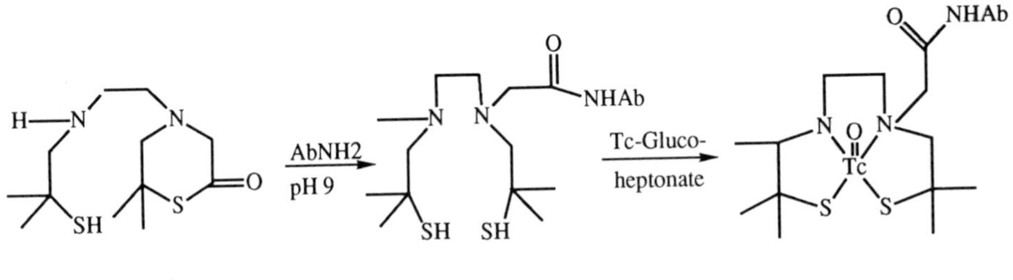

9

SCHEME 3. Post-formed radiolabeling of antibody using thiolactone BFC.

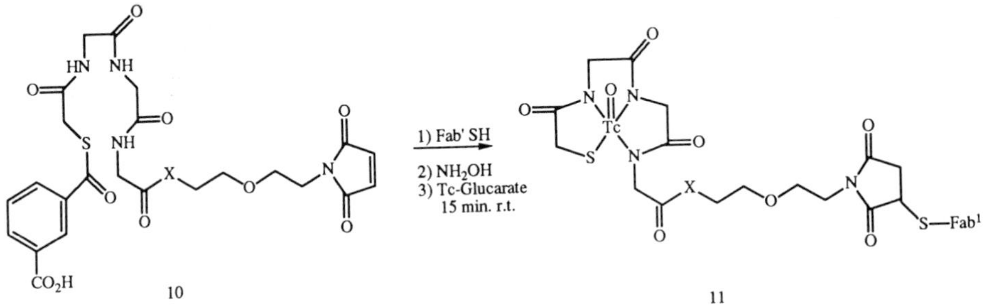

SCHEME 4. Post-formed radiolabeling of antibody using MAG$_3$ ester malemide BFC.

conjugation reaction is conducted at neutral pH to minimize possible lysine amino participation and thus limits the number of chelators/antibody to the number of available sulfhydryls. In designing the BFC, the MAG$_3$ chelating agent was chosen since ^{99m}Tc MAG$_3$ has been shown to be specifically and rapidly excreted through the kidneys.[64] The S-isophthaloyl ester protecting group increases the hydrophilicity of the BFC to facilitate conjugation to the protein under aqueous condition. Removal of the sulfur protecting group and room temperature ^{99m}Tc labeling by exchange with ^{99m}Tc glucarate was quantitative. The ester-linked conjugate **11** gave a twofold improvement in clearance from the kidney compared to the amide-linked conjugate.

The N$_2$S$_2$ BFC **12** (Scheme 5) was developed for preformed labeling of antibodies with Tc.[65] In the first step, ^{99m}Tc chelation is accomplished by heating for a few minutes at acidic pH. The ethoxyethyl sulfur protecting group is selectively cleaved during chelation via metal assistance, as the protecting group is otherwise stable under the acidic chelation conditions. The ^{99m}Tc N$_2$S$_2$ chelate active ester is conjugated to the antibody at basic pH and then purification of the conjugate by elution of only the ^{99m}Tc protein through anion exchange is done. The labeled conjugate showed retention of immunoreactivity, excellent *in vitro* and *in vivo* stability, as previously demonstrated for ^{99m}Tc chelate conjugates obtained by *in situ* chelation and esterification.[66] Clinical trials have been completed using the N$_2$S$_2$ formulation for ^{99m}Tc NR-ML-05 Fab in melanoma,[67,68] and NR-LU-10 for small cell, and non-small cell lung cancers.[69]

Biodistribution studies in mice that received ^{99m}Tc N$_2$S$_2$ NR-ML-05 Fab showed the organ of highest accumulation to be tumor, with good clearance from normal organs. After 4 h, 3 to 4% of injected dose per gram remained in blood and intestine and 17% was in the kidneys. The kidney accumulation is typical for Fab fragments and clears by later times. The accumulation in the intestines was found to be due to an antibody catabolite, the ε-aminolysine ^{99m}Tc N$_2$S$_2$ derivative **13** (Figure 3).[70] Since the small molecule generated by catabolism of the conjugate is responsible for nontarget organ retention, design of the BFC may alter the biodistribution. Second-generation BFCs containing one or two carboxylates, such as **14** and **15** (Figure 3), were prepared. Biodistribution studies in mice showed that introduction of a single carboxylic acid reduced the intestinal activity by 20 to 30% and when two carboxylic acid substituents were present, intestinal accumulation was reduced 40 to 60% compared to N$_2$S$_2$-BFCs lacking carboxylic acid substituents.[71] The use of a

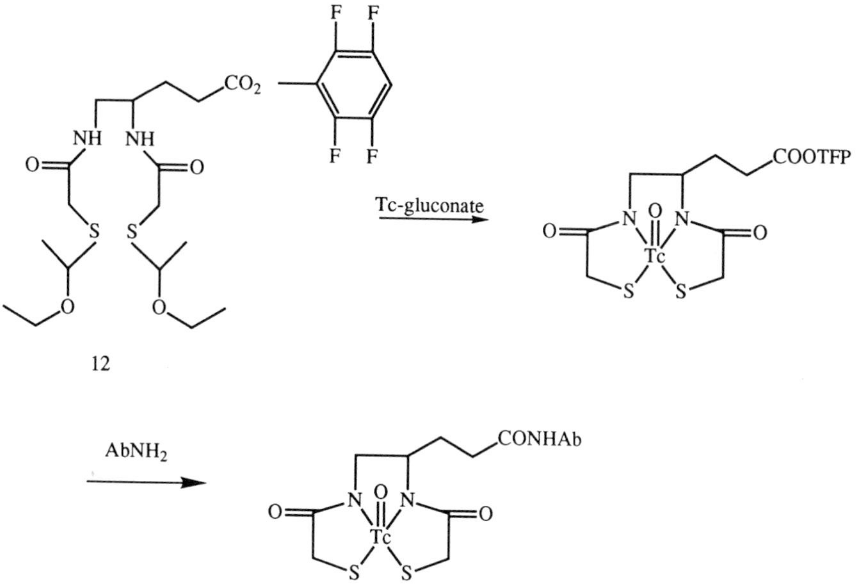

SCHEME 5. Preformed radiolabeling using C5-N2S2-TFP ester.

cleavable ester linkage between the chelate core and the active ester, as in BFC **16** (Figure 3), resulted in reduced accumulation in both kidney and intestines.[72] The preformed chelate approach thus has demonstrated rational design of chelate and correlation of biodistribution with chelate structure.

B. DESIGN OF RADIOTHERAPEUTIC AGENTS

An obvious extension to the delivery of radioactivity for the imaging of pathology is to replace diagnostic gamma ray-emitting radionuclides with those that emit particulate radiation for therapeutic intent. The diagnostic tracer can demonstrate the target to normal organ biodistribution properties and with a time course provide estimates of radiation dose that can be delivered to a tumor as well as radiation exposure to normal organs based on physical properties of selected therapeutic radionuclides. Thus, a pattern has developed for evaluation of the targeting of diagnostic imaging agents and the development of parallel chemistries for appropriate therapeutic radionuclides. A large number of potential, particle-emitting therapeutic radionuclides are available. Several reviews have appeared that provide an analysis of candidate therapeutic radionuclides in terms of physical properties, biological targeting pharmacokinetics parameter match considerations, and availability of appropriate amounts and specific activities.[10,73–75] The list of preferred radionuclides that have had significant developmental effort for targeting include phosphorous (^{32}P), copper (^{67}Cu), yttrium (^{90}Y), iodine (^{131}I), rhenium (^{186,188}Re), astatine (^{211}At), and bismuth (^{212}Bi). The chemistries associated with attachment of these radionuclides to targeting moieties, primarily antibodies, will be discussed in this section.

1. Phosphorous (^{32}P)

Phosphorous (^{32}P), with its 14.2-d half-life and high-energy betas (1.71 MeV max) has been in routine use for decades as ^{32}P phosphate for polycythemia vera[76] and as chromic phosphate for cystic neoplasms in the peritoneal cavity.[77] Recently, it has been proposed for use attached to tumor-targeting antibodies via the Kemptide peptide.[78–80]

2. Copper (^{67}Cu)

Copper (^{67}Cu) has attractive physical properties for targeted radiotherapy with a half-life of 2.58 d and a medium beta energy (0.57 MeV max). It emits gamma radiation (92 keV at 23%,

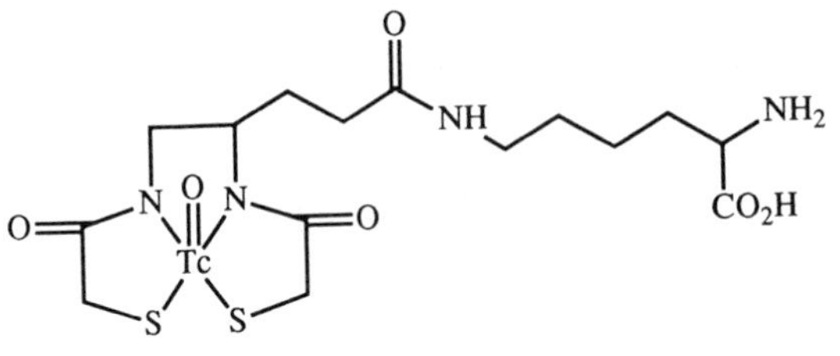

Lysine Adduct 13

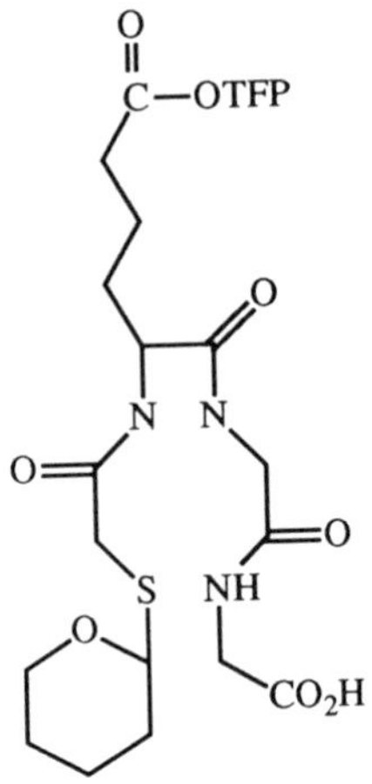

N₃S - Monocarboxylate 14

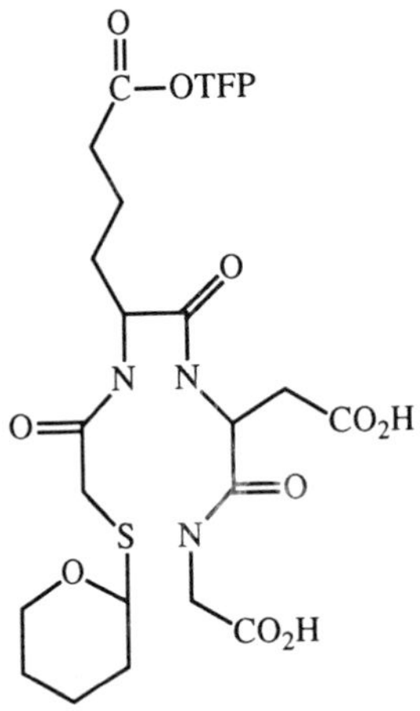

N₃S - dicarboxylate 15

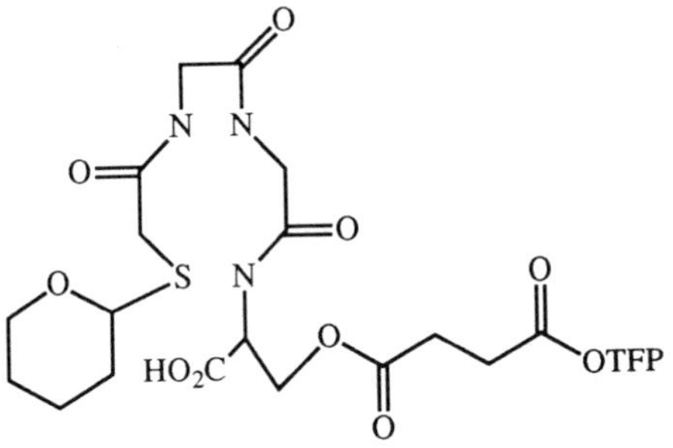

N₃S - ester linker 16

FIGURE 3. Lysine metabolite of C_5-N_2S_2-TFP ester and second-generation N₃S ligands.

184 keV at 40%), allowing imaging. Several chelating agents have been evaluated for stable attachment of ^{67}Cu to antibodies by Cole et al.[81] Serum stability studies indicated loss of ^{67}Cu by transfer from EDTA or DTPA antibody conjugates to serum albumin. The functionalized polyaza polycarboxylic acid macrocyclic bifunctional chelating agent, 1,4,8,11-tetraazacyclotetradecane, *N*, *N′, N″, N‴* tetracetic acid (TETA, **17**) (Figure 4), attached via a *p*-bromoacetamidobenzyl group, however, stably contained the ^{67}Cu with losses of only 4 and 6% at days 3 and 5 in serum.

3. Yttrium (^{90}Y)

Yttrium (^{90}Y) has had the most interest, following ^{131}I, for use in antibody-targeted radiotherapy. It has a high-energy beta maximum of 2.12 MeV and a half-life of 64 h suitable for antibody-targeting pharmacokinetics. As the chemistry of yttrium is similar in respect to the trivalent cation chemistry of In, initial immunoconjugates were based on the same EDTA- and DTPA-related chelating agents. However, the ionic radius of yttrium is 15% larger than indium, resulting in higher coordination numbers.[82] Thus, while EDTA and DTPA chelates were stable with ^{111}In, yttrium chelates did exchange. As ^{90}Y localizes in bone when released *in vivo*,[83] the resulting dose to bone

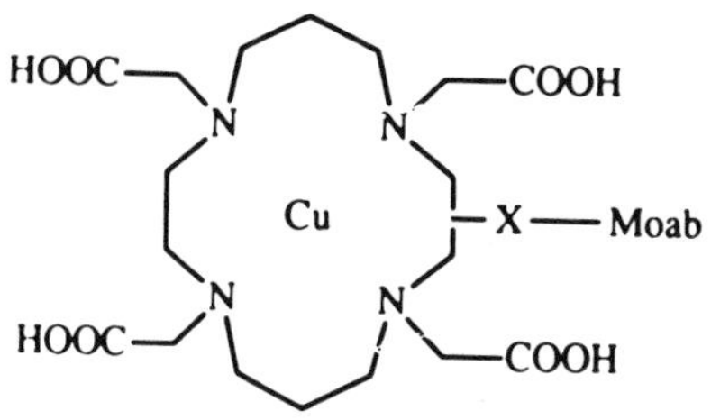

FIGURE 4. N_4 macrocycles.

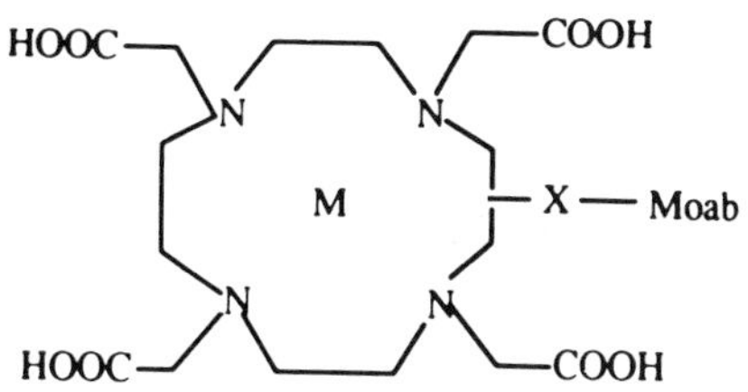

marrow has significantly limited administered doses in clinical studies.[84] *In vivo* stability for ^{90}Y immunoconjugates has been achieved with suitably benzyl backbone substituted DTPA,[82,85] and the 12-member cyclic tetraza tetracarboxylate, DOTA **18** (Figure 4).[86]

4. Radioiodines (^{125}I/^{131}I)

Radiotherapy can be delivered using either ^{125}I via low-energy Auger emissions if a process is available to target the ^{125}I intracellularly in the vicinity of the nucleus, or ^{131}I via its moderate energy (0.61 MeV max) beta particle emission. As ^{131}I emits 364-keV gamma rays in high abundance, significant nontarget irradiation inevitably occurs. Radioiodine has been predominantly used in animal and human studies due to its availability and ease of direct attachment via iodination of tyrosine moieties. As the relevant chemistry was discussed earlier under diagnostic imaging applications, the reader is referred to that section. The chemistries that result in improved stability should be used in therapeutic applications, as targeting peptides or proteins that undergo significant deiodination at therapeutic doses result in hypothyroidism, an avoidable toxicity.

5. Rhenium (^{186}Re/^{188}Re)

Rhenium (^{186}Re), with its moderate 1.07-MeV max beta emission and low-abundance 137-keV photons (9%), is attractive for antibody-based targeting with its 3.7-d half-life. ^{188}Re, with its 2.12-MeV max betas and 155-keV (15%) photons, has a 17-h half-life and thus requires more rapid targeting and normal organ clearance to develop a favorable therapeutic index. However, ^{188}Re is available in high specific activity from a W-188/Re-188 generator[87] and thus a convenient supply can be made available.

Rhenium lies below technetium in the periodic table and because of the lanthanide contraction, has a virtually identical ionic radius.[88] Diamide dithiolate MO (M = Tc, Re)$^{3+}$ structures are identical structurally[89] and are seen biologically the same.[90] Thus, the development of stable bifunctional chelate immunoconjugates with ^{99m}Tc allowed application of the same chelating agents for rhenium therapeutic radioisotopes. While Re is more difficult to reduce from ReO_4^- than Tc from TcO_4^-,[88] and chelate exchange occurs at slower rates with Re,[91,92] use of the preformed chelate approach allows adjustment of N,S amide thiolates chelate conditions to overcome these differences prior to protein labeling.

As discussed with the ^{99m}Tc N,S amide thiolate modifications to core chelate structure and cleavable linkages, improved biodistribution properties have been seen similarly with the ^{186}Re immunoconjugates. Using the N_3S ligand with serine as the terminal amino acid, mercaptoacetylglycylglycylserine, succinate was used to create a cleavable ester link with the metal complex and a stable amide link to an antibody lysine (**20**, Figure 5). When tested on a whole antibody adenocarcinoma targeting molecule, NR-LU-10, no difference in tumor uptake and retention of ^{186}Re compared to stable radioiodinated control (peak values of 40% injected dose per gram) or serum disappearance pharmacokinetics was seen in a xenograft LS-180 colon tumor model.[93]

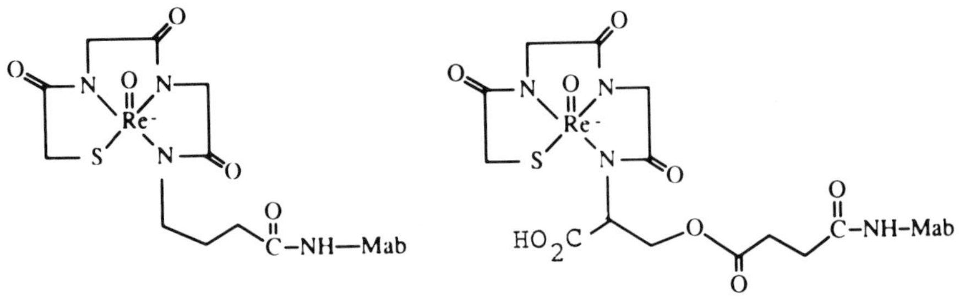

FIGURE 5. Antibody conjugates using N$_3$S chelates.

FIGURE 6. Preparation of *N*-succinimidyl-*p*-[^{211}At]astatobenzoate for protein labeling.

However, intestinal excretion of ^{186}Re was reduced from 34% for amide-linked ^{186}Re to 14% for ester-linked ^{186}Re in rats, while urinary excretion was correspondingly increased from 28 to 48% of the injected dose. Thus, the seryl succinate ester linkage was seen as stable in circulation and at tumor, but cleavable in liver and kidneys to release the core ^{186}Re N$_3$S complex, which then is efficiently cleared by the kidneys. Analysis of urine radioactivity indicated only the cleaved form of the ^{186}Re complex present.

6. Astatine (^{211}At)

Astatine (^{211}At) provides alpha particle emission with its attendant high potency at short range (5 to 6 cell diameters). It is of interest due to its 7.2-h half-life, which allows selective *in vivo* targeting, as well as its chemical similarities to iodine, which allow labeling via conditions and reagents developed for radioiodine.

The labeling chemistry of ^{211}At in general parallels the chemistry of radioiodine. However, its complex halogen and metallic properties allow it to react with sulfhydryl groups as well as undergo electrophilic aromatic substitution when oxidized.[94] Thus, while directly attached ^{211}At-labeled protein is unstable *in vivo*,[95,96] tumor targeting and *in vivo* stability have been shown for *p*-astatophenyl antibody conjugates, prepared from *n*-succinimodyl *p*-astatobenzoate **21** (Figure 6).[97,98]

7. Bismuth (^{212}Bi)

Bismuth (^{212}Bi) has a short half-life of 1.0 h, but is a conveniently available alpha emitter from a ^{212}Pb generator.[99] Although thermodynamic studies of the trivalent Bi metal indicate greater stability than those of In and Y, *in vivo* instability has been observed with EDTA- and DTPA-based chelating agents.[82] However, evaluation of cyclic polyaza macrocyclic ligands indicated that DOTA **18** chelating agents provide *in vivo* stability.[82]

C. LINKERS FOR DRUGS AND TOXIN CONJUGATES

A number of different linker systems have been utilized for conjugation of drugs or toxins to targeting proteins. These linkers, classified as either metabolically stable or cleavable, often contribute in directing the metabolic fate of the cytotoxic agent.

TABLE 1
Selection Criteria for Linkers

1. Retention of cytotoxicity after conjugation
2. Drug-linker reactivity
3. Stability of intermediate (reaction vs. hydrolysis)
4. Cleavability under physiological conditions
5. Reaction functionality (mono, bi, homo, hetero)
6. Reaction specificity (amino, thiol, etc.)
7. Molecular dimensions (chain length, steric)
8. Affinity (hydrophobic, hydrophilic, charge)
9. Presence of reporter groups (chromophores, metal complexes, etc.)

Stable linkers, based on amide, thioether, and secondary amine functionalities, are resistant to degradation or cleavage in most metabolic pathways. Cleavable linkers, including disulfides, esters, acid labile moieties, hydrazones, imines, and certain peptide spacer groups, may undergo metabolic cleavage as a result of hydrolysis, oxidation, nucleophilic displacement, or change in pH. The ultimate goal of delivering a toxic agent to a target site may require selection of a linker to accommodate the expected cellular processing of the conjugate.

In general, an effective linker should remain noncleavable (stable) while the conjugate is in circulation en route to its target. After endocytosis of the conjugate at the target site, the linker should then undergo cleavage to release the cytotoxic species, or a derivative thereof. For large-scale preparation, the cytotoxic agent/linker adduct should be easily prepared and exhibit reasonable stability. A multifunctional linker which utilizes different functional groups on the targeting protein and cytotoxic agent may be required for the eventual conjugation reaction. The functional groups present on the linker determine reaction specificity or sites of derivatization on the protein or cytotoxic agent. The affinity and molecular dimensions of the linker and presence of peripheral groups on the two conjugate components also must be considered when selecting a linker (Table 1).[8]

In this section, the chemistry involved in the preparation of drug and toxin conjugates with linkers based on different functionalities will be reviewed. A comprehensive survey and complete acknowledgment of all the research conducted in both the drug and toxin linker chemistry is beyond the scope of this chapter.

Proteins having reactive lysine amino, cysteine sulfhydryls, or carbohydrate residues are ideal for the conjugation of cytotoxic agents. Other reactive groups such as phenols, carboxylates, and imidazoles are also used.[100,101] In all cases, an important concern is the type and degree of protein derivatization carried out for the conjugation of cytotoxic agents. Overloading or modification of functional groups critical to activity or binding may result in altered biological activity. Likewise, derivation of the drug or toxin may also lead to altered biological potency. The ultimate goal of conjugate preparation is to ensure the biological activity of both targeted protein and cytotoxic agent be preserved.

Conjugation of toxins and drugs to targeting proteins has been performed by different methods, including direct covalent attachment, use of small spacer (linker) molecules, and initial attachment to a high-molecular-weight carrier followed by attachment to targeted proteins. In general, toxins have been covalently conjugated to targeted proteins using either disulfide or thioether linkers. These include holotoxins, such as *Pseudomonas* exotoxin A, ricin, diphtheria, abrin, and gelonin which possess a catalytic active A fragment and binding B fragment. Toxin fragments such as ricin A, diphtheria A, and abrin A and ribosome-inactivating proteins including saporin, barley toxin II, gelonin, and pokeweed antiviral protein, have also been conjugated.

Drug conjugates have been prepared by direct covalent attachment to protein but, more frequently, linkers containing a variety of different functional groups have been used. Examples of drugs which have been conjugated to targeting proteins include: methotrexate, vindesine, neocarzinostatin, *cis*-platinum, chlorambucil, cytosine arabinoside, 5-fluorouridine, melphalan,

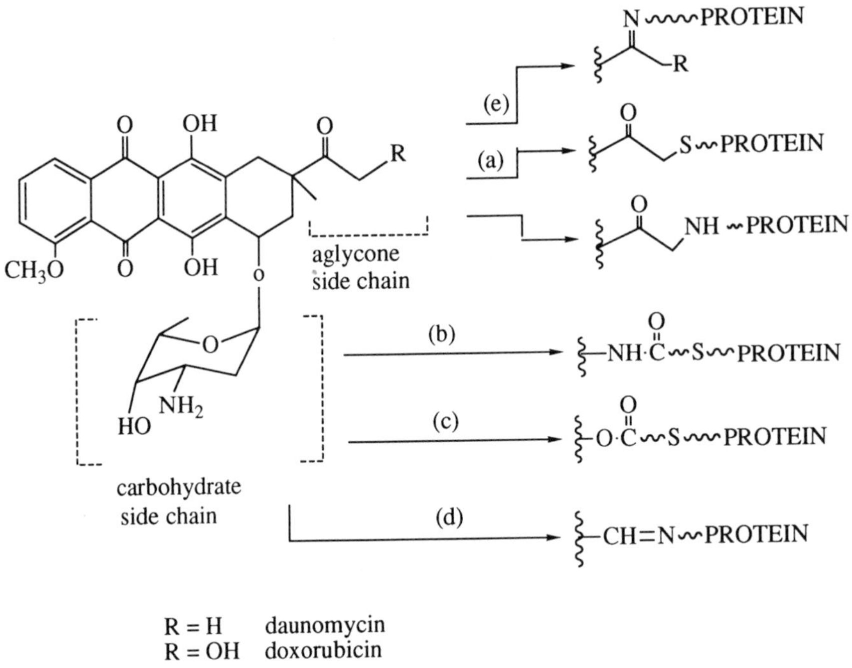

FIGURE 7. Major routes of conjugation of anthracyclines to proteins.

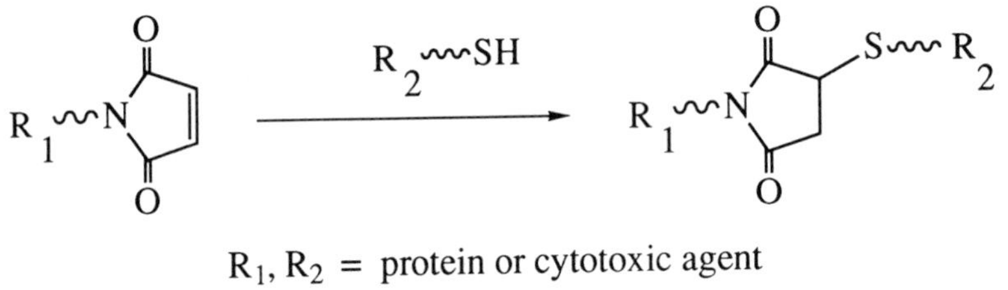

R_1, R_2 = protein or cytotoxic agent

FIGURE 8. Reaction of reactive thiols with maleimide-derivatized protein or cytotoxic agent to generate thioether conjugates.

trichothecenes, calicheamicin, mitomycin, daunomycin, and adriamycin as examples. In this review, anthracycline derivatives will be addressed extensively, since the type of linker chemistry used for their conjugation to proteins is diverse and representative.

Methods that have been used for covalent attachment of daunomycin to protein include: reaction at the carbohydrate amino group on daunomycin to generate amides, imines, and disulfides; reaction at the C-3 sugar hydroxyl to form an ester; oxidative cleavage of the carbohydrate moiety followed by reaction with an amine to form a Schiff base; conjugation through the methyl ketone of the aglycone side chain to form hydrazones; nucleophilic substitution on the 14-bromo derivatives to form thioethers and secondary amines (Figure 7).[102–105]

1. Thioethers

Toxin conjugates containing a thioether linker are usually prepared by reaction of a thiol-containing protein with a protein containing a reactive maleimide or haloacetamide at pH 6 to 7.5 (Figure 8).

Reactive thiols are generated on the first protein by several methods, including treatment with reducing agents such as dithiothreitol (DTT)[106] or mercaptoethanol (MCE); iminothiolane to produce reactive thiols via an amidation reaction;[107] N-succimidyl-3(2-pyridyldithio)-proprionate (SPDP) followed by mild reduction,[108,109] or N-succinimidyl S-acetylthioacetate (SATA)[109] (Figure 9).

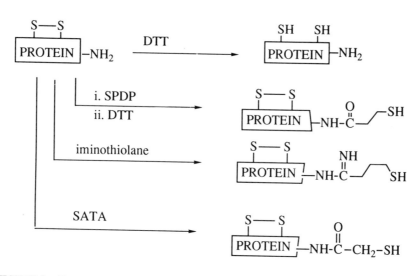

FIGURE 9.   Generation of reactive thiols on protein using DTT, SPDP iminothiolane, and SATA.

The complementary species (protein or drug) is derivatized with N-(4-carboxycyclohexyl-methyl)maleimide (SMCC),[110] m-maleimidobenzoyl-N-hydroxysuccinimide ester (MBS),[111] or N-succinimidyl(4-iodoacetyl)aminobenzoate (SIAB)[112] at pH 7.5 to 9.5. After the derivatized proteins have reacted, the desired thioether conjugate can be isolated. Careful reaction optimization is often required to minimize formation of cross-linked and aggregated proteins.[113]

Although immunotoxins prepared using a thioether linkage exhibit a high degree of stability, they generally show a substantial loss of potency as compared to the analogous immunotoxin prepared using a disulfide linker.[114,115] The exceptions to this include thioether conjugates of *Pseudomonas* exotoxin A (PE),[116–118] pokeweed antiviral protein,[119] and a ricin toxin/IgG conjugate prepared using a peptide spacer with thioether linkages.[113,120]

Conjugates of daunomycin, possessing thioether linkers, have been prepared using the aglycone side chain and carbohydrate portions of daunomycin. Preparation of the 14-bromodaunomycin followed by conjugation to a thiol-containing protein afforded the corresponding thioether conjugate (Figure 7a).[121,122] Another thioether conjugate of daunomycin was prepared by acylation of the carbohydrate amine with a linker containing both maleimide and active ester functionalities. The maleimide derivatized daunomycin was then reacted with protein thiols[123] (Figure 7b). Unfortunately, like the majority of toxin conjugates prepared with metabolically stable linkers, a substantial loss of potency for daunomycin conjugates is often observed.[124] A peptide derivative of methotrexate (MTX) was also conjugated to IgG antibody with a thioether linker by reaction of IgG thiols with its iodoacetyl derivative.[125] In this case, potent antibody-directed cytotoxicity was observed, most likely as a result of eventual lysosomal degradation of the tetrapeptide spacer.

Anthracyclines such as B-rhodomycin and its derivatives have also been conjugated to IgG by esterification of a carbohydrate hydroxyl moiety of rhodomycin with a maleimido active ester. The maleimido ester was then reacted with a thiol-containing protein[126] (Figure 7c). Unfortunately, the high metabolic stability of the thioether functionality is likely to be offset by the additional presence of the more metabolically labile ester bond.

2. Amide Linkers

The reaction of drugs containing active esters with protein amines has been used extensively for the preparation of drug conjugates containing an amide linkage. The procedure typically involves preparation of an N-hydroxysuccinimide ester (active ester) of a carboxylic acid present on the drug. The active ester is then reacted with protein at pH 8 to 10 to produce the amide-linked protein conjugate. This technology has been used to prepare IgG conjugates of methotrexate,[127] chlorambucil,[128] N-acetylmelphalan,[129] and trichothecenes.[130]

Amide linkages between drug amino and protein carboxylates have also been prepared using water-soluble carbodiimides,[131] which do not require preactivation of the carboxylate to an active ester. Carbodiimide-catalyzed amide formation appears to involve formation of an *O*-acylisourea. These reactive intermediates hydrolyze and condense with amines to yield amides or with other nucleophiles to produce different carboxylate derivatives. Despite the potential problems with polymerization,[132] protein conjugates of daunomycin[133] and methotrexate[127,134] prepared by this method have been reported.

3. Schiff Base Linkers

Schiff base (imine) linkers, resulting from the reaction of an amine with an aldehyde such as glutaraldehyde, have been used extensively to conjugate anthracyclines to proteins.[133,135] Although not well defined, it is likely that the chemistry involves Schiff base formation with both protein and drug by the reaction of aldehyde functionalities of glutaraldehyde with amino groups on the respective moieties. The resultant Schiff base adduct may be reduced to a secondary amine using sodium borohydride or sodium cyanoborohydride to increase stability. Conjugation reactions using glutaraldehyde often result in appreciable undesired protein-protein cross-linking.

Hydrazone linkers can also be prepared by reaction of hydrazides of vinblastine,[136] methotrexate,[137] and verrucarin A[138] with oxidized sugar residues on antibody. The potential advantage of this methodology is site-specific modification of the antibody. Use of hydrazone linkers for preparation of toxin conjugates was also demonstrated using ricin toxin.[139] In this case, periodate oxidation of terminal mannose residues on ricin produced aldehydes which were reacted with antibody amino groups. Unfortunately, greater than 90% loss in cytotoxicity was observed, again illustrating potential problems that may be confronted by chemical derivatization of proteins and drugs. In an analogous sequence of reactions, Schiff base coupling of the oxidized carbohydrate residue of daunomycin and protein amines has been conducted (Figure 7d).[140]

Preparation of hydrazones and semicarbazones of the aglycone side chain 13-keto group of daunomycin and morpholino doxorubicin have been reported using hydrazine and semicarbazide derivatives (Figure 7e). These adducts were subsequently conjugated to antibody via disulfide,[141,142] amide,[143] or macromolecular linkers.[121] The hydrazone conjugates obtained showed varying degrees of ability to undergo hydrolysis to release free drug under slightly acidic conditions. Similarly, calicheamicin conjugates were prepared which contained both disulfide and hydrazone functionalities.[144]

4. Disulfide Linkers

Metabolically labile disulfide linkers are used in preparation of the majority of chemically linked immunotoxins and, to a lesser extent, for the preparation of drug conjugates. Disulfides are generally formed by reaction of a thiol from one species with a disulfide on a second, forming a new disulfide bond. These reactions have been routinely carried out using a variety of reagents. Thiolation of the first protein is carried out by one of the methods described above for preparation of thioether conjugates (Figure 9).

Derivatization of the second protein or drug with SPDP is attractive, since pyridyldisulfides undergo a facile reaction with protein thiols to give mixed disulfides[108] (Figure 10). Generation of 2-thiopyridone, produced by the displacement reaction, can be conveniently monitored spectrophotometrically to assess the extent of the reaction. Similar to the preparation of thioether conjugates, undesirable side reactions such as production of protein dimers and aggregated species are often encountered.

The fact that disulfide linkers are usually required for maximal *in vitro* and *in vivo* potency for most immunotoxins, particularly using A-chain toxins, suggests that metabolic cleavage to release free toxin during the cellular processing plays a vital role. For example, a ricin-A chain disulfide conjugate is 2 logs more active than the corresponding thioether analog.[145] Unfortunately, the desired *in vivo* localization of the disulfide-linked toxin conjugate is seriously compromised by its rapid breakdown in serum, liberating free toxin. The poor potency observed for the corresponding thioether conjugate parallels its inability to release free toxin in the cell. In contrast, both thioether

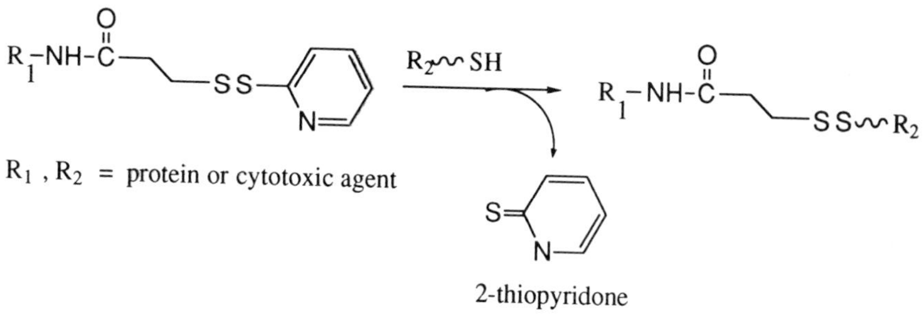

FIGURE 10. Preparation of disulfide conjugate by displacement of thiopyridone.

and disulfide conjugates of PE show similar *in vitro* cytotoxic properties, although the more stable thioether conjugate is required for lower toxicity *in vivo*.[116,117] Cellular proteases are likely to be responsible for breakdown of the PE conjugate into the highly potent enzymatic portion.[146]

Hindered disulfide linkers, methyl-substituted iminothiolanes or methyl-substituted disulfide derivatives, have also been prepared.[147–150] The objective was to improve conjugate stability towards metabolism and preserve biological potency by allowing eventual release of free toxin upon cellular endocytosis. Conjugation of ricin A chain to antibody using hindered disulfides resulted in more metabolically stable conjugates with similar potencies to the unhindered disulfide analogs.[147–149]

Disulfide-linked doxorubicin conjugates, which also contained hydrazone functionalities, were prepared in order to evaluate the pH sensitivity of the hydrazone.[133,142] No evidence regarding any cleavage of the disulfide under the experimental conditions employed was reported. The utility of disulfide linkers for drug conjugation is illustrated by the preparation of an antibody neocarzinostatin disulfide conjugate. *In vivo* studies showed adequate stability to allow a substantial level of drug to reach the tumor.[151]

5. Ester Linkers

Conjugates containing ester linkers have been prepared by several methods. Trichothecenes such as verrucarin A and roridin A have been conjugated to antibody by initial succinylation of the 13′ or 2′ secondary alcohol functionalities. Following preparation of the *N*-hydroxysuccinimide ester of the trichothecene succinate, aminolysis with protein lysine residues afforded the conjugate which contained both amide and ester functionalities in the linker. Interestingly, significant *in vitro* and *in vivo* differences in stability between the 2′ and 13′ ester conjugates were observed.[152–154] Conjugates of vericarin A containing an ester or hydrazone linkage showed good *in vivo* stability.[138] A rhodomycin conjugate was also prepared with a linker possessing thioether and ester functionalities.[126]

6. Acid Cleavable Linkers

The development of acid-labile linkers is based on the observation that immunoconjugates typically require release of free drug once internalization into the cell has occurred. If the mechanism of internalization involves a lysosomal vesicle, which has a characteristic lowered internal pH, hydrolysis of the linker could occur, allowing release of the cytotoxic agent. With the correct chemical design, release of underivatized drug or toxin is possible to ensure maximum potency.

Citraconate anhydride, representing a pH-sensitive linker, has been used for the conjugation of daunomycin[155–157] and gelonin toxin[158] to IgG. At pH 7 the linkage is chemically stable, but at pH 4 to 6 hydrolysis occurs, releasing the cytotoxic agent. The acetone ketal linker, bismaleimido-ethoxypropene (BMEP) has been used to conjugate diphtheria and ricin toxins to antibodies. The pH sensitivity was shown to be dependent on electron donation and bond angles at the ketal group. Depending on the ketal derivatives studied, the $t_{1/2}$ of hydrolysis at pH 5.5 varied between 0.1 and 130 h.[159,160]

Lysosomal-sensitive peptides, LeuAlaLeu, AlaLeuAlaLeu, were used to conjugate daunorubicin to bovine serum albumin (BSA). Release of free drug was observed when the conjugate was treated with lysosomal hydrolases.[161] Additionally, the hydrolyzable peptide, GlyPheLeuGly, has been used to conjugate methotrexate to antibody.[162]

7. Other Linkers

In order to achieve a higher drug:antibody ratio for an enhanced therapeutic effect, use of macromolecules such as dextran, polyamino acids, and human serum albumin as drug carriers have been investigated. Dextran, consisting of glucopyranosyl moieties, can be readily oxidized with sodium periodate to generate reactive aldehydes or treated with diamines to form amino dextran. A number of amine-containing drugs, including daunomycin,[163] mitomycin-C,[164] and drugs containing carboxylates, such as methotrexate,[165] have been used to form covalent complexes with dextrans. Following attachment of drug, the derivatized dextran can be conjugated to protein.

Oligopeptide spacers such as poly(L-lysine)[166] and poly(L-aspartic acid)[167] provide an additional mechanism to carry larger quantities of drug on a protein. One avenue to drug conjugation using a poly(L-aspartate) involved initial derivatization of the polyamino acid with the cross-linking reagent, SPDP, on the α-amino group. The carbohydrate amine of daunomycin was then covalently attached to the carboxyl groups of aspartate residues by an amide bond using carbodiimide. The antibody disulfide conjugate of SPDP-derivatized poly(L-aspartate) was finally obtained by a disulfide exchange reaction.[168] Branched polypeptide antibody conjugates of daunomycin have also been prepared and evaluated with the intention of increasing blood survival time without reducing the *in vitro* cytotoxicity of the drug.[169]

Other examples of synthetic carriers which have been investigated include poly[Glu(N_2H_3)],[170] poly(divinyl ether-co-maleic anhydride),[171] *N*-(2-hydroxypropyl)-methacrylamide,[172] and poly[acryloyl-2-amido-2-(hydroxymethyl)-1,3-propanediol].[173]

D. BIFUNCTIONAL ANTIBODIES

Bifunctional or bispecific antibodies were developed initially as agents for the diagnosis and therapy of cancer. It is now clear, however, that they have a much broader field of use in medicine. These antibodies have two distinct binding specificities. Typically one Fab is reactive with antigen on a tumor cell or other target cell and the other Fab is reactive with an effector molecule or an endogenous cytotoxic white cell. The antibody may or may not have an Fc which carries additional binding specificities. This definition excludes chemically cross-linked or otherwise aggregated whole antibody, which, in addition to their lack of defined structure, have much different pharmacokinetics and biodistribution which limits their usefulness.

It is clear that most monoclonal antibodies, regardless of specificity and inherent effector functions, do not have the ability to be therapeutic or diagnostic on their own. However, they still hold tremendous promise as conjugates, providing both specific binding and carrying capacity for a variety and number of effector agents, such as small-molecule drugs, protein toxins, and radioisotopes. Bifunctional antibodies, chemically or genetically constructed, offer several advantages over directly coupled antibody conjugates. Direct conjugation protocols typically subject both the antibody and effector agent to extremes of pH and/or organic solvents. Further they are topographically nonspecific, reacting with the target amino acid regardless of its location on the antibody. Although conjugations can be directed to cysteines or carbohydrate entities on the antibody these functionalities are not always available in sufficient quantity or only accessible under potentially denaturing conditions. The construction of bifunctional antibodies by cell fusion avoids all of these problems. The pharmacokinetics and biodistribution of directly coupled conjugates are altered in proportion to the degree of modification, whereas bifunctional antibodies have *in vivo* properties predictable for the construct, whether whole molecule, F(ab')$_2$, or chimera. Cytotoxic agents (toxins and drugs) need to gain an intracellular location through a defined sequence of cell trafficking. The binding of the effector agent to the bifunctional antibody is noncovalent and of a defined affinity. This avoids having to deal with chemistries (linkers) that are designed to release their cytotoxic agent under various conditions.

In conclusion, bifunctional or bispecific antibodies are uniquely engineered to provide specific and potent therapeutic or diagnostic agents that avoid the shortcomings of direct conjugation.

1. Applications

Although the bifunctional antibody can be precomplexed with its effector agent prior to administration, the real potential of these agents is maximized by pretargeting the bifunctional antibody with subsequent administration of the effector agent. Excellent and comprehensive reviews have been published.[174,175] Pretargeting of the "cold" or nontoxic bifunctional antibody by large bolus dosing, multiple dosing, or even continuous infusion potentially allows saturation of the tumor or target site. The monovalent binding is less likely to induce internalization and metabolism than the usual bivalent binding, thus allowing time for the circulating antibody to clear. Also specific clearing agents based on the antigenic or effector binding specificities could be designed to accelerate the clearance of circulating antibody. The effector drug, toxin, or radioisotope should be selected or designed to clear through an appropriate route with sufficient rapidity to minimize toxicity while allowing it to bind to the pretargeted antibody. Again several dosing options are available for the subsequent administration of the effector. Pretargeting in this manner should allow the concentration of even small administered amounts of effector at the tumor or target site.

A review of the recent literature has yielded examples where the effector is the therapeutic agent itself, for example a drug, toxin, or radioisotope, and where it is the initiating agent in a cascade of events leading to the therapeutic effect. An example of the latter is an activating antigen on a cytotoxic cell or an enzyme that converts a subsequently administered prodrug to an active drug. By far the most cited application of bispecific antibodies and effector pairs was for various cancers, but also cited were applications for thrombolysis, infectious diseases, and autoimmune indications. This review includes examples of the construction and application of bispecific antibodies with these effector agents and indications.

2. Effector Agents

a. Drugs

A bispecific antibody with specificity for carcinoembryonic antigen (CEA) and vinca alkaloids was used successfully to target the drugs to human tumor xenografts in nude mice.[176–180] Significant suppression of tumor growth was seen. An important finding was that the bispecific antibody did not interfere with the potency of the drugs. In these studies the investigators premixed the bispecific antibody with the drugs prior to administration. No attempt was made to pretarget the antibody. Another group prepared a bispecific antibody reactive with the tumor-associated antigen gp72 and methotrexate (MTX).[181] To increase potency these workers used MTX conjugated to human serum albumin (HSA). With up to 40 drug molecules per HSA molecule, the conjugates showed potency and antigen-positive selectivity in *in vitro* cytotoxicity assays using pre-exposure of the bispecific antibody followed by the MTX-HSA conjugate. They anticipate using the pretargeting approach in animal studies.

The approach of pretargeting to the tumor an enzyme immunoconjugate which can convert a relatively nontoxic drug into a potent therapeutic agent appears promising.[182–184] To overcome the problems associated with direct chemical conjugation of the enzyme to the antibody, a bispecific antibody was prepared with reactivity towards the Hodgkin's-associated CD30 antigen and alkaline phosphatase.[185] Alkaline phosphatase converted mitomycin phosphate, a relatively nontoxic drug, to mitomycin alcohol, a potent cytotoxic agent. The combination of pretargeted bispecific antibody and subsequently applied alkaline phosphatase showed potency and specificity to antigen-positive cells. Of particular interest are the data presented indicating that nearby antigen-negative cells can also be killed by release of the active drug. This is an important aspect in tumor therapy and prior to this work only radioimmunoconjugates had this attribute, which gave them a major advantage over drug and toxin conjugates.

b. Radioisotopes

Much effort has gone into the use of radioimmunoconjugates for cancer therapy, but with limited success. The pharmacokinetics and biodistribution of the directly radiolabeled conjugates results in a dose-limiting toxicity to the bone marrow cells and uptake into tumor of less than 1% of the injected dose. However, the potent cytotoxicity of radioisotopes and their inherent ability to kill cells within a defined radius obviating the need to target every cell, make radioisotopes an attractive effector to harness. Reardan and coworkers prepared monoclonal antibodies with specificity for metal chelates.[186] These antibodies, which possess no antigenic specificity to tumor, passively localized in tumor and captured significant amounts of subsequently administered ^{111}In-EDTA chelates.[187] In theory, bispecific antibodies should enhance tumor uptake of radiolabel over directly conjugated antibodies.

Bispecific antibodies prepared by cell fusion have not been used with radioisotopes. LeDoussal et al. have produced chemically linked Fab' molecules which react with melanoma and ^{111}In-DTPA haptens.[188] The bispecific construct had low immunoreactivity, but bound to cells with appropriate specificity and was able to concentrate the radiolabeled hapten in tumor in tumor-bearing mice. Gridley and coworkers also used chemically cross-linked Fab' fragments (supplied by Hybritech, Inc.) reactive toward CEA and ^{111}In-benzyl-EDTA.[189] Using pretargeting of the antibody fragments, specific localization of the radiolabel in tumored mice was enhanced by increasing the body temperature of the mice. Neither of these groups used any agent to accelerate the blood clearance of the antibody prior to injection of the radiolabeled chelate hapten. Although these studies used a gamma-emitting radioisotope suitable for tumor imaging it is most likely that both imaging and therapy with a bispecific antibody will be enhanced by the use of an intervening clearance step, as stated by Goodwin et al.[187]

c. Toxins

Bispecific antibodies have been prepared with specificity for either ricin A-chain (RTA) or saporin and a tumor-associated antigen. Ricin is dimeric in structure and the ribosome-inhibiting A-chain can be separated from the cell-binding B-chain. Raso and Griffin were the earliest to exploit bispecific antibodies binding to RTA.[190] They used chemically cross-linked rabbit polyclonal fragments with reactivity to RTA and human Ig. Webb and coworkers produced bispecific antibodies reactive with RTA and prostate-restricted antigen by hybridoma cell fusion.[191] The specificity of binding was confirmed with very elegant studies using cultured cells which also showed them to be potent inhibitors of protein synthesis when combined with RTA. These studies indicated that immune complexes of RTA can be internalized, transit through the appropriate intracellular pathway and inhibit ribosomal function. RTA is normally cleared very rapidly in mice by the kidneys and liver. However, RTA complexed at 1:1 ratios with an anti-RTA monoclonal antibody has an extended circulation time and is not cleared by the recticuloendothelial cells in the liver.[192] Robins et al.[193] and Embleton et al.[194] prepared a bispecific antibody using cell fusion which reacts with CEA and RTA. The antibody complexed with RTA was at least as potent as a chemical RTA-anti-CEA conjugate and also specific for antigen-positive cells.

The ribosome-inactivating protein saporin was initially used by Glennie and coworkers.[195,196] They prepared bispecific antibodies by chemically cross-linking Fab' fragments with specificity to saporin (polyclonal and monoclonal) and the guinea pig lymphoblastic leukemia, L_2C (monoclonal). They were able to show complete tumor regression in the animal model, although most relapsed. Most of the relapses were due to the escape of antigen-negative cells, which points out one of the shortcomings of using the protein toxins. Another bispecific antibody has been prepared by Flavell and coworkers using the same method but with reactivity towards CD7.[197] It also has shown potent and selective cytotoxicity in studies with a CD7-positive human T cell leukemia cell line.

d. Effector Cell Targeting

The focusing and concomitant activation of endogenous T lymphocytes to lyse tumor cells that have previously escaped recognition by the immune system is dependent upon specific antigen-induced cross-linking of the target and effector cells. Once activated the effector cell can induce lysis of surrounding cells not directly attached to it. Applications of this technology to viral infections have also been published. Although most work completed to date incorporated chemically aggregated whole antibodies, simply because they are the most convenient to make, truly bispecific antibodies have also been used and are required for the successful application of cell targeting for human therapy. Reviews detailing the history and current status of this technology have recently been published.[198,199]

Bispecific antibodies reactive with a target cell antigen and the TcR/CD3 complex on cloned T cells have been shown to lyse the target cells. The bispecific antibodies both cross-link the cells as a first requirement and secondly activate the T cells to lyse the target cell. It is also known that T cells freshly isolated from the peripheral blood have poor cytolytic activity. What activity is seen is due to a small percentage of the cells. To overcome this limitation, recent work has been focused on the identification of additional signals (antigens) which together with the TcR/CD3 signal will activate a larger percentage of the endogenous T cell population.[200–202] Trispecific antibodies therefore have been created through chemical cross-linking to simultaneously bind to the target cell and two antigens on the effector cell. These constructs may prove to be more efficacious than bispecific antibodies. Other efforts involve identifying other additional cell types available in the peripheral blood that can be selectively activated.[203]

3. Noncancer Applications
a. Antivirals

Chemically aggregated bispecific antibodies have been used to induce cytotoxic T cell lysis of cells infected with influenza virus,[204] herpesvirus,[205] and human immunodeficiency virus (HIV).[206] The hallmarks of specificity are the telltale antigens left on the surface of infected cells by the invading virus. Specificity for these viral targets by the antibody allow the selective lysis of infected cells. In theory this approach will target infected cells well before progeny infectious virus are produced. However, it still remains to be demonstrated that lysis of late-stage infected cells will not exacerbate the infection.

In a particularly elegant approach to an HIV therapy a bispecific antibody was genetically engineered with one arm reactive with CD3 on a cytotoxic T cell and the other arm the actual CD4 domain to bind to gp120 of HIV of any strain.[207] Although not fully purified, studies *in vitro* showed that the bispecific antibody was very effective at mediating T cell killing of HIV-infected cells. Another approach uses bispecific antibodies to target HIV virus directly to human monocytes and macrophages through Fc receptors where it is internalized and deactivated.[208] Although these cleverly designed constructs work well in cell culture only testing in an appropriate animal model or clinical trial can determine their full therapeutic value.

b. Thrombolysis

Bispecific antibodies have been used to enhance the specificity and the potency of tissue plasminogen activator (tPA) and urokinase. Both thrombolytic agents mediate clot dissolution through enzymatic action. However, urokinase does not bind specifically to clots and although tPA does bind to clots its affinity is low. Therefore, the use of a specific, high-affinity antibody should prove beneficial. These constructs have one Fab reactive with fibrin and the other Fab reactive with tPA[209–213] or urokinase.[214] Presumably, streptokinase would work also. Constructs have been prepared through chemical aggregation,[209,210] recombined $F(ab')_2$,[211] and cell fusion.[212,213] Using a bispecific antibody reactive with both fibrin and tPA, Bode and coworkers found that in an *in vitro* clot lysis assay the bispecific antibody together with tPA resulted in substantially more clot lysis than the equivalent amount of tPA alone.[210] Further, using a human thrombus in a rabbit jugular vein, the bispecific antibody and a small amount of tPA caused significant clot lysis, whereas the same amount of tPA alone gave no significant lysis. It is interesting that the amount of tPA used is less then would be found in normal human plasma, suggesting that the bispecific antibody could be therapeutic alone by capturing endogenous tPA and concentrating it at the site of thrombus. This may avoid the potential complications associated with the direct infusion of tPA.[215] These results have been extended to bispecific antibodies created by cell fusion by Branscomb et al.[212] and Kurokawa et al.[213]

Evaluation of these constructs has mostly involved *in vitro* biochemical analyses. Although some data from work in rabbits are available, a more extensive evaluation even in the same model is necessary to establish a basis for human clinical studies. Although not truly a bispecific antibody, an interesting construct to evaluate would be a fusion protein between an anti-fibrin single-chain antibody or Fab fragment and the whole molecule or enzymatic portion of tPA, urokinase, or streptokinase.

4. Production

If bispecific antibodies are going to have an impact on human health care, efficient methods of providing these drugs need to be devised. Although overall cost dominates this analysis, some of the items that impact cost are yield (activity and mass), the reproducibility of the process, the ability to use conventional equipment, the ability to apply conventional quality control assays and procedures, and the shelf-life stability of the product. Methods to make bispecific antibodies have been reviewed by Paulus.[216] They include somatic cell fusion as described by Milstein and Cuello,[217] the chemical recombination method of Brennan et al.,[218] and various genetic engineering technologies and combinations of these.[219] Although these methods are not necessarily optimal they are capable of producing chemically defined, homogeneous product after suitable purification. For example, cell fusion of two hybridomas will result in multiple products since the association of heavy and light chains appears to be a random process.[220] Since the identity of the impurities is known, however, it should be possible to devise efficient purification schemes. Other methods include protein A-mediated aggregation,[221] formation of immune complexes using rat anti-mouse Ig,[222,223] and various chemical aggregation methods usually involving glutaraldehyde or *N*-succinimidyl-3-(1-pyridyldithio)propionate (SPDP). These methods appear to be limited in application. Certainly they are not acceptable for human use.

IV. CLINICAL STATUS OF MOLECULAR TARGETING

Much effort has been carried out to use targeting agents for increasing the efficiency of delivery of pharmaceutical agents to their targets. Clinical success would mean greater efficacy if more drug is delivered without raising toxicity or, alternatively, with lower toxicity while maintaining efficacy. As targeted delivery results in alternative mechanisms of intracellular delivery, potential exists for overcoming drug resistance, a serious problem in cancer chemotherapy.

The most advanced of targeted modalities is diagnostic radiation for imaging. While more years than anticipated have been required for commercialization, several agents for imaging mela-noma,[67,68] colon cancer,[224,225] and myocardial infarct[226] are at the doorstep of approval. Despite the ability of radiolabeled antibodies to target lesions successfully in patients, issues of manufacturing using mammalian cell culture, regulatory criteria requirements for approval, and immunogenicity of murine-derived monoclonal antibodies needed to be dealt with. A number of radiolabeled antibodies and fragments, as well as peptides such as the somatostatin analog octreotide[227,228] and melanocyte stimulating hormone[229] are undergoing clinical studies that may provide the advantages of second-generation products.

The clinical status of targeted radiotherapy is at a much earlier stage than radioimaging. As discussed in a recent review by Langmuir,[84] the use of antibody targeting or radiotherapy treatment of cancer suffers from problems of low fraction of dose delivered, heterogeneity of dose distribution, low dose rate, and systemic irradiation that results in toxicity to radiosensitive bone marrow cells. Significant responses have been seen in radiation-sensitive leukemias and lymphomas[230] while response in solid tumors have been anecdotal from systemic administration.[231] More success with solid tumors has resulted from intraperitoneal administration in which a compartmental targeting advantage is achieved.[232]

Several clinical studies with enzymatic toxin conjugates have been carried out.[233–235] As with targeted radiotherapy, studies in solid tumors have shown only minimal and anecdotal responses. Rapid clearance from the circulation limiting amount of toxin conjugate delivered, immunogenicity limiting numbers of doses that can be given, and potency are factors that need to be overcome for solid tumor efficacy. Some success in lymphomas and leukemias has been achieved.[234,235] In these diseases, the malignant cells are more accessible, such that targeting is effective within the short time in circulation and some patients are immunocompromised allowing multiple dosing. Clinical potential has also been shown for use of anti-CD5-ricin A in patients with steroid-resistant graft-versus-host disease.[236] Unfortunately, severe neuropathy toxicities were seen from administration of an anti-breast cancer ricin A chain immunotoxin in which cross-reactivity with Schwann cells was thought to be responsible.[233]

The most limited clinical evaluation of immunoconjugates for cancer treatment is in those utilizing drugs. Drug immunoconjugates have suffered from both potency as well as, until recently, selectivity.[8] A phase I clinical trial of the vinca conjugate, KS1/4-DAVLB, has been performed in patients with lung adenocarcinoma. Although significant delivery of the drug DAVLB to tumor cells was shown, dose-dependent toxicity due to cross-reactivity of the antibody with duodenal epithelium was observed and studies were discontinued.[237] Antiglobulin responses were observed both to the antibody KS1/4 and to the vinca-alkaloids in most of the patients, indicating a limiting factor in repeat administration of these murine antibody-based drug conjugates.[238] The same antibody KS1/4 as a delivery agent for methotrexate was evaluated in a Phase I trial involving non-small cell lung cancer patients.[239] A variety of mild-to-moderate side effects and anti-mouse antibody response was seen. Of the eleven patients studied, one possible response was reported. Finally, colon cancer targeting of neocarcinostatin via antibody A7 was reported in a limited number of patients.[240] The conjugate was given intraarterially. Of eight patients with liver metastases, three showed evidence of tumor reduction on computed tomography scans and three reported relief of pain. No benefit was seen in cases of multiple lung or peritoneal metastases. Follow-up of these encouraging results in a larger study would be of interest.

A great deal of effort has been expended in attempts to exploit antibodies as agents of targeted delivery for diagnosis and therapy. Although progress has been slow and problems have been greater than perhaps anticipated, improvements are being seen. Many issues, including cross-reactivity, pharmacokinetics, tumor uptake and penetration kinetics, molecular form, immunogenicity, linkage chemistry, effector potency and mechanism of action, molecular dose requirements for therapeutic effect, and so on have been dealt with in some fashion. As the extent of rational design of molecular conjugates increases, the data accumulated should facilitate application of the best combination of the chemistries available.

REFERENCES

1. **Woolfenden, J.M. and Barber, H.B.,** Radiation detection probes for tumor localization using tumor-seeking radioactive tracers, *Am. J. Radiol.,* 153, 35, 1989.
2. **Brodsky, F.M.,** Monoclonal antibodies as magic bullets. *Pharm. Res.,* 5, 1, 1988.
3. **Vaickus, L. and Foon, K.A.,** Overview of monoclonal antibodies in the diagnosis and therapy of cancer, *Cancer Invest.,* 9, 195, 1991.
4. **Schally, A.V.,** Oncological applications of somatostatin analogs, *Cancer Res.,* 48, 6977, 1988.
5. **Cuttitta, F., Carney, D.N., Mulshine, J., Moody, T.W., Fedorko, J., Fischler, A., and Minna, J.D.,** Bombesin-like peptides can function as autocrine growth factors in human small-cell lung cancer, *Nature,* 316, 823, 1985.
6. **Hruby, V.J.,** Designing molecules: specific peptides for specific receptors, *Epilepsia,* 30 (Suppl. 1), 542, 1989.
7. **Moore, M.A.S.,** Hematopoietic growth factor interactions: in vitro and in vivo preclinical evaluation, *Cancer Surv.,* 9, 8, 1990.
8. **Reisfeld, R.A., Mueller, B.M., Schrappe, M., Wargalla, U.C., Yang, H.M., and Wrasidlo, W.,** Antibody-drug conjugates for cancer therapy — promises and problems, *Immunol. Allergy Clin. North Am.,* 11, 341, 1991.
9. **Matteucei, M.D. and Bischofberger, N.,** Sequence-defined oligonucleotides as potential therapeutics, *Annu. Rep. Med. Chem.,* 26, 287, 1991.
10. **Humm, J.L.,** Dosimetric aspects of radiolabeled antibodies for tumor therapy, *J. Nucl. Med.,* 27, 1490, 1986.

11. **Yorke, E.D., Beaumier, P.L., Wessels, B.W., Fritzberg, A.R., and Morgan, A.C., Jr.,** Optimal antibody-radionuclide combinations for clinical radioimmunotherapy: a predictive model based on mouse pharmacokinetics, *Nucl. Med. Biol.,* 18, 827, 1991.

12. **Wheldon, T.E., O'Donoghue, J.A., Barett, A., and Michalowski, A.S.,** The curability of tumors of differing size by targeted radiotherapy using ^{131}I or ^{90}Y, *Radiother. Oncol.,* 21, 91, 1991.

13. **Oeltmann, T.N. and Frankel, A.E.,** Advances in immunotoxins, *FASEB J.,* 5, 2334, 1991.

14. **Rybak, S.M. and Youle, R.J.,** Clinical use of immunotoxins — monoclonal antibodies conjugated to protein toxins, *Immunol. Allergy Clin. North Am.,* 11, 359, 1991.

15. **Fritzberg, A.R., Berninger, R.W., Hadley, S.W., and Wester, D.W.,** Approaches to radiolabeling of antibodies for diagnosis and therapy of cancer, *Pharm. Res.,* 5 (6), 325, 1988.

16. **Hnatowich, D.J.,** Recent developments in the radiolabeling with iodine, indium, and technetium, *Semin. Nucl. Med.,* 20, 80, 1990.

17. **Abrams, M.J., Juweid, M., tenKate, C.I., Schwartz, D.A., Hauser, M.M., Gaul, F.E., Fuccello, A.J., Rubin, R.H., Strauss, H.W., and Fischman, A.J.,** Technetium-99m-human polyclonal IgG radiolabeled via the hydrazino nicotinamide derivative for imaging focal sites of infection in rats, *J. Nucl. Med.,* 31 (12), 2022, 1990.

18. **Verbruggen, A.M.,** Radiopharmaceuticals: state of the art, *Eur. J. Nucl. Med.,* 17, 346, 1990.

19. **Dykes, P.W., Bradwell, A.R., Chapman, E.C., and Vaughan, A.T.M.,** Radioimmunotherapy of cancer: clinical studies and limiting factors, *Cancer Treat. Rev.,* 14, 87, 1987.

20. **Engler, D. and Burger, A.G.,** The deiodination of the iodothyronines and of their derivatives in man, *Endocr. Rev.,* 5, 151, 1984.

21. **Vaidyanathan, G. and Zalutsky, M.,** Protein radiohalogenation: observations on the design of N-succinimidyl ester acylation agents, *Bioconjugate Chem.,* 1, 269, 1990.

22. **Wilbur, D.S., Hadley, S.W., Grant, L.M., and Hylarides, M.D.,** Radioiodinated iodobenzoyl conjugates of a monoclonal antibody, *Bioconjugate Chem.,* 30, 111, 1991.

23. **Wilbur, D.S., Hadley, S.W., Hylarides, M.D., Abrams, P.A., Beaumier, P.A., Morgan, A.C., Reno, J.M., and Fritzberg, A.R.,** Development of a stable radioiodinating reagent to label monoclonal antibodies for radiotherapy of cancer, *J. Nucl. Med.,* 30, 216, 1989.

24. **Zalutsky, M.R. and Narula, A.S.,** Radiohalogenation of a monoclonal antibody using an N-succinimidyl 3-(tri-n-butylstannyl) benzoate intermediate, *Cancer Res.,* 48, 1446, 1988.

25. **Srivastava, P.C., Knapp, F.F., Allred, J.F., and Buchsbaum, D.J.,** Evaluation of N-(p-I-125-iodophenyl) maleimide for labeling monoclonal antibodies, *J. Labelled Compd.,* 26, 296, 1989.

26. **Wilbur, D.S., Hylarides, M.D., Hadley, S.W., Schroeder, J., and Fritzberg, A.R.,** A general approach to radiohalogenation of proteins, radiohalogenation of organometallic intermediates containing protein reactive substituents, *J. Labelled Compd.,* 26, 316, 1989.

27. **Hylarides, M.D., Wilbur, D.S., Reed, M.W., Hadley, S.W., Schroeder, J.R., and Grant, L.M.,** Preparation and in vivo evaluation of an N-(p-[^{125}I]iodophenethyl)maleimide-antibody conjugate, *Bioconjugate Chem.,* 2, 435, 1991.

28. **Krejcarek, G.E. and Tucker, K.L.,** Covalent attachment of chelating groups of macromolecules, *Biochem. Biophys. Res. Commun.,* 77, 581, 1977.

29. **Hnatowich, D.J., Layne, W.W., and Childs, R.L.,** Radioactive labeling of antibody: a simple and efficient method, *Science,* 220, 613, 1983.

30. **Paik, C.H., Murphy, P.R., Eckelman, W.C., Volkert, W.A., and Reba, R.C.,** Optimization of the DTPA mixed-anhydride reaction with antibodies at low concentration, *J. Nucl. Med.,* 24, 932, 1983.

31. **Hnatowich, D.J., Childs, R.C., Lanteigne, D., and Najafi, A.,** The preparation of DTPA-coupled antibodies radiolabeled with metallic radionuclides: an improved method, *Immunol. Methods,* 65, 147, 1983.

32. **Paik, C.H., Yoloyama, K., Reynolds, J.C., Quadri, S.M., Min, C.Y., Shin, S.Y., Maloney, P.J., Larson, S.M., and Reba, R.C.,** Reduction of background activities by introduction of a diester linkage between antibody and a chelate in radioimmunodetection of tumor, *J. Nucl. Med.,* 30, 1693, 1987; and references cited therein.

33. **Colcher, D. and Gansow, D.,** Biodistribution of indium-111-labeled monoclonal antibodies, *J. Nucl. Med.,* 28, 1924, 1987.

34. **Esteban, J.M., Schlom, J., Gansow, A., Atcher, R.W., Martin, W.B., Simpson, D.E., and Colcher, D.,** New methods for the chelation of indium-111 to monoclonal antibodies: biodistribution and imaging of athymic mice bearing human colon carcinoma xenografts, *J. Nucl. Med.,* 28, 861, 1987.

35. **Deshpande, S.V., Ramaswamy, S., McCall, M.J., DeNardo, S.J., DeNardo, G.L., and Meares, C.F.,** Metabolism of indium chelates attached to monoclonal antibody: minimal transchelation of indium from benzyl-EDTA chelate in vivo, *J. Nucl. Med.,* 31, 218, 1990.

36. **Blend, M.J., Greager, J.A., Atcher, R.W., Brown, J.M., Brechbiel, M.W., Gansow, O.A., and Das Gupta, T.K.,** Improved sarcoma imaging and reduced hepatic activity with indium-111-SCN-Bz-DTPA linked to MoAb 19–24, *J. Nucl. Med.,* 29, 1810, 1988.

37. **Adams, G.P., DeNardo, S.J., Deshpande, S.V., DeNardo, G.L., Meares, C.F., McCall, M.J., and Epstein, A.L.,** Effect of mass of ^{111}In-benzyl-EDTA monoclonal antibody on hepatic uptake and processing in mice, *Cancer Res.,* 49, 1701, 1989.

38. **Rodwell, J.D., Alvarez, V.L., Lee, C., Lopes, A.D., Goers, W.F., King, H.D., Powsner, H.J., and McKearn, T.J.,** Site specific covalent modification of monoclonal antibodies: in vitro and in vivo evaluations, *Proc. Natl. Acad. Sci. U.S.A.,* 83, 2632, 1986.

39. **Haseman, M.K., Goodwin, D.A., Meares, C.F., Kaminski, M.S., Wensel, T.G., McCall, M.J., and Levy, R.,** Metabolizable ^{111}In chelate conjugated anti-idiotype monoclonal antibody for radioimmunodetection of lymphoma in mice, *Eur. J. Nucl. Med.,* 12, 455, 1986.

40. **Deshpande, S.V., DeNardo, S.J., Meares, C.F., McCall, M.J., Adams, G.P., and DeNardo, G.L.,** Effect of different linkages between chelates and monoclonal antibodies on levels of radioactivity in the liver, *Nucl. Med. Biol.,* 16, 587, 1991.

41. **Khaw, B.A., Yasuda, T., and Gold, H.K.,** Acute myocardial infarct imaging with In-111 labeled monoclonal antimyosin Fab, *J. Nucl. Med.,* 28, 1671, 1987.

42. **Johnson, L.L. and Seldin, D.W.,** The role of antimyosin antibodies in acute myocardial infarction, *Semin. Nucl. Med.,* 19, 238, 1989.

43. **Rhodes, B.A., Zamora, P.O., Newell, K.D., and Valdez, E.F.,** Technetium-99m labeling of murine monoclonal antibody fragments, *J. Nucl. Med.,* 27, 685, 1986.

44. **Siccardi, A.G., Callegaro, L., Mariani, G., Natali, P.G., Abbati, A., Bestagno, M., Caputo, L.V., Mansi, L., Masi, R., Paganelli, G., Riva, P., Salvatore, M., Sanguineti, M., Troncone, L., Turco, G.L., Scassellati, A., and Ferrone, S.,** Multicenter study of immunoscintigraphy with radiolabeled monoclonal antibodies in patients with melanoma, *Cancer Res.,* 46, 4817, 1986.

45. **Som, P., Oster, Z.H., Zamora, P.O., Yamamoto, K., Sacker, D.F., Brill, A.B., Newell, K.D., and Rhodes, B.A.,** Radioimmunoimaging of experimental thrombi in dogs using technetium-99m-labeled monoclonal antibody fragments reactive with human platelet, *J. Nucl. Med.,* 27, 1315, 1986.

46. **Reno, J.M., Bottino, B.J., and Wilbur, D.S.,** Improved antibody coupling, U.S. Patent No. 4, 877, 868, 1989.

47. **Schwarz, A. and Steinstrasser, A.,** A novel approach to Tc-99m-labeled monoclonal antibodies, *J. Nucl. Med.,* 28, 721, 1987.

48. **Mather, S.J. and Ellison, D.** Reduction-mediated Tc-99m labeling of monoclonal antibodies, *J. Nucl. Med.,* 31, 692, 1990.

49. **Pimm, M.V., Rajput, R.S., Frier, M., and Gribben, S.J.,** Anomalies in reduction-mediated technetium-99m labeling of monoclonal antibodies, *Eur. J. Nucl. Med.,* 18, 973, 1991.

50. **Thakur, M.L., De Fulvio, J., Richard, M.D., and Park, C.H.,** Technetium-99m labeled monoclonal antibodies: evaluation of reducing agents, *Nucl. Med. Biol.,* 18, 227, 1990.

51. **Pak, K.Y., Nedelman, M.A., Fogler, W.E., Tam, S.H., Wilson, E., Van Haarlem, L.J.M., Colognola, R., Warnaar, S.O., and Daddona, P.E.,** Evaluation of the 323/A3 monoclonal antibody and the use of technetium-99m-labeled 323/A3 Fab' for the detection of pan adenocarcinoma, *Nucl. Med. Biol.* 18, 1491, 5, 483–497, *Int. J. Radiat. Appl. Instru. B.,* 18, 483, 1991.

52. **Hansen, H.J., Jones, A.L., Sharkey, R.M., Grebenau, R., Blazejewski, N., Kunz, A., Buckley, M.J., Newman, E.S., Ostella, F., and Goldenberg, D.M.,** Preclinical evaluation of an "instant" [99mTc]-labeling kit for antibody imaging, *Cancer Res. (Suppl.),* 50, 7945, 1990.

53. **Childs, R.L. and Hnatowich, D.J.,** Optimum conditions for labeling of DTPA-coupled antibodies with Tc-99m, *J. Nucl. Med.,* 26, 293, 1985.

54. **Lanteigne, D. and Hnatowich, D.J.,** The labeling of DTPA-coupled proteins, *Int. J. Appl. Radiat. Isot.,* 35, 617, 1984.

55. **Arano, Y., Ykoyama, A., Furukawa, T., Horiuchi, K., Yahata, T., Saji, H., Sakahara, H., Nakashima, T., Koizumi, M., Endo, K., and Torizuka, K.,** Tc-99m-labeled monoclonal antibody with preserved immunoreactivity and high in vivo stability, *J. Nucl. Med.,* 28, 1027, 1987.

56. **Misra, H.K., Hnatowich, D.J., and Wright, G.,** Synthesis of a novel diaminodithiol ligand for labeling proteins and small molecules with Tc-99m, *Tetrahedron Lett.,* 30, 1885, 1988.

57. **Abrams, M.J., Juweid, M., Ten Kate, C.I., Schwartz, D.A., Hauser, M.M., Gaul, F.E., Fuccello, A.J., Rubin, R.H., Strauss, H.W., and Fischman, A.J.,** Technetium-99m-human polyclonal IgG radiolabeled via the hydrazino nicotinamide derivative for imaging focal sites of infection in rats, *J. Nucl. Med.,* 31, 2022, 1990.

58. **Schwartz, D.A., Abrams, M.J., Hauser, M.M., Gaul, F.E., Larsen, S.K., Rauh, D., and Zubieta, J.A.,** Preparation of hydrazino-modified proteins and their use for the synthesis of [99mTc]-protein conjugates, *Bioconjugate Chem.,* 2, 333, 1991.

59. **Kopicka, K.K., Ketring, A.R., Volkert, W.A., Singh, P.R., and Katti, K.V.,** New bifunctional frameworks derived from phosphorous hydrazide chelating agents for labeling proteins with Tc-99m and Re-188, *J. Nucl. Med.,* 33 (Abstr.), 910, 1992.

60. **Lever, S.Z., Baidoo, K.E., Kramer, A.V., and Burns, D.H.,** Synthesis of a novel bifunctional chelate designed for labeling proteins with technetium-99m, *Tetrahedron Lett.,* 29, 3219, 1988.

61. **Baidoo, K.E. and Lever, S.Z.,** Synthesis of a diaminedithiol bifunctional chelating agent for incorporation of Tc-99m into biomolecules, *Bioconjugate Chem.,* 1, 132, 1990.

62. **Baidoo, K.E., Scheffel, U., and Lever, S.Z.,** Tc-99m labeling of proteins: initial evaluation of a novel diaminedithiol bifunctional chelating agent, *Cancer Res. (Suppl.),* 50, 799, 1990.

63. **Weber, R.W., Boutin, R.H., Nedelman, M.A., Lister-James, J., and Dean, R.T.,** Enhanced kidney clearance with an ester-linked ^{99m}Tc-radiolabeled antibody Fab′-chelator conjugate, *Bioconjugate Chem.,* 1, 431, 1990.

64. **Fritzberg, A.R., Kasina, S., Eshima, D., and Johnson, D.L.,** Synthesis and biological evaluation of technetium-99m MAG$_3$ as a hippuran replacement, *J. Nucl. Med.,* 27, 111, 1986.

65. **Kasina, S., Rao, T.N., Srinivasan, A., Sanderson, J.A., Fitzner, J.N., Reno, J.M., Beaumier, P.L., and Fritzberg, A.R.,** Development and biologic evaluation of a kit for preformed chelate technetium-99m radiolabeling of an antibody Fab fragment using a diamide dimercaptide chelating agent, *J. Nucl. Med.,* 32, 1445, 1991.

66. **Fritzberg, A.R., Abrams, P.G., Beaumier, P.L., Kasina, S., Morgan, A.C., Rao, T.N., Reno, J.M., Sanderson, J.A., Srinivasan, A., Wilbur, D.S., and Vanderheyden, J.-L.,** Specific and stable labeling of antibodies with technetium-99m with a diamide dithiolate chelating agent, *Proc. Natl. Acad. Sci. U.S.A.,* 85, 4025–4029, 1988.

67. **Salk, D.,** Tc-labeled monoclonal antibodies for imaging metastic melanoma: results of a multicenter clinical study, *Semin. Oncol.,* 15, 608, 1988.

68. **Eary, J.F., Schroff, R.W., Abrams, P.G., Fritzberg, A.R., Morgan, A.C., Kasina, S., Reno, J.M., Srinivasan, A., Woodhouse, C.S., Wilbur, D.S., Natale, R.B., Collins, C., Stehlin, J.S., Mitchell, M., and Nelp, W.B.,** Successful imaging of malignant melanoma with technetium-99m-labeled monoclonal antibodies, *J. Nucl. Med.,* 30, 25, 1989.

69. **Friedman, S., Sullivan, K., Salk, D., Nelp, W.B., Griep, R.J., Johnson, D.H., Blend, M.J., Aye, R., Suppers, V., and Abrams, P.G.,** Staging non-small cell carcinoma of the lung using technetium-99m-labeled monoclonal antibodies, *Hematol. Oncol. Clin. North Am.,* 4, 1069, 1990.

70. **Axworthy, D.B., Kasina, S., Rao, T.N., Srinivasan, A., and Fritzberg, A.R.,** Metabolite analysis, biliary and urinary excretion of Tc-99m-N$_2$S$_2$ preformed chelate labeled anti-melanoma monoclonal antibody Fab fragment in mice and rats, *J. Nucl. Med.,* 32 (Abstr.), 915, 1991.

71. **Srinivasan, A., Kasina, S., Fitzner, J.N., Gustavson, L.G., Reno, J.M., Rao, T.N., Sanderson, J.A., Gray, M.A., Axworthy, D., and Fritzberg, A.R.,** Modified bifunctional amide thiolate ligands: enhanced utility of Tc-99m Fab for radioimmunodetection, *J. Nucl. Med.,* 31 (Abstr.), 747, 1990.

72. **Srinivasan, A., Kasina, S., Fitzner, J.N., Axworthy, D., Rao, T.N., Reno, J.M., Sanderson, J.A., and Fritzberg, A.R.,** Enhanced targeting specificity of Tc-99m amide thiolate-Fab conjugates utilizing (3-amido) alkyl succinate cleavable linkers, *J. Nucl. Med.,* 32 (Abstr.), 1017, 1991.

73. **Troutner, D.E.,** Chemical and physical properties of radionuclides, *Nucl. Med. Biol.,* 14, 171, 1987.

74. **Volkert, W.A., Goeckeler, W.F., Ehrhardt, G.J., and Ketring, A.R.,** Therapeutic radionuclides: production and decay property considerations, *J. Nucl. Med.,* 32, 174, 1991.

75. **Wessels, B.W. and Rogus, R.D.,** Radionuclide selection and model absorbed dose calculations for radiolabeled tumor associated antibodies, *Med. Phys.,* 11, 638, 1984.

76. **Chaudhuri, T.K.,** Role of P-32 in polycythemia vera and leukemia, in *Therapy in Nuclear Medicine,* Spencer, R.P., Ed., Grune & Stratton, New York, 1978, 223.

77. **Currie, J.L., Bagne, F., Harris, C., Sullivan, D.L., Surwit, E.A., Willonsin, R.A., and Creasman, W.T.,** Radioactive chromic phosphate absorption and effective radiation in phantoms, dogs, and man, *Gynaecol. Oncol.,* 12, 193, 1981.

78. **Britton, K.E.,** Overview of radioimmunotherapy: a European perspective, *Antibody Immunoconj. Radiopharm.,* 4, 133, 1991.

79. **Kemp, B.E., Graves, D.J., and Benjami, E.,** Synthetic peptide substrates of the cAMP-dependent protein kinase, *Fed. Proc.,* 35, 1384, 1976.

80. **Foxwell, B.M.J., Band, H.A., Jeffrey, W.A., Snook, D., Thorpe, P.E., Watson, G., Parker, P.J., Epenetos, A.A., and Creighton, A.M.,** Conjugation of monoclonal antibodies to a synthetic peptide substrate for protein kinase with 32-P, *Br. J. Cancer,* 57, 489, 1988.

81. **Cole, W.C., DeNardo, S.J., Meares, C.F., McCall, M.J., DeNardo, G.L., Epstein, A.L., O'Brien, H.A., and Moi, M.K.,** Comparative serum stability of radiochelates for antibody radiopharmaceuticals, *J. Nucl. Med.,* 28, 83, 1987.

82. **Gansow, O.A.,** Newer approaches to the radiolabeling of monoclonal antibodies by use of metal chelates, *Nucl. Med. Biol.,* 18, 369, 1991.

83. **Durbin, P.W.,** Metabolic characteristics within a chemical family, *Health Phys.,* 2, 225, 1960.

84. **Langmuir, V.K.,** Radioimmunotherapy, clinical results and dosimetric considerations, *Nucl. Med. Biol.,* 19, 213, 1992.

85. **Brechbiel, M.W., Gansow, O.A., Atcher, R.W., Schlom, J., Esteban, J., Simpson, D.E., and Colcher, D.,** Synthesis of 1-(*p*-isothiocyanatobenzyl) derivatives of DTPA and EDTA. Antibody labeling and tumor imaging studies, *Inorg. Chem.,* 25, 2772, 1986.

86. **Meares, C.F., Moi, M.D., Diril, H., Kukis, D.L., McCall, M.J., Deshpande, S.V., DeNardo, S.J., Snook, D., and Epenetos, A.A.,** Macrocyclic chelates of radiometals for diagnosis and therapy, *Br. J. Cancer,* 62 (Suppl. X), 21, 1990.

87. **Ehrhardt, G.J., Ketring, A.R., Turpin, T.A., Razavi, M.-S., Vanderheyden, J.-L., Su, F.-M., and Fritzberg, A.R.,** A convenient tungsten-188/rhenium-188 generator for radiotherapeutic applications using low specific activity tungsten-188, in *Technetium and Rhenium in Chemistry and Nuclear Medicine,* 3rd ed., Nicolini, M., Bandoli, G., and Maggi, U., Eds., Raven Press, New York, 1990, 631.

88. **Deutsch, E., Libson, K., Vanderheyden, J.-L., Ketring, A.R., and Maxon, H.R.,** The chemistry of rhenium and technetium as related to the use of isotopes of these elements in therapeutic and diagnostic nuclear medicine, *Nucl. Med. Biol.,* 13, 465, 1986.

89. **Rao, T.N., Adhikesavalu, D., Camerman, A., and Fritzberg, A.R.,** Technetium (V) and rhenium (V) complexes of 2,3-bis (mercaptoacetamido) proponate. Chelate ring stereochemistry and influence on chemical and biological properties, *J. Am. Chem. Soc.,* 412, 5798, 1990.

90. **Fritzberg, A.R., Vanderheyden, J.-L., Rao, T.N., Kasina, S., Eshima, D., and Taylor, A.T.,** Comparative renal handling of Tc-99m and Re-186 CO$_2$DADS: implications for diagnostic/therapeutic pair applications, *J. Nucl. Med.,* 30 (Abstr.), 743, 1989.

91. **Johnson, D.L., Fritzberg, A.R., Hawkins, B.L., Kasina, S., and Eshima, D.,** Stereochemical studies in the development of technetium and radiopharmaceuticals. I. Fluxional racemization of technetium and rhenium penicillamine complexes, *Inorg. Chem.,* 23, 4204, 1984.

92. **Helm, L., Deutsch, K., Deutsch, E.A., and Merbach, A.E.,** Multinuclear NMR studies of ligand-exchange reactions on analogous technetium (V) and rhenium (V) complexes. Relevance to nuclear medicine, *Helv. Chem. Acta,* 75, 210, 1992.

93. **Su, F.-M., Axworthy, D.B., Vanderheyden, J.-L., Srinivasan, A., Fitzner, J., Galster, J., Kasina, S., Reno, J., Beaumier, P., and Fritzberg, A.R.,** Evaluation of a cleavable N$_3$S-hydroxyl ester ligand for improved targeting of rhenium-186 labeled antibodies, *J. Nucl. Med.,* 32 (Abstr.), 1020, 1991.

94. **Wilbur, D.S.,** Potential use of alpha emitting radionuclides in the treatment of cancer, *Antibody Immunoconj. Radiopharm.,* 4, 85, 1991.

95. **Aaij, C., Tschroots, W.R.J.M., Linder, L., and Feltkamp, T.G.W.,** The preparation of astatine labeled proteins, *Int. J. Appl. Radiat. Isot.,* 26, 25, 1978.

96. **Vaughn, A.T.M. and Fremlin, J.H.,** The preparation of astatine labeled proteins using an electrophilic reaction, *Int. J. Nucl. Med. Biol.,* 5, 229, 1978.

97. **Zalutsky, M.R., Garg, P.K., Friedman, H.S., and Bigner, D.D.,** Labeling monoclonal antibodies and F (ab')$_2$ with the α-particle emitting nuclide astatine-211: preservation of immunoreactivity and in vivo localizing capacity, *Proc. Natl. Acad. Sci. U.S.A.,* 86, 7149, 1989.

98. **Hadley, S.W., Wilbur, D.S., Gray, M.A., and Atcher, R.W.,** Astatine-211 labeling of an antimelanoma antibody and its Fab fragment using N-succinimidly *p*-astatatobenzoate: comparisons in vivo with the *p*-[^{125}I] iodobenzoylconjugate, *Bioconjugate Chem.,* 2, 171, 1991.

99. **Atcher, R.W., Friedman, A.M., and Hines, J.J.,** An improved generator for the production of ^{212}Pb and ^{212}Bi from ^{224}Ra, *Appl. Radiat. Isot.,* 39, 283, 1988.

100. **Means, G.E. and Feeney, R.E.,** Chemical modality of proteins: history and applications, *Bioconjugate Chem.,* 1, 2, 1990.

101. **Brinkley, M.,** A brief survey of methods for preparing protein conjugates with dyes, haptens and cross-linking reagents, *Bioconjugate Chem.,* 3, 2, 1992.

102. **Ghose, T. and Blair, A.H.,** The design of cytotoxic agent-antibody conjugates, *Crit. Rev. Ther. Drug. Carrier Sys.,* 3, 263, 1987.

103. **Ghose, T., Blair, A.H., and Kulkarni, P.N.,** Preparation of antibody-linked cytoxic agents, *Methods Enzymol.,* 93, 280, 1983.

104. **Pietersz, C.A.,** The linkage of cytotoxic drugs to monoclonal antibodies for the treatment of cancer, *Bioconjugate Chem.,* 1, 89, 1990.

105. **Hermentin, P. and Seiler, F.R.,** Investigations with monoclonal antibody drug (anthracycline) conjugates, *Behring Inst. Mitt.,* 82, 197, 1988.

106. **Cleland, W.W.,** Dithiothreitol, a new protective reagent for SH groups, *Biochemistry,* 3, 480, 1964.

107. **Schramm, H.J. and Dulffer, T.,** The use of 2-iminothiolane as a protein cross-linking reagent, *Hoppe-Seyler's Z. Physiol. Chem.,* 358, 137, 1977.

108. **Carlsson, J., Drevin, H., and Axen, R.,** Protein thiolation and reversible protein-protein conjugation. N-succinimidyl-3-(2-pyridyldithio) propionate, a new heterobifunctional reagent, *Biochem. J.,* 173, 723, 1978.

109. **Duncan, J.S., Weston, P.D., and Wrigglesworth, R.,** A new reagent which may be used to introduce sulfhydryl groups into proteins, and its use in preparation of conjugates for immunoassay, *Anal. Biochem.,* 132, 68, 1983.

110. **Yoshitake, S., Yamada, Y., Ishikawa, E., and Masseyeff, R.,** Conjugation of glucose oxidase for Aspergillus niger and rabbit antibodies using N-hydroxysuccinimide ester of N-(4-carboxycyclohexylmethyl) maleimide, *Eur. J. Biochem.,* 101, 395, 1979.

111. **Kitagawa, T. and Aikawa, T.,** Enzyme coupled immunoassay of insulin using a novel coupling reagent, *J. Biochem. (Tokyo),* 79, 233, 1976.

112. **Weltman, J.K., Johnson, S.-A., Langevin, J., and Riester, E.F.,** N-Succinimidyl (4-iodoacetyl) aminobenzoate: a new heterobifunctional crosslinker, *Biotechniques,* 1, 148, 1983.

113. **Marsh, J.W. and Neville, D.M., Jr.,** in *Protein Tailoring for Food and Medical Uses,* Feeney, R.E. and Whitaker, J.R., Eds., Marcel Dekker, New York, 1986, 291.

114. **Jansen, F.K., Blythman, H.E., Carriere, D., et al.,** Immunotoxins: hybrid molecules combining high specificity and potent cytotoxicity, *Immunol. Rev.,* 62, 185, 1982.

115. **Masuho, Y., Kishida, K., Saifo, M., Umenofo, N., and Hara, T.,** Importance of the antigen binding valency and the nature of the cross-linking bond in ricin-A chain conjugates with antibody, *J. Biochem.,* 91, 1583, 1982.

116. **Bjorn, M.J., Groetsema, G., and Scalapino, L.,** Antibody-Pseudomonas exotoxin A. Conjugates cytotoxic to human breast cancer cells in vitro, *Cancer Res.,* 46, 3262, 1986.

117. **Fitzgerald, D., Idziorek, T., Batra, J.K., Willingham, M., and Pastan, I.,** Antitumor activity of a thioether-linked immunotoxin: OVB3-PE, *Bioconjugate Chem.,* 1, 264, 1990.

118. **Morgan, A.C., Jr., Sivam, G., Beaumier, P., McIntyre, R., Bjorn, M., and Abrams, P.G.,** Immunotoxins of Pseudomonas exotoxin A (PE): effect of linkage on conjugate yield, potency, selectivity and toxicity, *Mol. Immunol.,* 27, 273, 1990.

119. **Letvin, N.L., Goldmacher, V.S., Ritz, J., Vetz, J.M., Schlossman, S.F., and Lambert, J.M.,** In vivo administration of lymphocyte specific monoclonal antibodies in non-human primates. In vivo stability of disulfide linked immunotoxin conjugates, *J. Clin. Invest.,* 77, 977, 1986.

120. **Marsh, J.W., Srinivasachar, K., and Nevill, D.M., Jr.,** Antibody toxin conjugation, in *Immunotoxins,* Frankel, A.E., Ed., Kluwer Academic Publishers, Boston, 1988, 213.

121. **Hurwitz, E., Wilchek, M., and Pitha, J.,** Soluble macromolecules as carriers for daunorubicin, *J. Appl. Biochem.,* 2, 25, 1980.

122. **Zunino, F., Gambetta, R., Vigevani, A., Penco, S., Geroni, C., and DiMarco, A.,** Biological activity of daunorubicin linked to proteins via the methylketone side chain, *Tumoric,* 67, 521, 1981.

123. **Fujiwara, K., Yasuno, M., and Kitagawa, T.,** Novel preparation method of immunogen for hydrophobic hapten, enzyme immuno-assay for daunomycin and adriamycin, *J. Immunol. Meth.,* 45, 511, 1985.

124. **Gallego, J., Price, M.R., and Baldwin, R.W.,** Preparation of four daunomycin-monoclonal antibody 76IT/36 conjugates with antitumor activity, *Int. J. Cancer,* 33, 737, 1984.

125. **Umemoto, N., Kato, Y., Endo, N., Takeda, Y., and Hara, T.,** Preparation and in vitro cytotoxicity of a methotrexate-anti-MM 46 monoclonal antibody conjugate via an oligopeptide spacer, *Int. J. Cancer,* 43, 677, 1989.

126. **Hermentin, P., Doenges, R., Gronski, P., Bosslet, H., Kraemer, H.P., Hoffman, D., Zilg, H., Steinstraesser, A., Schwarz, A., Kuhlmann, L., Luben, G., and Seiler, F.R.,** Attachment of rhodosaminyl anthracyclinone-type anthracyclines to the hinge region of monoclonal antibodies, *Bioconjugate Chem.,* 1, 100, 1990.

127. **Kulkari, P.N., Blair, A.H., and Ghose, T.I.,** Covalent binding of MTX to immunoglobulins and the effect of antibody-linked drug or tumor growth in vivo, *Cancer Res.,* 41, 2700, 1981.

128. **Smyth, M.J., Pietersz, G.A., Classor, B.J., and McKenzie, I.F.C.,** Specific targeting of chlorambucil to tumors with use of monoclonal antibodies, *J. Natl. Cancer Inst.,* 76, 503, 1986.

129. **Smyth, M.J., Pietersz, G.A., and McKenzie, I.F.C.,** Selective enhancement of anti-tumor activity of N-acetyl melphalan upon conjugation to monoclonal antibodies, *Cancer Res.,* 47, 62, 1987.

130. **Vrudhula, V., Srinivasan, A., and Comezoglu, T.F.,** Selective Synthetic Transformations with Roridin A, Poster presentation #50, American Chemical Society Meeting, Boston, 1990.

131. **Lundblad, R.L. and Meyers, C.M.,** *Chemical Reagents for Protein Modification,* Vol. 2, CRC Press, Boca Raton, 1984, chap. 4.

132. **Kovich, T.R.,** Polymerization site reactions during protein modification with carbodiimide, *Biochem. Biophys. Res. Commun.,* 74, 1463, 1977.

133. **Hurwitz, E., Levy, R., Maron, R., Wilchek, M., Arnon, R., and Sela, M.,** The covalent building of daunomycin and adriamycin to antibodies with retention of both drug and antibody activities, *Cancer Res.,* 35, 1175, 1975.

134. **Burstein, S. and Knapp, S.,** Chemotherapy of murine ovarian carcinoma by methotrexate antibody conjugate, *J. Med. Chem.,* 20, 950, 1977.

135. **Belles-Isleo, M. and Page, M.,** Anti-onco foetal proteins for targeting cytotoxic drugs, *Int. J. Immunopharmacol.,* 3, 97, 1981.

136. **Laguzza, B.C., Nichols, C.L., Briggs, S.L., Cullman, G.J., Johnson, D.A., Starling, J.J., Baker, A.L., Bumol, T.F., and Corvolan, J.R.F.,** New antitumor monoclonal antibody vinca conjugates LY203725 and related compounds: design, preparation and representative in vivo activity, *J. Med. Chem.,* 32, 548, 1989.

137. **Ghose, T., Blair, A.H., Krapovec, J., Uadia, P.O., and Mammen, M.,** Synthesis and testing of antibody-antifolate conjugates for drug targeting, in *Target Diagnosis and Therapy,* Vol. 1, Rodwell, J.D., Ed., Marcell Dekker, New York, 1988, 81.

138. **Sivam, G.P., Comezoglu, T., Manger, R., Gray, M.A., Jarvis, B.B., and Morgan, A.C.,** Immunoconjugates of trichothecenes and monoclonal antibody, presented at 4th Int. Conference of Monoclonal Immunoconjugates for Cancer, Univ. of California, San Diego, 1989.

139. **Thorpe, D.E., Detre, S.J., Foxwell, B.M., Braun, A.N., Skilleter, D.N., Wilson, G., Forrester, J.A., and Stirpe, F.,** Modification of the carbohydrate in ricin with metaperiodate-cyano borohydrate mixtures, *Eur. J. Biochem.,* 47, 197, 1985.

140. **Hurwitz, E., Levy, R., Maron, R., Wilchek, M., Aran, R., and Sela, M.,** The covalent binding of daunomycin and adriamycin to antibodies with retention of both drug and antibody activities, *Cancer Res.,* 35, 1175, 1975.

141. **Kaneka, T., Willner, D., Mankovic, I., Knipe, J.O., Braslawsky, G.R., Greenfield, R.S., and Vyas, D.M.,** New hydrazone derivatives of adriamycin and their conjugates — a correlation between acid stability and cytotoxicity, *Bioconjugate Chem.,* 2, 133, 1991.

142. **Greenfield, R.S., Kaneko, T., Daves, A., Edson, M.A., Fitzgerald, K.A., Olech, L.J., Grattan, J.A., Spitalny, G.L., and Braslawsky, G.R.,** Evaluation in vitro of adriamycin immunoconjugates synthesized using an acid sensitive hydrazone linker, *Cancer Res.,* 50, 6600, 1990.

143. **Mueller, B.M., Wrasidlo, W.A., and Reisfeld, R.A.,** Antibody conjugates with morpholino doxorubicin and acid cleavable linkers, *Bioconjugate Chem.,* 1, 325, 1990.

144. **Hamann, P.R., Hinman, L.M., and Upeslacis, J.,** Monoclonal antibody conjugates prepared from the calicheamicin family of highly potent antitumor antibiotics. Presented at 5th Int. Conference of Monoclonal Immunoconjugates for Cancer, San Diego, 1990.

145. **Bjorn, M.J., Ring, D., and Frankel, A.,** Evaluation of monoclonal antibodies for development of breast cancer immunotoxins, *Cancer Res.,* 45, 1214, 1985.

146. **Ogata, M., Chaudhary, V.K., Pastan, I., and Fitzgerald, D.J.,** Processing of *Pseudomonas* exotoxin by a cellular protease results in the generation of a 37,000-Da toxin fragment that is translocated to the ceptosol, *J. Biol. Chem.,* 265, 20678, 1990.

147. **Thorpe, P.E., Wallace, P.M., Knowles, P.P., Relf, M.G., Brown, A.F., Watson, G.J., Knyba, R.E., Wawtzenyczak, E.J., and Blakely, D.C.,** New coupling agents for the synthesis of immunotoxins containing a hindered disulfide bond with improved stability in vivo, *Cancer Res.,* 47, 5924, 1987.

148. **Worrell, N.R., Camber, A.J., Parnell, G.D., Mirza, A., Forrester, J.A., and Ross, W.C.J.,** Effect of linkage variation on pharmacokinetics of ricin A chain-antibody conjugates in normal rats, *Anti-Cancer Drug Des.,* 1, 179, 1986.

149. **Greenfield, L., Bloch, W., and Moreland, M.,** Thiol-containing cross-linking agent with enhanced steric hindrance, *Bioconjugate Chem.,* 1, 400, 1990.

150. **Goff, D.A. and Carroll, S.F.,** Substituted 2-iminothiolanes: reagents for preparation of disulfide cross-linked conjugates with increased stability, *Bioconjugate Chem.,* 1, 381, 1990.

151. **Tsurumi, H., Takahasi, T., Yamaguchi, T., Kitomura, K., Noguchi, A., Chumori, Y., Kamiguchi, M., Honda, M., Noguchi, A., Takashima, K., Yamaoka, N., and Mayagaki, T.,** In vivo stability of the disulfide linkage between the monoclonal antibody A7 and anticancer agent — Neocarzinostatin (A7-NCS), Poster presentation #103, 6th Int. Conf. of Monoclonal Immunoconjugates for Cancer, 1991.

152. **Sivam, G.,** Trichothecene conjugates and methods of use, U.S. Patent 4,906,452, 1990.

153. **Vrudhula, V., Srinivasan, A., and Comezoglu, T.F.,** Selective Synthetic Transformations with Roridin A, Poster presentation #50, American Chemical Society meeting, Boston, 1990.

154. **Comezoglu, T.F., Manger, R., Woodle, D., Jackson, T., Priest, J., Morgan, A.C., and Sivam, G.P.,** Comparative cytotoxicities of Verrucarin A and standard chemotherapeutic agents, Poster presentation #71, 4th Int. Conference of Monoclonal Immunoconjugates for Cancer, San Diego, 1989.

155. **Shen, W.C. and Ryser, J.P.,** Cis-aconityl spacer between daunomycin and macro-molecular carriers: a model of pH-sensitive linkage releasing drug from a lysosomotropic conjugate, *Biochem. Biophys. Res. Commun.,* 102, 1048, 1981.

156. **Blattler, W.A., Kuenzi, B.S., Lambert, J.M., and Senter, P.D.,** New heterobifunctional protein cross-linking reagent that forms an acid-labile link, *Biochemistry,* 24, 1517, 1985.

157. **Ming Yang, H. and Reisfeld, R.A.,** Pharmacokinetics and mechanism of action of a doxorubicin monoclonal antibody 9.2.27 conjugate directed to a human melanoma proteoglycan, *J. Natl. Cancer Inst.,* 80, 2154, 1988.

158. **Lambert, J.M., Blattler, W.A., McIntyre, G.D., Goldmacher, V.S., and Scott, C.F.,** Immunotoxins containing single chain ribosome-inactivating proteins, in *Immunotoxin,* Frankel, A.E., Ed., Kluwer Academic Publishers, Boston, 1988, 175.

159. **Srinivasachar, K. and Neville, D.M., Jr.,** New protein cross-linking reagents that are cleaved by mild acid, *Biochemistry,* 28, 2501, 1989.

160. **Neville, D.M., Jr., Srinivasachar, K., Stone, R., and Scharff, J.,** Enhancement of immunotoxin efficacy by acid-cleavable cross linking agents utilizing diptheria toxin and toxin mutants, *J. Biol. Chem.,* 264, 14653, 1989.

161. **Troutner, A., Masquelier, M., Baurain, R., and Deprez-deCampeneere, D.,** A covalent linkage between daunorubicin and proteins that is stable in serum and reversible by lysosomal hydrolases, as required for a lysosomotropic drug-carrier conjugate: in vitro and in vivo studies, *Proc. Natl. Acad. Sci. U.S.A.,* 79, 626, 1982.

162. **Umemoto, N., Kato, Y., Endo, N., Takeda, Y., and Hara, T.,** Preparation and in vitro cytotoxicity of a methotrexate-anti MM46 monoclonal antibody conjugate via an oligopeptide spacer, *Int. J. Cancer.* 42, 1665, 1989.

163. **Tsukada, Y., Hurwitz, E., Kashi, R., Sela, M., Hibi, N., Hara, A., and Hirai, H.,** Chemotherapy by intravenous administration of conjugates of daunomycin with monoclonal and conventional anti-rat α-fetoprotein antibodies, *Proc. Natl. Acad. Sci. U.S.A.,* 79, 7896, 1982.

164. **Noguchi, A., Takuhashi, T., Yamaguchi, T., Kitamura, K., Takakura, Y., Hashida, M., and Sezaki, H.,** Preparation and properties of the immunoconjugate composed of anti-human colon cancer monoclonal antibody and mitomycin-C dextran conjugate, *Bioconjugate Chem.,* 3, 132, 1992.

165. **Shih, L., Sharkey, R.M., Princes, F.J., and Goldenberg, D.M.,** Site specific linkage of methotrexate to monoclonal antibodies using an intermediate carrier, *Int. J. Cancer,* 4, 832, 1988.

166. **Arnold, L.J., Jr.,** Polylysine-drug conjugates, *Meth. Enzymol.,* 112, 270, 1985.

167. **Zunino, F., Guiliani, F., Sovi, G., Dasdia, T., and Gambetta, R.,** Antitumour activity of daunomycin linked to poly-L-aspartic acid, *Int. J. Cancer,* 30, 465, 1982.

168. **Tsukada, Y., Umemoto, N., Takeda, Y., Hara, T., and Hirai, H.,** An anti–α–feto protein antibody-daunorubicin conjugate with a novel poly-L-glutamic acid derivative as intermediate drug carrier, *J. Natl. Cancer Inst.,* 73, 721, 1984.

169. **Hudecz, F., Clegg, J.A., Kajtar, J., Embleton, M.J., Szekerke, M., and Baldwin, R.W.,** Synthesis conformation, biodistribution and in vitro cytotoxicity of daunomycin-branched polypeptide conjugates, *Bioconjugate Chem.,* 3, 49, 1992.

170. **Hurwitz, E., Wilchek, M., and Pitha, J.,** Soluble macromolecules as carriers for daunorubicin, *J. Appl. Biochem.,* 2, 25, 1980.

171. **Hirano, T., Ohaski, S., Morimoto, S., Tsukada, K., Kobayashi, T., and Tsukagoshi, S.,** Synthesis of antitumor conjugates of adriamycin or daunomycine with the copolymer of divinyl ether and maleic anhydride, *Macromol. Chem.,* 189, 2815, 1986.

172. **Duncan, R., Kopeckova, P., Rejmanova, P., Strohalm, J., Hume, I., Cable, H.C., Pohl, J., Lloyd, J.B., and Kopecek, J.,** Anticancer agents coupled to N-(2-hydroxypropyl) methacrylamide copolymers. I. Evaluation of daunomycin and puromycin in vitro, *Br. J. Cancer,* 55, 165, 1987.

173. **Daussin, F., Boschetti, E., Delmotte, F., and Monsigny, M.,** p-Benzylthio carbonyl-aspartyl-daunomycine-substituted poly-trisacryl. A new drug acid-labile arm-carrier conjugate, *Eur. J. Biochem.,* 176, 625, 1988.

174. **Songsivilai, S. and Lachmann, P.J.,** Bispecific antibody: a tool for diagnosis and treatment of disease, *Clin. Exp. Immunol.,* 79, 315, 1990.

175. **Nolan, O. and O'Kennedy, R.,** Bifunctional antibodies: concept, production and applications, *Biochim. Biophys. Acta,* 1040, 1, 1990.

176. **Corvalan, J.R.F. and Smith, W.,** Construction and characterization of a hybrid-hybrid monoclonal antibody recognizing both carcinoembryonic antigen (CEA) and vinca alkaloids, *Cancer Immunol. Immunother.,* 24, 127, 1987.

177. **Corvalan, J.R.F., Smith, W., Gore, V.A., and Brandon, D.R.,** Specific *in vitro* and *in vivo* drug localization to tumor cells using a hybrid-hybrid monoclonal antibody recognizing both carcinoembryonic antigen and vinca alkaloids, *Cancer Immunol. Immunother.,* 24, 133, 1987.

178. **Corvalan, J.R.F., Smith, W., Gore, V.A., Brandon, D.R., and Ryde, P.J.,** Increased therapeutic effect of vinca alkaloids targeted to tumour by hybrid-hybrid monoclonal antibody, *Cancer Immunol. Immunother.,* 24, 138, 1987.

179. **Corvalan, J.R.F., Smith, W., and Gore, V.A.,** Tumour therapy with vinca alkaloids targeted by a hybrid-hybrid monoclonal antibody recognising both CEA and vinca alkaloids, *Int. J. Cancer,* 2 (Suppl.), 22, 1988.

180. **Smith, W., Gore, V.A., Brandon, D.R., Lynch, D.N., Cranstone, S.A., and Corvalan, J.R.F.,** Suppression of well-established tumour xenografts by a hybrid-hybrid monoclonal antibody and vinblastine, *Cancer Immunol. Immunother.,* 31, 157, 1990.

181. **Pimm, M.V., Robins, R.A., Embleton, M.J., Jacobs, E., Markham, A.J., Charleston, A., and Baldwin, R.W.,** A bispecific monoclonal antibody against methotrexate and a human tumour associated antigen augments cytotoxicity of methotrexate-carrier conjugate, *Br. J. Cancer,* 61, 508, 1990.

182. **Senter, P.D., Saulnier, M.G., Schreiber, G.J., Hirschberg, D.L., Brown, J.P., Hellstrom, I., and Hellstrom, K.E.,** Anti-tumor effects of antibody-alkaline phosphatase conjugates in combination with etoposide phosphate, *Proc. Natl. Acad. Sci. U.S.A.,* 85, 4842, 1988.

183. **Bagshawe, K.D., Springer, C.J., Searle, F., Antoniw, P., Sharma, S.K., Melton, R.G., and Sherwood, R.F.,** A cytotoxic agent can be generated selectively at cancer sites, *Br. J. Cancer,* 58, 700, 1988.

184. **Senter, P.D., Schreiber, G.J., Hirschberg, D.L., Ashe, S.A., Hellstrom, K.E., and Hellstrom, I.,** Enhancement of the *in vitro* and *in vivo* antitumor activities of phosphorylated mitomycin and etoposide derivatives by monoclonal antibody-alkaline phosphatase conjugates, *Cancer Res.,* 49, 5789, 1988.

185. **Sahin, U., Hartmann, F., Senter, P., Pohl, C., Engert, A., Diehl, V., and Pfreundschuh, M.,** Specific activation of the prodrug mitomycin phosphate by a bispecific anti-CD30/anti-alkaline phosphatase monoclonal antibody, *Cancer Res.,* 50, 6944, 1990.

186. **Reardan, D.T., Meares, C.F., Goodwin, D.A., McTigue, M., David, G.S., Stone, M.R., Leung, J.P., Bartholomew, R.M., and Frincke, J.M.,** Antibodies against metal chelates, *Nature,* 316, 265, 1985.

187. **Goodwin, D.A., Meares, C.F., McCall, M.J., McTigue, M., and Chaovapong, W.,** Pre-targeted immunoscintigraphy of murine tumors with indium-111-labeled bifunctional haptens, *J. Nucl. Med.,* 29, 226, 1988.

188. **LeDoussal, J.-M., Gruaz-Guyon, A., Martin, M., Gautherot, E., Delaage, M., and Barbet, J.,** Targeting of indium 111-labeled bivalent hapten to human melanoma mediated by bispecific monoclonal antibody conjugates: imaging of tumors hosted in nude mice, *Cancer Res.,* 50, 3445, 1990.

189. **Gridley, D.S., Ewart, K.L., Cao, J.D., and Stickney, D.R.,** Hyperthermia enhances localization of ^{111}In-labeled hapten to bifunctional antibody in human colon tumor xenografts, *Cancer Res.,* 51, 1515, 1991.

190. **Raso, V. and Griffin, T.,** Hybrid antibodies with dual specificity for the delivery of ricin to immunoglobulin-bearing target cells, *Cancer Res.,* 41, 2073, 1981.

191. **Webb, K.S., Ware, J.L., Parks, S.F., Walther, P.J., and Paulson, D.F.,** Evidence for a novel hybrid immunotoxin recognizing ricin A-chain by one antigen-combining site and a prostate-restricted antigen by the remaining antigen-combining site: potential for immunotherapy, *Cancer Treat. Rep.,* 69, 663, 1985.

192. **Pimm, M.V., Gunn, B., Lord, J.M., and Baldwin, R.W.,** The influence of anti-(ricin toxin A chain) monoclonal antibodies on the pharmacokinetics of ricin toxin A chain and recombinant ricin A chain in mice, *Cancer Immunol. Immunother.*, 32, 235, 1990.

193. **Robins, R.A., Embleton, M.J., Pimm, M.V., Betfs, D.S., Charleston, A., Markham, A.J., and Baldwin, R.W.,** Bispecific antibody that binds carcinoembryonic antigen and ricin A chain cytotoxic for gastrointestinal tract tumor cells, *J. Natl. Cancer Inst.*, 82, 1295, 1990.

194. **Embleton, M.J., Charleston, A., Robins, R.A., Pimm, M.V., and Baldwin, R.W.,** Recombinant ricin toxin A chain cytotoxicity against carcinoembryonic antigen expressing tumour cells mediated by a bispecific monoclonal antibody and its potentiation by ricin toxin B chain, *Br. J. Cancer*, 63, 670, 1991.

195. **Glennie, M.J., Brennand, D.M., Bryden, F., McBride, H.M., Stirpe, F., Worth, A.T., and Stevenson, G.T.,** Bispecific F(ab'γ)₂ antibody for the delivery of saporin in the treatment of lymphoma, *J. Immunol.*, 141, 3662, 1988.

196. **French, R.R., Courtenay, A.E., Ingamells, S., Stevenson, G.T., and Glennie, M.J.,** Cooperative mixtures of bispecific F(ab')2 antibodies for delivering saporin to lymphoma *in vitro* and *in vivo*, *Cancer Res.*, 51, 2353, 1991.

197. **Flavell, D.J., Cooper, S., Morland, B., and Flavell, S.U.,** Characteristics and performance of a bispecific F(ab')2 antibody for delivering saporin to a CD7⁺ human acute T-cell leukaemia cell line, *Br. J. Cancer*, 64, 274, 1991.

198. **Nelson, H.,** Targeted cellular immunotherapy with bifunctional antibodies, *Cancer Cells*, 3, 163, 1991.

199. **Bolhuis, R.L.H., Sturm, E., and Braakman, E.,** T cell targeting in cancer therapy, *Cancer Immunol. Immunother.*, 34, 1, 1991.

200. **Jung, G., Freimann, U., Marschall, Z.V., Reisfeld, R.A., and Wilmanns, W.,** Target cell-induced T cell activation with bi- and trispecific antibody fragments, *Eur. J. Immunol.*, 21, 2431, 1991.

201. **Tutt, A., Stevenson, G.T., and Glennie, M.J.,** Trispecific F(ab'γ)₃ derivatives that use cooperative signaling via the TcR/CD3 complex and CD2 to activate and redirect resting cytotoxic T cells, *J. Immunol.*, 147, 60, 1991.

202. **Tutt, A., Greenman, J., Stevenson, G.T., and Glennie, M.J.,** Bispecific F(ab'γ)₃ antibody derivatives for redirecting unprimed cytotoxic T cells, *Eur. J. Immunol.*, 21, 1351, 1991.

203. **Segal, D.M., Garrido, M.A., Qian, J.-H., Mezzananica, D., Andrew, S., Perez, P., Kurucz, I., Valdayo, M.J., Titus, J.A., Winkler, D.F., and Wunderlich, J.R.,** Effectors of targeted cellular cytotoxicity, *Mol. Immunol.*, 27, 1339, 1990.

204. **Staerz, U., Yewdell, J.W., and Bevan, M.J.,** Hybrid antibody-mediated lysis of virus-infected cells, *Eur. J. Immunol.*, 17, 571, 1987.

205. **Paya, C.V., McKean, D.J., Segal, D.M., Schoon, R.A., Schowalter, S.D., and Leibson, P.J.,** Heteroconjugate antibodies enhance cell-mediated anti-herpes simplex virus immunity, *J. Immunol.*, 142, 666, 1989.

206. **Zarling, J.M., Moran, P.A., Grosmarie, L.S., McClure, J., Shriver, K., and Ledbetter, J.A.,** Lysis of cells infected with HIV-1 by human lymphocytes targeted with monoclonal antibody heteroconjugates, *J. Immunol.*, 140, 2609, 1988.

207. **Berg, J., Lotscher, E., Steimer, K.S., Capon, D.J., Baenziger, J., Jack, H.-M., and Wabl, M.,** Bispecific antibodies that mediate killing of cells infected with human immunodeficiency virus of any strain, *Proc. Natl. Acad. Sci. U.S.A.*, 88, 4723, 1991.

208. **Conner, R.I., Dinces, N.B., Howell, A.L., Romet-Lemonne, J.-L., Pasquali, J.-L., and Fanger, M.W.,** Fc receptors for IgG (FcγRs) on human monocytes and macrophages are not infectivity receptors for human immunodeficiency virus type 1 (HIV-1): studies using bispecific antibodies to target HIV-1 to various myeloid cell surface molecules, including the FcγR, *Proc. Natl. Acad. Sci. U.S.A.*, 88, 9593, 1991.

209. **Runge, M.S., Bode, C., Matsueda, G.R., and Haber, E.,** Antibody-enhanced thrombolysis: capture of tissue plasminogen activator by a bispecific antibody and direct targeting by an antifibrin-tissue plasminogen activator conjugate *in vivo*, *Trans. Assoc. Am. Physicians*, 100, 250, 1987.

210. **Bode, C., Runge, M.S., Branscomb, E.E., Newell, J.B., Matsueda, G.R., and Haber, E.,** Antibody directed fibrinolysis: an antibody specific for both fibrin and tissue plasminogen activator, *J. Biol. Chem.*, 264, 944, 1989.

211. **Runge, M.S., Bode, C., Savard, C.E., Matsueda, G.R., and Haber, E.,** Antibody-directed fibrinolysis: a bispecific (Fab')₂ that binds to fibrin and tissue plasminogen activator, *Bioconjugate Chem.*, 1, 274, 1990.

212. **Branscomb, E.E., Runge, M.S., Savard, C.E., Adams, K.M., Matsueda, G.R., and Haber, E.,** Bispecific monoclonal antibodies produced by somatic cell fusion increase the potency of tissue plasminogen activator, *Thrombo. Haemost.*, 64, 260, 1990.

213. **Kurokawa, T., Iwasa, S., and Kakinuma, A.,** Enhancement of fibrinolysis by bispecific monoclonal antibodies reactive to fibrin and tissue plasminogen activators, *Thrombo. Res.*, 10 (Suppl.), 83, 1990.

214. **Charpie, J.R., Runge, M.S., Matsueda, G.R., and Haber, E.,** A bispecific antibody enhances the fibrinolytic potency of single-chain urokinase, *Biochemistry*, 29, 6374, 1990.

215. The TIMI Study Group, The thrombolysis in myocardial infarction (TIMI) trial. Phase I findings, *N. Engl. J. Med.*, 312, 932, 1985.

216. **Paulus, H.,** Preparation and biomedical applications of bispecific antibodies, *Behring Inst. Mitt.*, 78, 118, 1985.

217. **Milstein, C. and Cuello, A.C.,** Hybrid hybridomas and their use in immunohistochemistry, *Nature*, 305, 537, 1983.

218. **Brennan, M., Davison, P.F., and Paulus, H.,** Preparation of bispecific antibodies by chemical recombination of monoclonal immunoglobulin G₁ fragments, *Science*, 229, 81, 1985.

219. **De Monte, L., Nistico, P., Tecce, R., Dellabona, P., Momo, M., Tarditi, L., Natali, P.G., Mariani, M., and Malavasi, F.,** Gene transfer by retrovirus-derived shuttle vectors in the generation of murine bispecific MAbs, *Dev. Biol. Stand.,* 71, 15, 1990.

220. **De Lau, W.B.M., Heije, K., Neefjes, J.J., Oosterwegel, M., Rozemuller, E., and Bast, B.J.E.G.,** Absence of preferential homologous H/L chain association in hybrid hybridomas, *J. Immunol.,* 146, 906, 1991.

221. **Ghetie, V. and Mota, G.,** Multivalent hybrid antibody, *Mol. Immunol.,* 17, 395, 1980.

222. **Lansdorp, P.M., Aalberse, R.C., Bos, R., Schutter, W.G., and Van Bruggen, E.F.J.,** Cyclic tetramolecular complexes of monoclonal antibodies: a new type of cross-linking reagent, *Eur. J. Immunol.,* 16, 679, 1986.

223. **Lansdorf, P.M., and Thomas, T.E.,** Purification and analysis of bispecific tetrameric antibody complexes, *Mol. Immunol.,* 27, 659, 1990.

224. **Abdel-Nabi, H., Doerr, R.J., Chan, H.-W., Balu, D., Schmelter, R.F., and Maguire, R.T.,** In-111-labeled monoclonal antibody immunoscintigraphy in colorectal carcinoma: safety, sensitivity and preliminary clinical results, *Radiology,* 175 (163), 171, 1990.

225. **Doerr, R.J., Nabi, H.A., and Merchant, B.M.,** In-111-ZOE-025 immunoscintigraphy in occult recurrent colorectal cancer with deviated CEA, *Arch. Surg.,* 125, 226, 1990.

226. **Khaw, B.A. and Haber, E.,** Imaging necrotic myocardia detection with ^{99m}Tc-pyrophosphate and radiolabeled antimyosin, *Cardiovasc. Clin.,* 7, 577, 1989.

227. **Kwekkeboom, D.J., Krenning, E.P., Bakker, W.H., Oei, Y., Splinter, T.A.W., Kho, G.S., and Lamberts, S.W.J.,** Radioiodinated somatostatin analog scintigraphy in small-cell lung cancer, *J. Nucl. Med.,* 32, 1845, 1991.

228. **Krenning, E.P., Breeman, W.A.P., Kooij, P.P.M., Lameris, J.S., Bakker, W.H., Kaper, J.W., Ausema, L., Reubi, J.C., and Lamberts, S.W.J.,** Localization of endocrine-related tumors with radioiodinated analogue of somatostatin, *Lancet,* 1, 242, 1989.

229. **Wraight, E.P., Bard, D.R., Maughan, T.S., Knight, C.G., and Page-Thomas, D.P.,** The use of a chelating derivative of alpha melanocyte stimulating hormone for the clinical imaging of malignant melanoma, *Br. J. Radiol.,* 65, 112, 1992.

230. **DeNardo, S.J., Denardo, G.L., O'Grady, L.F., Levy, N.B., Mills, S.L., Macey, D.J., McGahan, J.P., Miller, C.H., and Epstein, A.L.,** Pilot studies of radioimmunotherapy of B cell lymphoma and leukemia using I-131 Lym-1 monoclonal antibody, *Antibody Immunoconj. Radiopharm.,* 1, 17, 1988.

231. **Lenhard, R.E., Order, S.E., Spunberg, J.J., Asbell, S.O., and Leibel, S.A.,** Isotopic immunoglobulin: a new systemic therapy for advanced Hodgkin's disease, *J. Clin. Oncol.,* 3, 1296, 1985.

232. **Epenetos, A.A., Munro, A.J., Stewart, S., Rampling, R., Lambert, H.E., McKenzie, C.G., Soutter, P., Rahentulla, A., Hooker, G., Sivolapenko, G.B., Snook, D., Courteney-Luck, N., Dhokia, B., Krausc, T., Taylor-Papa Dimitriou, J., Durbin, H., and Bodmer, W.F.,** Antibody-guided irradiation of advanced ovarian cancer with intraperitoneally administered radiolabeled monoclonal antibodies, *J. Clin. Oncol.,* 5, 1890, 1987.

233. **Gould, B.J., Borowitz, M.J., Graves, E.S., Carter, P.W., Anthony, D., Weiner, L.M., and Frankel, A.E.,** Phase I study of an anti-breast cancer immunotoxin by continuous infusion: report of a targeted toxic effect not predicted by animal studies, *J. Natl. Cancer Inst.,* 81, 775, 1989.

234. **Mendelsohn, J.,** Immunotoxins: prospects and problems, *J. Clin. Oncol.,* 9, 2088, 1991.

235. **Vitetta, E.S., Stone, M., Amlot, P., Fay, J., May, R., Till, M., Newman, J., Clark, P., Collins, R., Cunningham, D., Ghetie, V., Uhr, J.W., and Thorpe, P.E.,** Phase I immunotoxin trial in patients with B-cell lymphoma, *Cancer Res.,* 51, 4052, 1991.

236. **Hertler, A.A. and Frankel, A.E.,** Immunotoxins in the therapy of leukemias and lymphomas, *Cancer Invest.,* 9, 211, 1991.

237. **Schneck, D., Butler, F., Dugan, W., Littrell, D., Petersen, B., Bowsher, R., DeLong, A., and Dorrbecker, S.,** Disposition of a murine monoclonal antibody vinca conjugate (KS1/4-DAVLB) in patients with adenocarcinoma, *Clin. Pharmacol. Ther.,* 47, 36, 1990.

238. **Petersen, B.H., Detterdt, S.V., Schneck, D.W., and Bumol, T.F.,** The immune response to KS1/4-desacetylvinblastine (LY256787) and KS1/4-desacetylvinblastine hydrazide (LY203728) in single and multiple dose clinical studies, *Cancer Res.,* 51, 2286, 1991.

239. **Elias, D.J., Hirschowitz, L., Kline, L.E., Kroener, J.F., Dillman, R.O., Walker, L.E., Robb, J.A., and Timms,** Phase I clinical comparative study of monoclonal antibody KS1/4 and KS1/4-methotrexate immunoconjugate in patients with non-small cell lung carcinoma, *Cancer Res.,* 50, 4154, 1990.

240. **Takahashi, T., Yamaguchi, T., Kitamura, K., Suzuyama, H., Honda, M., Yokota, T., Kotanagi, H., Takahashi, M., and Hashimoto, Y.,** Clinical application of monoclonal antibody-drug conjugates for immunotargeting chemotherapy of colorectal carcinoma, *Cancer,* 61, 881, 1988.

Section II:
Structural Approaches to Rational Drug Design

Chapter 6

THE ROLE OF X-RAY CRYSTALLOGRAPHY IN STRUCTURE-BASED RATIONAL DRUG DESIGN

Alexander McPherson

CONTENTS

I. OVERVIEW

Prior to the emergence of powerful analytical methods for the determination and delineation of macromolecular structure, drug discovery was based predominantly on the identification of lead compounds and their development into useful pharmacological agents by mass screening for desired activities. Generally this required enormous commitments of time, effort, and financial support with extremely small probabilities of success to drive the enterprise. With the advent of modern structural analytical approaches, the target macromolecules of a directed drug development program could be examined in extraordinary chemical detail, and rational approaches to the design of specific pharmacological agents applied. While still in its infancy, this new science of systematic and directed drug design has already assumed an important place in the biotechnology revolution that has emerged over the past 15 years.

The use of macromolecular structure, principally proteins, and specifically enzymes, but also including nucleic acids as well, can be roughly divided into three different approaches: (1) rational design of small molecules to interact with the target macromolecule in specified and controlled ways; (2) the genetic engineering of protein drugs, generally natural polypeptide products normally found in low abundance, their amplification, and their mass production; and (3) the newest and most novel approach, that of antibody-directed drug discovery based on the mimicry of induced antibody combining sites.

To understand fully how these methods of drug design and discovery are utilized in practice, it is necessary to understand the scientific techniques in which they are grounded. In principle, this means an understanding of at least the kinds of results that are obtained from high-resolution structural analyses, how they may be used by researchers, and what problems are involved in their

0-8493-7818-4/95/$0.00+$.50
© 1995 by CRC Press Inc.

routine application. In practice, this means at least a passing comprehension of the process known as X-ray crystallography or X-ray diffraction analysis.

II. INTRODUCTION

The discovery of new drugs, such as antibiotics or antitumor agents, as well as herbicides, insecticides, and other pharmacologically active compounds has traditionally depended on the, more or less, trial and error synthesis of potential lead compounds and their testing for effect in some physiological system. Such lead compounds are suggested through knowledge of biochemical pathways and inferred antimetabolites, or through successful trials in either receptor- or cellular-based assays. Following identification of such lead compounds, the chemical structure and physical properties of the drug are gradually optimized in a very painstaking and time-consuming manner by synthesis of a succession of variants with subsequent testing for physiological effect and behavior.

This entire process is long and difficult, and extremely demanding of resources because a correlation between the chemical structure of a drug and its physiological effect is often very difficult to discern. In addition, success at the molecular level is often submerged in side effects and transport problems. In general, the development of a new drug, a rare event indeed, may require the synthesis and testing of literally thousands of new compounds.

The targets of most drugs are proteins, usually enzymes, and nucleic acids to a much lesser extent. Drugs act at the molecular level by binding to either the active site of a specific target enzyme and inhibiting its activity, or by binding to some allosteric or regulatory site on the target protein and altering in a directed fashion the physiological activity of the macromolecule.

In the case of nucleic acids, the binding may alter the manner by which specific regions of the polynucleotide are recognized and acted upon by systems of enzymes. Many tumor suppressor drugs, for example, bind to nucleic acids and induce the generation of mistakes by accessory enzymes during the course of replication or transcription. In other cases the drugs, such as bromo dexory uridine, may actually be incorporated into daughter nucleic acid strands because of their similarity to natural components, and concomitantly confuse and disrupt subsequent essential processes.

Enzymes, however, are the usual targets of drugs, and it is thus proteins that are of particular interest to pharmacologists. Enzymes are highly specific biological catalysts that promote certain specified chemical reactions in cells with extraordinary efficiency and exquisite accuracy (see Reference 6 for an excellent discussion). The fastest enzymes, such as carbonic anhydrase or catalase for example, can carry out their specified task from 250,000 to 400,000 times per second with no byproducts formed. Other enzymes have the amazing ability to choose among substrates of such close chemical similarity that we are still at a loss to fully explain the basis of their discrimination.

Enzymes act by combining with their specific substrates at a unique, confined region on their surface (see Reference 12 for a survey of X-ray results), known as the active site. Once this enzyme-substrate complex is formed, certain catalytic amino acid residues provided by the enzyme focus their catalytic effects on the substrate, induce it to form a transition state, and ultimately see that it is converted chemically to product. The enzyme in no case alters the thermodynamic potential between substrates and products, but simply provides an alternate pathway for the transition. That is, it lowers the activation barrier for the reaction.

The extraordinary specificity for the binding and fixing of the substrate to the active site of the enzyme is a consequence of an exceptional stereochemical complementarity between the macromolecule and the conventional molecule, or substrate, that it complexes with. An example[19] of an extensive complex between enzyme, substrate analog, and a drug is shown in Plate 1* and a second example, that of α-cyclodextrin binding at different sites on pancreatic α-amylase, is shown in Plate 2.* This high degree of complementarity is both physical and chemical. The substrate must occupy

*Plates 1 and 2 follow page 162.

A.

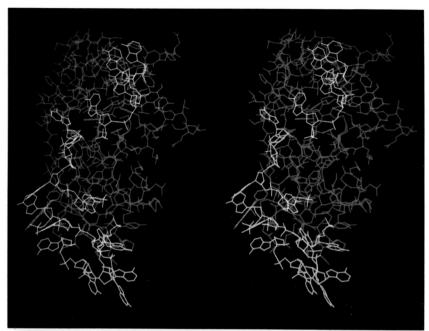

B.

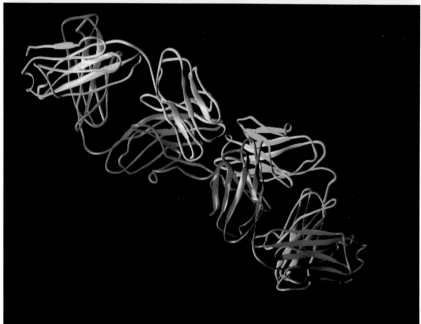

PLATE 1. The elucidation of complexes between proteins and conventional small molecules, as well as with other types of macromolecules is illustrated in (A) by the triple complex between pancreatic ribonuclease (in blue), four oligomers (in yellow) of d(pA)$_4$, and two molecules of a drug, propidium iodide (in red). The first two components were co-crystallized and their structure solved, while the locations and orientations of the drug molecules were determined by difference Fourier methods from crystals of the binary complex suffused with the drug. In (B) is shown the backbone structure of an idiotypic-antiidiotypic Fab complex, where the Abl was against feline infectious peritonitis virus. Complexes of such complementary antibodies yield images of the complementarity determining regions that may lead to new classes of drugs based on peptidomemetic designs.

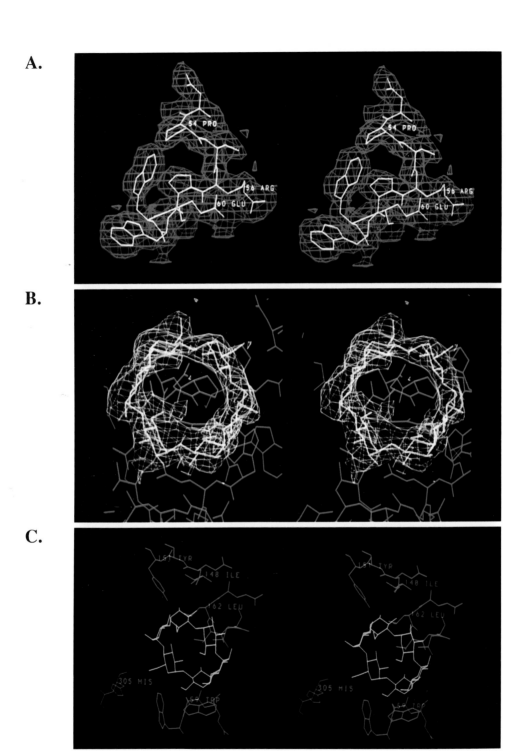

PLATE 2. In (A) is shown a computer graphics display of electron density (in blue) calculated from X-ray diffraction data, with a segment of the corresponding model for pig pancreatic α-amylase superimposed. In (B) is shown the electron density (in purple) corresponding to a molecule that has been bound at the active site of α-amylase (in blue). A skeletal model of the bound cyclodextrin is superimposed (in yellow). In (C) is shown the binding of a second molecule of α-cyclodextrin to α-amylase, which binds three in total, illustrating the molecular interactions by which the ring of sugars is bound.

a volume of very complex shape at the active site of the enzyme, a consequence of the disposition and orientation of amino acid side chains extending from the polypeptide, as well as by main chain atoms themselves. This "lock and key" relationship is extremely demanding in that supplementary or improperly placed atoms on the substrate, like an additional tooth on a key, immediately excludes its binding at the active site. A substrate too small leaves the active site exposed to invasion by water, imperfect bonding distances, and multiple nonproductive orientations. These, in the case of "induced fit" mechanisms, fail to trigger the optimal conformational change required of the macromolecule for catalysis to occur. The fit must be exactly right, for that is what the evolution of the enzyme structure over hundreds of centuries has come to demand.

In addition to perfect fit in a physical sense, the match must be exactly right in a chemical sense as well if the strength, or binding energy, of the interaction is to be maximized. This binding energy is of course the crucial determinant in terms of substrate affinity. The dissociation constant for the enzyme-substrate complex is roughly an exponential function of that energy.

The negative binding energy is provided by the chemical interactions that are formed between the substrate and the amino acids that compose the surface of the active site. These are in general electrostatic interactions which include both salt bridge and hydrogen bonds, and in those cases of metalloenzymes may involve more complex coordination bonds. Hydrophobic interactions may also play some role, though many hydrophobic interactions must in general contribute to constitute a meaningful force of attraction. The bonds between substrate and enzyme are almost always comprised of what we refer to as ensembles of weak secondary interactions. Only infrequently do they involve true covalent bond formation and a covalent intermediate between enzyme and substrate.

Clearly, even if the fit is superb, juxtaposition of a negative charge on the enzyme surface with a negative charge on that of the substrate, particularly at the active site where there is usually no water and the bulk dielectric constant is about one, would immediately disrupt enzyme-substrate complex formation. The same would of course be true for two positive charges, two hydrogen bond donors, or any two repulsive species. Thus, not only must the shape of the substrate be acceptable, but its inclinations as well; it must have charged groups and hydrogen bond donors and acceptors at the right positions. Not only must they be at the correct locations on the substrate molecules, but they must have exactly the right geometry as well in order to form linear hydrogen bonds or salt bridges of maximum strength. All of these requirements are rigorously imposed by the enzyme on the chemical structure of the substrate molecule.

While the discussion has been confined, so far, to the requirements for substrate binding to the active site of an enzyme, the identical arguments hold as well for the affinity of effector molecules for regulatory sites on proteins and to the binding of conventional small molecules such as antibiotics, to specific sequences of bases on a nucleic acid. The binding interactions are both highly cooperative and strictly defined, and this is the basis for many of the important regulatory mechanisms by which living systems maintain their biochemical integrity and insure their metabolic security.

It follows from the "lock and key" hypothesis of Fischer that there must be instances where the inviolable specificity of enzyme substrate interaction breaks down and other molecules that closely resemble the natural substrate or effector molecules bind in their place. Such competition is now, of course, well known, as are a host of other modes of interaction between enzymes and various small molecules or metabolites that we refer to as allosteric interactions. Cells in fact even utilize such mechanisms to regulate the rates at which certain metabolic pathways proceed. We have the classic examples of feedback inhibition in the synthesis of threonine and the regulation of expression of genes, such as that for β-galactaosidase, by the products of their activity.

We have also come to appreciate that many natural toxins are in fact inhibitors of crucial enzymes and act by mimicking true substrates, and that many natural processes such as fermentation can be controlled by addition of appropriate inhibitory compounds. Most importantly, beginning with Erlich,[7] it has been demonstrated that what we commonly know as drugs express their physiological activity by acting as specific inhibitors of target enzymes. They compete for active or regulatory sites on particular enzymes because of their close similarity to an intended substrate or effector and, because of their nonperforming assets, bring critical biochemical processes to a halt.

III. STRUCTURE-BASED DESIGN OF CONVENTIONAL DRUGS

It is not a recent idea that if one knew, in accurate atomic detail, the structure of the active site, or some other important regions of the enzyme, that one might then be in a position to manufacture conventional small molecules that would bind to those sites and interfere with the activity of the biochemical catalyst (see Reference 26 for a discussion). Such molecules are properly called drugs. This is exactly the concept underpinning what we now commonly refer to as rational or directed drug design, to fabricate bioactive molecules in the chemical laboratory based on a detailed knowledge of the geometry and chemical character of the active site, its hydrophobic valleys and electrostatic landmarks, the chemical hooks and eyes that endow it with its unique biochemical properties.

Plate 1A illustrates an example of a drug molecule, propidium iodide, bound to an enzyme whose activity it affects,[15,16] pancreatic ribonuclease. Two representations of the interaction between drug and protein are shown. A host of other visualization forms and techniques, one of which is seen in Plate 2, are available to the computer graphics analyst to assist in the identification of important contacts and the delineation of a comprehensive binding mechanism.

This idea of rational drug design based on a clear picture of the target's structure, is not particularly novel or clever; what is exciting and intriguing is that we are now in a position to actually apply the procedure in practice (see Appelt et al.[1] and Baldwin et al.[3] for fine examples). We now have the analytical tools at our disposal to determine at the atomic level of detail the structures of enzymes and their active and regulatory sites, and to exploit this knowledge in the design of new and potentially useful drugs. Through a careful delineation of active site architecture, lead compounds emerge directly from molecular structure, not from vast arrays and sequences of low probability trials. In principle, the long traditional process of drug discovery and optimization can be dramatically shortened. In addition, because drug design is based on examination of a known macromolecular structure, new and otherwise improbable possibilities come into view that would not likely have been previously considered.

In the application of structure-based rational drug design (see for example, Kuyper et al.[13]), analogs of natural substrates are modeled by computer graphics to the active site of the enzyme. The structure of the protein is generally well established at the atomic level through the use of X-ray diffraction techniques. Those analogs that show particular promise in such simulated binding studies can then be synthesized and examined for their physiological activity. Frequently, the degree of physical and chemical complementarity can be precisely quantitated by considering the overall energy of the interaction of the ligand with the macromolecule.

The enzyme α-amylase, synthesized in the pancreas and active in the small intestine, is a target for drug design. By regulating its activity in humans, dental carries may be minimized and blood glucose levels moderated following starch ingestion. This may have benefits in dentistry in the first case and in the treatment of diabetes in the second. Plate 2 illustrates the level of detail that can be obtained using X-ray diffraction methods, and illustrates the structural basis that may be obtained for the design of useful drugs.

The approach of rational drug design based on knowledge of target macromolecules is made all the more powerful by the ability to actually visualize the complex that is formed between the enzyme and the synthetic drug. This is, like the original structure determination of the native macromolecule, dependent on successful application of the X-ray diffraction technique to reveal the complex structure. In so doing, there is the added bonus that one visualizes not only the structure of the bound molecule and its intimate relationship with the target, but one sees as well those changes that occurred in the enzyme through conformational changes brought about by its interaction with the drug.

Understanding the conformational alterations associated with drug binding are of crucial importance in drug design and optimization. The interaction between drug and enzyme is a mutually

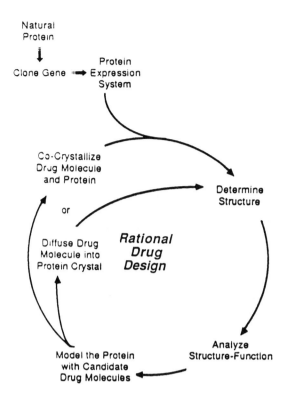

FIGURE 1.   The cyclic pathway of rational drug design showing the analytic-synthetic relationship and the role of X-ray crystallography.

cooperative process, and it is essential to understand the dynamics of the macromolecular response and interaction if one is to correlate the ultimate biochemical effect with the properties of the drug. Thus, in practice, the application of rational design to drug discovery and development requires not only determination of the native enzyme structure, but many cycles of drug synthesis, review by X-ray crystallography of the consequences, further synthesis and review, and ultimately optimization of the stereochemical parameters that govern the event. This is now possible in some favorable cases using current technology. It is widely expected that it will be even more broadly applicable as our facility with the techniques improves. The drug design process based on known macromolecular structure is illustrated in Figure 1.

IV. THE DEVELOPMENT OF PROTEIN DRUGS

To this point we have for the most part considered drugs as conventional, low-molecular-weight compounds that can be synthesized, granted with some difficulty, in the traditional chemical laboratory using well-understood organic synthetic techniques. Macromolecular X-ray crystallographers commonly refer to these simply as "small molecules", somewhat disdainfully at that. These are what the average man thinks of as drugs. They come in a bottle or vial, they are swallowed or injected, they are expensive, and they are generally quite specific in their effects.

More recently, we have come to understand that there exists another class of drugs, though we have in fact utilized such molecules for several generations; these are the peptides and proteins manufactured by natural sources and used internally to promote, inhibit, or otherwise regulate important biochemical activities. These range from pituitary and other glandular polypeptide hormones to immunologically active proteins such as TPA, erythropoietin, interleukin and interferon, to synthetic insulin, to bioactive peptides of enormous promise, such as endothelin. In a sense, the power of most of our traditional drugs pales in comparison with these almost hyper-potent natural products.

The fundamental reason that this newer class of drugs, these proteinaceous drugs, are so impressive is that they do not exert their power by forming simple one-to-one stoichiometric complexes with target enzymes. They act by forming complexes with specific receptor molecules, generally on the surfaces of target cells. When the appropriate complex is formed, a whole cascade of subsequent responses is triggered and subsequently amplified within the cell, then successively throughout an entire tissue. These polypeptide drugs function as signals and switches to ignite a vast series of cellular events that would otherwise require a virtual bank of conventional drugs to attain, if it were even possible at all.

Because of their inherent power, the extent and scope of their response, and the enormous potential and promise of their utilization, they have become the focus of great interest to the pharmacological community. They provide an unsurpassed research tool for cell biologists and pharmacologists and the most promising source of our drugs for tomorrow. The question then is what restrains their exploitation, what inhibits the fulfillment of their promise.

These proteinaceous effectors are, in general, extraordinarily expensive to isolate from natural sources, if they can even be purified at all in significant amounts. They occur at very low abundance in organisms because of their great inherent potency. They are generally very labile and sensitive to their environment once purified, are difficult to store for long periods, denature and alter their structure, elicit immune responses with use, survive for only a short time in the body before they are cleared or degraded, often cannot cross transport barriers, and frequently present problems in administration. Clearly we would like to avail ourselves of their properties, but ideally we would like to manufacture them on a commercially viable scale and in a form that maintains quality, potency, structure and has, if possible, improvements over the naturally available molecules.

Again the idea of directed drug design has application, but in a somewhat different way than that described previously. The objective in this case is to improve on the natural product, not to find and fabricate a conventional molecule that will bind to it (see Oxender and Fox[25] and McPherson[21]). The synthetic capability is again present, but is of course different than in the case already described. Instead of a conventional laboratory based on organic chemical synthesis, we have the modern genetic engineering laboratory based on recombinant DNA techniques. Instead of synthesizing variants of drugs by introducing and varying chemical groups through the use of organic chemistry, we reshape and chemically modify target proteins and peptides by altering the base sequence of their genes, and hence their amino acid sequences. Because structure is a direct function of the order of the amino acids, we alter their form. With a few changes in the broader genetic structure we can overexpress and amplify the protein product a thousandfold. What more is required?

What is needed is a way of directly visualizing the results of our recombinant DNA handiwork to see directly the structural consequences of the genetic changes. If our goal is to stabilize the protein to make it more resistant to hydrolysis or heat denaturation, we must evaluate directly the structural changes produced by, for example, introduction of a new disulfide bridge. If we wish to increase solubility or the number of glycosylation sites, then the sites for such changes must be rigorously determined in advance and the effects determined subsequently. If an abbreviated protein or peptide is sought that might be less antigenic, cross barriers more easily, or exhibit a more robust character, then we must be able to directly visualize the natural architecture of the molecule, consider how it might be improved upon, and finally to measure the success or failure of our attempt in structural terms.

Again the analytical technique of X-ray diffraction as applied to single crystals of the proteins provides the necessary description for intelligent synthesis. The method serves as a visual guide to the genetic engineer, allowing him to examine the protein of interest, determine targets for directed change, and finally permit detailed review of the changes introduced. An image of protein with its associated electron density as it appears to the drug investigator on a computer graphics screen is shown in Plates 2A and 2B.

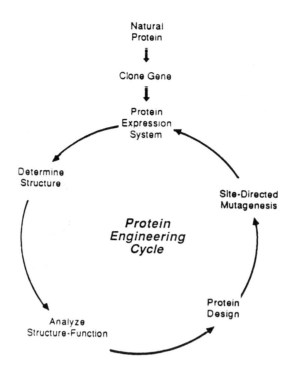

FIGURE 2. This diagram shows the integration of recombinant DNA technology and X-ray diffraction analysis in the process of genetic engineering a protein drug such as interferon or interleukin.

Once a protein or peptide structure has been determined in a particular crystalline environment, visualization of variant, recombinant proteins can be readily achieved with relatively little effort and expense through application of the difference Fourier technique. Again, as in the case described above for enzyme substrate complexes, one sees not only the exchange of one amino acid side chain for another, but any associated structural alterations or localized conformational changes induced by the modification. Thus, the genetic engineer intent on exploiting the pharmacological value of a particular protein or peptide can perform a succession of experiments, altering one or more amino acids at a time, viewing the results each time, until his desired result is achieved. Much as the synthetic and analytic chemists work in tandem, alternating cycles, the recombinant DNA practitioner and the X-ray crystallographer do the same, but on a macromolecular level. The protein engineering cycle as it is applied to drug development is illustrated in Figure 2.

V. ANTIBODY-DIRECTED DRUG DESIGN

The most recent approach to utilizing the properties of macromolecules for directed drug design, and certainly the most novel, is that employing the recognition feature of monoclonal antibodies. For targets that are fundamentally inaccessible to most biochemical techniques in terms of isolation and purification, and hence direct investigation by X-ray crystallography, targets at such low abundance that purification is impracticable, or for targets of such lability that removal from their native environment is intolerable, methods based on monoclonal antibodies may well represent the only effective tool.[2,10,27] Receptors embedded in membranes, viruses, and enzymes of a particularly elusive character all fall into this category.

The approach is based on the idea that when a monoclonal antibody, such as that one seen in Figure 3,[11] is raised against an antigen, a stereochemically negative image of the antigen is embedded in the structure and chemistry of the antibody's hypervariable region, a relatively small portion of the antibody comprised of peptides contributed in part by both the heavy and light chains

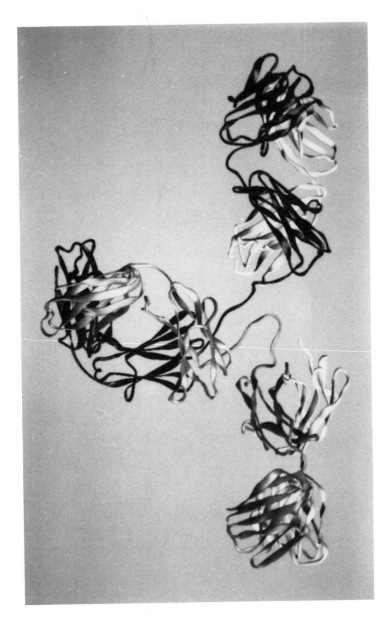

FIGURE 3. Backbone diagram of an intact monoclonal antibody against canine lymphoma showing the inherent asymmetry and flexibility of the molecule. Antibody based drug design using the intricate stereospecificity of the complementarity determining region provides a new and unique approach to discovery of new therapeutic agents.

of the immunoglobulin. The antigen could be the active site of an enzyme, a viral antigen, or the recognition or combining site of a receptor. The monoclonal antibody, called an Ab-1, that has the desired properties can be selected for its desired immunogenic properties through appropriate assays or on the basis of its observable immunological responses.

From the antibody (Ab-1) containing the mirror image of the target macromolecule a positive image can be extracted. This is accomplished by raising a second antibody (Ab-2) against the hypervariable region of the Ab-1. By virtue of its stereocomplementarity to the combining region of the Ab-1, a feature it shares with the original target antigen, it must therefore possess common structural and electrostatic characteristics. Though it may not be perfectly homologous with the target antigen, it must incorporate at least the essential physical disposition of key chemical groups and a similar distribution of salt bridge and hydrogen bonding elements.

In principle, the Ab-1 could serve as an antagonist to the receptor or act as a molecular sponge to clear, for example, virus from a system. Such antibody drugs have been shown to be of only marginal value at best and have not provided a great number of useful therapeutic agents. The Ab-2 could also be useful, for example as a vaccine to induce an immune response in an animal against the original antigen. Often this has advantages over the antigen itself as the immunizing agent and such vaccines have been at least partially successful. Other direct applications for Ab-2 molecules could be enumerated, but again for the same reasons as for the Ab-1, the successes are limited. Antibodies, at least in their present manifestations and formulations, simply do not make very good therapeutic agents.

A much more useful and effective drug could in principle be made if some small molecule, perhaps a short polypeptide, even one using nonstandard amino acids or unusual linkages, could be made that exhibited the structural and chemical features of the hypervariable region, or combining site, of an immunoglobulin. Because the hypervariable region is of rather limited extent in terms of amino acid content and spatial distribution, such a small molecule is well within the realm of possibility. Such molecules are called peptidomemetics, and there are now several biotechnology companies pursuing their development.

The crucial requirement in the construction of a useful peptidomemetic drug is first to understand as accurately as possible the structural and chemical features of the hypervariable combining site. This includes its amino acid sequence, the degree of homology in the case of an Ab-2 with the target antigen sequence, and the spatial distribution of the amino acid side chains. In practice, this means knowing the detailed three-dimensional structure of the antibody combining site.

The exact amino acid sequence of hypervariable regions can now be accurately and quickly determined by polymerase chain reaction (PCR) amplification of the gene region coding for the hypervariable chains followed by conventional DNA sequencing. Such direct sequencing can now be carried out in the course of a week. The spatial information is considerably more problematical. There are three approaches that may be used to some avail. Certainly the most accurate, detailed, and comprehensive method is to determine the structure of the hypervariable region on the antibody (or Fab fragment) of interest using X-ray crystallographic techniques. This is without doubt the preferred method because one obtains an unequivocal description at near atomic level. All the information one needs is abundantly provided by an X-ray structure determination.

The difficulty with X-ray crystallography in this case is that although one desires only the structure of the hypervariable region, in order to obtain it one must solve the structure of the entire protein of which it is a part, generally an entire Fab fragment but conceivably an entire antibody molecule (see Ban et al.[4] or Larson et al.[14] for examples). This is a time-consuming and usually demanding task, even with the availability of rather powerful computing tools such as molecular replacement. In addition, this is in spite of the fact that we already essentially know the structure of all of the rest of the molecule except the hypervariable region. Thus, even if the Fab or antibody of interest can be crystallized, a difficult problem in itself, the structure determination remains far from routine.

Two-dimensional nuclear magnetic resonance (NMR) techniques can be of some value in deducing structural information if the polypeptide or antigen is relatively small. In the case of the 22-amino acid peptide endothelin, an important vasoconstrictor, the NMR data yield many important clues, indeed a model structure has been advanced, as to the native three-dimensional structure.

In addition to X-ray diffraction and NMR, which are direct techniques, methods based on the calculation of predicted three-dimensional structures of molecules in the range of 3 to 50 amino acids based on energy considerations are under rapid development. These approaches use what are commonly called molecular dynamics and energy minimization equations to specify the most probable conformation of polypeptides and small proteins. Often when combined with information from other sources, such as X-ray or NMR studies, they have been demonstrated to be quite useful.

The extent of their power and the accuracy of their predictive capability remain to be seen, however, when standing alone.

Actual design and synthesis of a peptidomemetic drug from known antibody structure still has not been achieved, but it is becoming increasingly likely that breakthroughs are simply a matter of time. This rather involved and complex approach to drug discovery and development is in its early stages even so far as rational drug design is concerned, but it shows great promise in terms of harnessing the enormous power of the mammalian immune system.

VI. THE ROLE OF X-RAY CRYSTALLOGRAPHY IN DRUG DESIGN

It is evident from the above discussion that structure determination by the use of X-ray crystallography, or diffraction analysis, is an essential element in all of the three forms of rational drug design. In the case of ligand design and evaluation, it is necessary first to determine the detailed architecture of the drug binding site and delineate the amino acid residues and their dispositions that determine the macromolecule-drug interaction. Following that, the method is essential for evaluating the successes and failures of each synthetic attempt by visualizing directly the complex and the fundamental chemical interactions involved.

Figures 4, 5, and 6 are examples of protein crystals, the most essential components of the entire process of X-ray diffraction analysis. Growth of such crystals of target enzymes is now a major goal of the biotechnology revolution, precisely because they make possible the approaches described here.

In the cases where the directed drug is based on the redesign of a naturally occurring peptide or protein, the method of X-ray diffraction is again crucial. Crystallography must be used to provide the underlying native protein structure that will be the target for specific change, bring into focus those regions of the macromolecule that can tolerate change in a structural sense, or provide grounds for desired changes through recombinant DNA techniques. Subsequently, crystallography is the tool that allows the genetic engineer to examine the altered polypeptides or protein and see directly the structural consequences of the changes he has introduced. By sequentially examining the molecule, changing its sequence, and then reviewing visually again, ultimately the slow course of natural evolution can be accelerated by orders of magnitude.

Finally, in the application of rational drug design based on the use of antibodies, X-ray crystallography becomes again an almost essential instrument for mapping the structural and chemical geography of the hypervariable combining sites of the immunoglobulins on which drug development is directly dependent. Only through an accurate description of the natural structure can peptidomemetic substitutes be designed and fabricated in the chemistry laboratory. As with the other forms of directed drug design, X-ray crystallography is the eye and ear, the senses, that are linchpins for success.

It is useful for anyone interested in rational drug design to have at least a passing understanding of the technique of X-ray crystallography and some understanding of what it is able to provide the pharmacologist, the genetic engineer, or the immunologist. While it is well beyond the scope of this chapter to provide a fundamental physical and mathematical basis for the method, there is one aspect that is of particular value and interest to the topic of rational drug design, and is deserving of some attention here. This is the procedure known in X-ray crystallography as the difference Fourier technique. For more complete descriptions of the X-ray crystallographic approach it is recommended the reader see McPherson,[18,20,22–24] Blundel and Johnson,[5] Glusker and Trueblood,[9] and Pines.[8]

VII. THE DIFFERENCE FOURIER TECHNIQUE

The difference Fourier technique is not a method for determining the structures of macromolecules. The difference Fourier procedure is a method of comparing similar structures when both can be obtained in essentially the same crystalline form (for a detailed discussion, see Reference

20). This clearly is of enormous value to the drug designer of any persuasion, for it permits one to compare the differences between an enzyme or receptor in its native state with its structure when complexed to ligand, the drug of interest. It permits the genetic engineer to compare his recombinant protein to the native, or to other recombinants, and it allows the immunologist to see the subtle changes that occur in the hypervariable regions of immunoglobulins as the antigen is varied. It is really the key crystallographic technique that drives the entire field of rational drug design.

Most important to the application of directed drug design, the difference Fourier procedure is relatively simple in concept, frequently rapid and straightforward to perform, and in most cases requires a minimum of time, effort, and other precious resources. It can be readily made a useful adjunct to the search for useful drugs. It is appropriate, therefore, that the remainder of this chapter be devoted to a brief description of the basis, application, and possibilities of the difference Fourier procedure.

The structures of protein and nucleic acid molecules are extremely interesting in themselves, each being a representative member of some architectural class of macromolecule shaped by evolutionary time and process toward the optimal completion of a specific cellular or metabolic task. They are, nevertheless, static objects. Because the catalytic functions they perform are dynamic events involving the interaction of the macromolecule with substrates, effectors, inhibitors, and other cellular components, we are constantly searching for techniques that will allow us to visualize the macromolecule in some intermediate states of the catalytic or physiological event.

By observing different static images representative of different points in the course of catalysis or functional activity, we stand a far better chance of working out the actual sequence of microevents that reflect the molecule's unique structure and chemical capability. In general, the most important images, other than that of the native macromolecule itself, are those of the macromolecule intimately engaged with its substrate, its ligand, or with another macromolecule.

By observing directly the distribution of bonds and juxtaposition of chemical groups, the structural and steric complementation, and the alterations from the native structure required by the event, a great deal can be learned of the purpose of the individual amino acid residues in the structure and what catalytic or binding responsibility governs their placement. From such images the molecular scheme of things, in precise chemical terms, begins to emerge and the intentions and principles underlying function are revealed.

The principal crystallographic tool currently employed to directly visualize complexes between macromolecules and ligands is the difference Fourier synthesis. The great attraction and advantage of difference Fourier experiments is that much of the data required is already in hand when the experiment begins. This is in the form of a native crystal X-ray diffraction data set obtained in the course of the structure solution and a highly accurate set of corresponding phases. Two examples of X-ray diffraction patterns from protein crystals showing the distribution of intensities unique to each specimen are shown in Figures 7 and 8.

These latter components, the phases, are obtained by back-calculation of the Fourier transform, or diffraction pattern, from the refined protein structure (see McPherson[18]). Such phases are by every measure far more accurate than the phases that were experimentally estimated and used to solve the structure in the first place. The only missing component is a complete set of diffraction intensities collected from a crystal of the protein complexed with the ligand under investigation. If such a set of intensities can be obtained, then it is a simple matter to subtract the Fourier transform of the native crystals, its diffraction pattern, from the Fourier transform of the complex crystals and employ the resultant differences as the coefficients in a Fourier synthesis. Because the phases of these differences are well approximated by the computed phases from the native structure, the calculation is straightforward. No multiple isomorphous replacement technique need be applied to find new phases, and the entire experiment is, by crystallographic standards, relatively simple. The only problem in conducting such an experiment is obtaining a crystal of the macromolecule complexed to the ligand of interest. It is at this point that the otherwise perverse and difficult character of protein crystals becomes an advantage.

FIGURE 4A.

FIGURE 4B.

FIGURE 4.   Some typical protein crystals are seen here. In (A) is the enzyme peroxidase from the common peanut, in (B) fructose, 1,6-diphosphatase from chicken liver, in (C) a protease called bromelin from pineapple, and in (D) a crystal of the Fab fragment of an anti-idiotypic antibody.

FIGURE 4C.

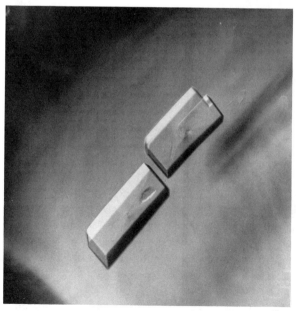

FIGURE 4D.

Protein crystals (for methods utilized to grow protein crystals, see McPherson[18,23,24]) and indeed all macromolecule crystals, are composed of approximately 50% solvent, though this amount may vary anywhere from 30 to 90% depending on the particular macromolecule. The protein or nucleic acid occupies the remaining volume so that the entire crystal is in many ways only an ordered gel

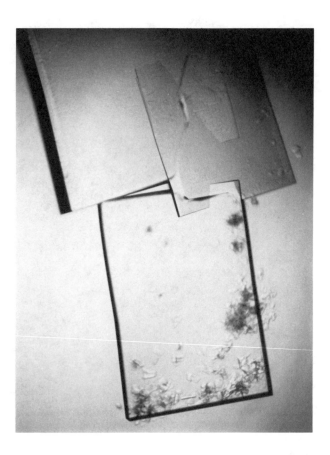

FIGURE 5.   Large crystals of the plant seed protein canavalin from the jack bean. The protein of $M_r = 150,000$ readily forms crystals like these that are greater than 2 mm in their longest dimension.

with vast interstitial spaces through which solvent and other small molecules may freely diffuse. This is shown clearly by electron micrographs of microcrystals and from X-ray-derived packing diagrams.

Shown in Figure 9 is an electron micrograph of a negatively stained microcrystal of the enzyme α-amylase from pig pancreas.[17] Apparent in the photo is the vast network of interstices and channels that permeate the crystal. Through these channels drugs and other ligands can be suffused into protein crystals where, in general, they will bind to the active site, or an effector site, of the enzyme.

Protein crystals are always grown from solution since complete hydration is essential for the maintenance of structure and, therefore, they are always, even during X-ray data collection, bathed in a liquid medium that we refer to as the mother liquor. The open channels and solvent cavities that characterize macromolecular crystals are primarily responsible for the severely limited resolution of most diffraction patterns which serve as accurate measures of inherent order. They also cause the crystals to be mechanically fragile and somewhat difficult to manipulate. The tight solvent content, however, proves invaluable for the formation of crystalline complexes between macromolecules and their relevant ligands.

There are basically two approaches for forming crystals of a protein ligand complex. One technique is to simply mix the protein with the ligand in solution at a ratio which ensures that essentially all protein molecules will have ligand bound to it. This will be a function of the equilibrium constant, or binding affinity, of the protein for the ligand. In the presence of ligand crystallization conditions are established by conventional techniques such as vapor diffusion or dialysis against highly concentrated salt solutions or polyethylene glycol solutions. If good fortune prevails, the protein-ligand complex will crystallize in a manner identical to the native protein alone. The crystal will have the identical unit cell dimensions and symmetry properties; that is, it will crystallize isomorphously with the native crystal. If this occurs, the X-ray data can be collected in a straightforward manner from the complex crystal, a difference Fourier image can be calculated, and one can visualize directly the ligand as it is bound to the protein.

Any parts of the protein or individual amino acids that have altered position during the binding process will ideally appear as pairs of closely associated negative and positive areas. The negative density represents the position occupied in the native structure, and the positive area that occupied in the complex. Thus, in carrying out a difference Fourier experiment, one sees not only how the ligand fits to the surface or active site of the protein, but how the protein alters its local conformation to accommodate and optimize its interactions with the ligand.

Unfortunately, difference Fourier studies predicated on successful "co-crystallization" as described above are seldom successful. This is because the binding of the ligand to the protein frequently perturbs either the conformation of the protein or some of those chemical groups on the protein's surface essential for forming the lattice contacts required for isomorphous crystallization. As a consequence, even very minor alterations often result in the macromolecule failing to crystallize at all, or as is more often the case, it crystallizes in a unit cell that is different from that found in the native crystal. That is, it crystallizes nonisomorphously with the native crystal. Because the crystal lattices are different, the diffraction patterns from the two crystals are not comparable in terms of either the amplitudes or the phases.

A more common experimental procedure utilizes the porous character of macromolecular crystals and their internal channels which are occupied by aqueous mother liquor. In these "diffusion" experiments, the ligand of interest is added directly to the mother liquor of the protein crystal, generally in incremental amounts over an extended period of time, until the concentration of ligand in the solvent reaches a concentration sufficient to saturate the binding site or sites. Such exposures may vary from, in some cases, no more than a day to several months in others. Longer times may be essential when the binding site for the ligand on the macromolecule is occluded by neighboring molecules in the crystal lattice; that is, by intermolecular contacts, or when the solvent content of the crystal is particularly low.

For a number of crystalline enzymes whose structures have been solved by X-ray diffraction analysis, such obstruction has been so severe that the ligand simply cannot be made to bind even when its concentration is far in excess of saturation and very long time periods are employed. In such instances, the only recourse is to solve the structure of the macromolecule in a second crystalline unit cell that is more accommodative toward ligand binding. Fortunately, once the structure of a protein is known in one crystalline unit cell, its structure determination in a second crystal form is appreciably simplified through the use of techniques collectively referred to as "molecular replacement methods".

Data are collected from the macromolecule-ligand complex crystals as for any other crystal generally using a multiwire detector with a high intensity source system. The complex crystal data set must be placed on a common numerical scale with the native crystal data so that relative differences between individual reflections can be formed by subtraction of the two.

If an unusually large degree of change is observed in the diffraction pattern of the complex crystal when compared with native when a relatively small ligand is diffused into a large protein, then one must assume that a considerable degree of conformational change has occurred in the protein as a consequence of binding. Careful analysis of the resultant difference electron density map, however, can allow delineation, in some cases very precisely, of the conformational changes that occurred.

Since one generally knows the structure of the ligand that has been diffused into the crystals, the problem is really reduced to one of fitting a small familiar structure into an irregular, but hopefully suggestive, mass of difference electron density. In the best cases, those produced with very accurate, high-resolution (2.0 Å) native protein phases from well-refined structures and of ligands that

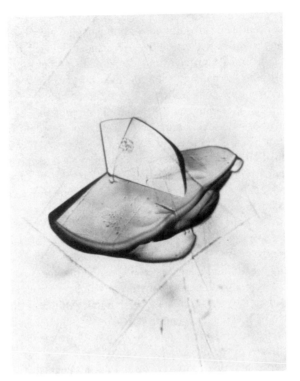

FIGURE 6A.

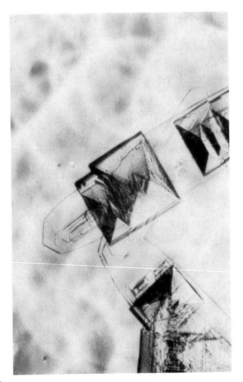

FIGURE 6B.

FIGURE 6. Shown here are three examples of protein crystals that grow as twins or as disordered aggregates, a very common and discouraging occurrence in macromolecular X-ray crystallography and the focus of much current research. In (a) is a cluster of crystals of an intact monoclonal antibody against canine lymphoma, in (b) two different forms of crystals of pancreatic α-amylase growing from one another, and in (c) the hypertensive polypeptide endothelin.

FIGURE 6C.

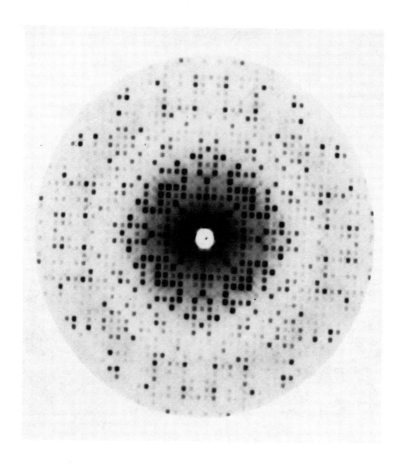

FIGURE 7. An X-ray diffraction pattern from a tetragonal crystal of the enzyme lactate dehydrogenase showing the fourfold symmetry of the crystal.

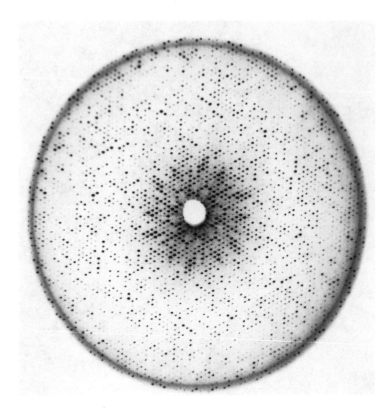

FIGURE 8. An X-ray diffraction photo from a hexagonal crystal of the plant protein concanavalin B showing its characteristic sixfold symmetry.

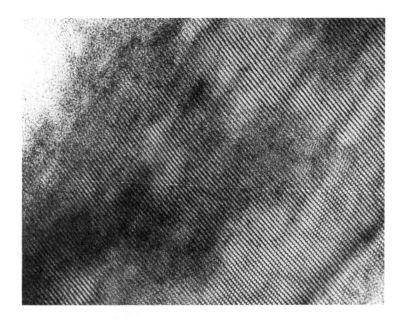

FIGURE 9. An electron micrograph of a microcrystal of pig pancreatic α-amylase negatively stained with uranyl acetate to reveal the cavities and paths that wind between the protein molecules as they are packed in the crystalline lattice.

produce little nonisomorphism, the individual chemical groups comprising the ligand may be readily fixed in the map.

The range of possible inhibitors, substrates, substrate analogs, drugs, coenzymes, ions, and other effectors that can be studied in the crystalline state using the difference Fourier approach is almost limitless. The method is most ideal for investigating ligands having a large complement of electrons concentrated in a small volume, such as metal ions, large anions, or molecules containing ring systems. Because the magnitude of difference electron density peaks will be proportional to the electron density of the ligand, such ligands can often be easily located even in relatively low-resolution difference Fourier maps, even when there is a substantial degree of nonisomorphism.

REFERENCES

1. **Appelt, K., Bacquet, R.J., Bartlett, C.A., Booth, C.L.J., Freer, S.T., Fuhry, M.A.M., Gehring, M.R., Herrmann, S.M., Howland, E.F., Janson, C.A., Jones, T.R., Kan, C.-C., Kathardekar, V., Lewis, K.K., Marzoni, G.P., Matthews, D.A., Mohr, C., Moomaw, E.W., Morse, C.A., Oatley, S.J., Ogden, R.C., Reddy, M.R., Reich, S.H., Schoettlin, W.S., Smith, W.W., Varney, M.D., Villafranca, J.E., Ward, R.W., Webber, S., Webber, S.E., Welsh, K.M., and White, J.,** *J. Med. Chem.,* 34(7), 1925, 1991.
2. **Linthicum, D.S. and Farid, N.R., Eds.,** *Anti-Idiotypes, Receptors and Molecular Mimicry,* Springer, New York, 1988.
3. **Baldwin, J.J., Ponticello, G.S., Anderson, P.S., Christy, M.E., Murcko, M.A., Randall, W.C., Schwam, H., Sugrue, M.F., Springer, J.P., Gautheron, P., Grove, J., Mallorga, P., Viader, M.P., McKeever, B.M., and Navia, M.A.,** *J. Med. Chem.,* 32, 2510, 1989.
4. **Ban, N., Escobar, C., Day, J., Greenwood, A., Larson, S., and McPherson, A.,** *J. Mol. Biol.,* 222, 445, 1991.
5. **Blundell, T.L. and Johnson, L.N.,** *Protein Crystallography,* Academic Press, New York, 1976.
6. **Dressler, D. and Potter, H.,** *Discovering Enzymes,* Scientific American Library, New York, 1991.
7. **Erlich, P.,** *Ber. Dtsch. Chem. Ges.,* 42, 17, 1909.
8. **Pines, M., Ed.,** *Finding the Critical Shapes,* Howard Hughes Medical Inst., Bethesda, MD, 1990.
9. **Glusker, J.P. and Trueblood, K.N.,** *Crystal Structure Analysis: A Primer,* Oxford University Press, Oxford, 1972.
10. **Goodman, M.,** *Biopolymers,* 24, 137, 1987.
11. **Harris, L.J., Larson, S.B., Hasel, K.W. et al.,** *Nature* 360, 369, 1992.
12. **Jurnak, F. and McPherson, A., Eds.,** *Biological Macromolecules and Assemblies,* Vol. 3, *The Active Sites of Enzymes,* John Wiley & Sons, New York, 1987.
13. **Kuyper, L.F., Roth, B., Baccanari, D.P., Ferone, R., Beddell, C.R., Champness, J.N., Stammers, D.K., Dann, J.G., Norrington, F.E., Baker, D.J., and Goodford, P.J.,** *J. Med. Chem.,* 28, 303, 1985.
14. **Larson, S., Day, J., Greenwood, A., Skaletsky, E., and McPherson, A.,** *J. Mol. Biol.,* 222, 17, 1991.
15. **McGrath, M.E., Cascio, D., Williams, R., Johnson, D., Greene, M., and McPherson, A.,** *Mol. Pharmacol.,* 32(5), 600, 1987.
16. **McGrath, M., Koszelak, S., Brayer, G., Williams, R., Cascio, D., Greene, M., Axelrod, H., and McPherson, A.,** in *From Proteins to Ribosomes, Structure and Expression,* Vol. 1, Sarma, M.H. and Sarma, R.H., Eds., Adenine Press, Guiderland, NY, 1988, 181.
17. **McPherson, A. and Rich, A.,** *J. Ultrastruct. Res.,* 44, 75, 1973.
18. **McPherson, A.,** *Preparation and Analysis of Protein Crystals,* John Wiley & Sons, New York, 1982.
19. **McPherson, A., Brayer, G., Cascio, D., and Williams, R.,** *Science,* 232, 765, 1986.
20. **McPherson, A.,** in *Crystallography Reviews,* Vol. 1, Moore, M., Ed., 1987, 191.
21. **McPherson, A.,** *Proceedings of the Conference on Biotechnology, Biology Pesticides and Novel Plant-Pest Resistance for Insect Pest Management,* July 18–20, Cornell University Press, 1988, 115.
22. **McPherson, A.,** *Sci. Am.,* Vol. 260 (No. 3), March, 62, 1989.
23. **McPherson, A.,** *Eur. J. Biochem.,* 1, 1, 1990.
24. **McPherson, A.,** *CRC Review: Crystallization of Membrane Proteins,* Michel, H., Ed., CRC Press, Boca Raton, FL, 1991, chap. 1.
25. **Oxender, D.L. and Fox, C.F.,** *Protein Engineering,* Alan R. Liss, New York, 1987.
26. **Walsh, C.T.,** *Trends Biochem. Sci.,* 8, 254, 1988.
27. **Wolff, M.E. and McPherson, A.,** *Nature,* 345, 365, 1990.

Chapter 7

BIOCOMPUTATIONAL APPROACHES IN PROTEIN-BASED DRUG DESIGN

Katherine V. Prammer, Matthew Wiener, and Thomas Kieber-Emmons

CONTENTS

I. INTRODUCTION

The goal in drug design is to produce a compound that will interact optimally with a unique endogenous molecule and which may also have enhanced metabolic stability, oral bioactivity, and decreased side effects. Biocomputational approaches in the rational design of bioactive proteins and peptides are driven by a general understanding of tertiary structure, electrostatic interactions, hydrogen bonding, hydrophobic interactions, desolvation effects, and cooperative motions of interacting ligands and receptors, backed up by data base searches and simulation techniques. With knowledge of the electrostatic and structural requirements to activate or inhibit a receptor, pharmacophores or mimetics can be designed or identified in data bases containing three-dimensional coordinates of small bioorganic compounds.[1]

Although observations extracted from the data bases of experimentally known structures are an extremely useful empirical guide to the design of modified proteins and peptides, current biocomputational approaches to drug design are limited by the small subset of proteins structurally characterized by X-ray crystallography, neutron diffraction, and, or, nuclear magnetic resonance (NMR) spectroscopy. The number of known protein sequences is one and a half orders of magnitude greater than the number of structurally characterized proteins. Fortunately, proteins belong to a limited number of structural families[2] whose members have very similar three-dimensional structures,[3] thus allowing comparative modeling approaches to fill the gap in the size of the two data bases.[4] Various protein design software tools have been and are being developed to evaluate secondary or tertiary structures from protein sequences and to design proteins with predetermined functions and, or, physical properties.

The procedure to predict protein structure has evolved from using probabilistic information on the preferential localization of residues, to relying heavily on protein structure data bases to define topological organization and three-dimensional relationships among proteins. The prediction of protein structure is now cast in a knowledge-based approach that identifies analogies in secondary structures, motifs, domains, or ligand interactions between a protein being modeled with homologous proteins whose structures are available.[5-8] This comparative model building or modeling by homology approach[9-13] combines knowledge-based systems involving crystallographic, NMR, and sequence data bases, experimentally determined constraints, conformational energy analysis, and molecular modeling.

This multidisciplinary approach has been used to study protein folding phenomena and protein structure-function relationships for protein engineering endeavors[1,14,15] and for protein-based and pharmacophore drug design.[16-24] There is no consensus, however, of the best approach to predict protein structures. The validity of the predicted structure depends on many factors, including the accuracy of the structures determined by crystallography and NMR, the percentage of homology in proteins used to establish the templates, the sequence alignment weights, the rules to translate alignments into geometric relationships between the template and the protein to be modeled, and the energy refinement of the resultant conformation. Nevertheless, many successful drug designs[25,26] and protein engineering projects,[27-31] without the benefit of X-ray or NMR structures, have been achieved by combining molecular biology techniques with judicious analysis of protein sequence information. New discoveries about receptor structure, coupled with rational strategies in molecular biology and biochemical experiments, promise to speed the development of new and better drugs. Presented in this chapter is a brief summary of various protein prediction strategies commonly used in computational biology, including computational approaches for predicting and energetically analyzing protein structures and for analyzing NMR data.

II. PROCEDURAL OVERVIEW OF PROTEIN-BASED DRUG DESIGN

Since Anfinsen et al. demonstrated that amino acid sequence directs protein folding,[32] much research has focused on understanding the molecular basis of the protein folding process and on extracting the inherent information in protein primary structures to predict protein tertiary structure. The naive approach to predict the tertiary structure of a protein would be to solve the protein folding problem;[33-37] start with a linear peptide chain, and let molecular dynamical or other computational chemistry calculations generate a structure.[38,39] For short peptides (<20 residues), a modification of this, the build-up method, can solve the structure by folding portions at a time, a numerical analog to certain protein folding models.[33,39-41] For longer peptides, selecting the native structure from the formidable protein conformational space, where interactions between distant residues stabilize the structure, might prove to be intractable with current approaches.[42] Another complication in *ab initio* protein folding is that folding can be dependent on external factors, such as ribosomes or enzymes.

Since no efficient methodology exists for *ab initio* structure prediction from sequence data alone, higher-order biochemical relationships are used. The challenge is to see how much information can be extracted from the protein data bases and how this information can be applied to a given protein. Computational researchers have made progress on a number of subproblems and alternative approaches to the protein folding puzzle that has enhanced structure prediction and drug design. Computational approaches to protein structure prediction include core domain and tertiary topology analysis,[43-56] amino acid packing and residue-residue contact studies,[57-67] side chain conformation studies,[68-70] mutational analysis and the contribution of single amino acids to protein stability,[71-73] multiple alignment or homology-derived techniques,[74-82] combined comparative modeling, restrained molecular dynamics techniques,[83] lattice folding models,[84-86] statistical or machine learning neural networks and multistrategy systems,[87-91] distinguishing native from incorrectly folded models,[92-95] and synthetic or simulated energetics or potentials.[96-99]

The general approaches to the design of bioactive proteins, peptides, and other pharmacophores are similar, depending on the type of structure-function information that is available about ligand-receptor interactions. Several potential caveats in the design process are that the binding site may influence the conformation of the bound ligand and that proteins and peptides may occupy one of several equally stable conformers requiring the description of an equilibrium population distribution of active conformers. To avoid these problems, conformational constraints are incorporated in the design to reduce the entropic contribution to binding through: main chain-main chain modifications using cyclization, peptide bond isosteres, or chain linkers such as CH_2 or ($-C=C-$); by side chain-side chain modifications using cyclization, disulfide bridge formation, or substitutions of methyl, aromatic, and heteroaromatic groups; and by main chain-side chain modifications such as proline or lactam ring analogs.[25,61,100-107] This section will briefly describe the biocomputational procedures commonly used to characterize protein and peptide conformations.

Case A — Ideally, high-resolution tertiary structures of the receptor with different ligands bound to the receptor are known from NMR or X-ray studies. Hence the function of cofactors, water, and the structure-dependent biological activities of the ligands can be considered in defining the required steric and electronic characteristics of the drug template.[108] Bioactive ligands can then be designed in several ways,[16-22] such as by modifying known ligands, by selecting structurally characterized molecules from data bases then measuring the fidelity of fit to the template,[23,24,109-113] and/or by *de novo* design in which the ligand is algorithmically constructed in the reaction site of the receptor.[19,114-117] Free energy perturbation approaches may also prove useful.[118] Entropy and the cooperative motions of ligand and receptor should be considered in the peptide design. One *de novo* design approach[119] is to iteratively piece together in a model of the target receptor a template set of amino acid conformations; after each amino acid addition, the structures are evaluated using a molecular mechanics-based energy function that considers van der Waals (VDW) and coulombic interactions, internal strain energy of the lengthening ligand, and desolvation of both ligand and receptor,[120] followed by a determination of the approximate binding enthalpy using the Metropolis method (simulated annealing).[121]

Case B — If the structures of the ligand and the receptor are known but not the exact position of the ligand in the reaction site, then the known ligands can be positioned using an automated docking program[117,122-129] that takes into account steric, van der Waals, and electrostatic interactions. The specific electostatic and structural requirements of the receptor can be inferred from a conformational analysis and the IC_{50} data of the known ligands. Mutation studies that pinpoint functionally important amino acids can also be used to position the ligands. The positions that generate the strongest interactions can then be used in drug design by implementing the data base search procedures or the direct method as outlined in Case A, or an indirect approach as outlined in Case C. Limitations in these approaches are that the effects of structural changes upon binding, water, and cofactors are difficult to include.

Case C — If the receptor active site conformation is unknown but the activities of a group of conformationally constrained ligands are known, then new ligands can be designed indirectly by inferring the structural and electronic requirements of activation from the structure and activity of the known ligands.[114,130-133] After a template is defined with the desired features, drugs can be designed through either structural modifications of known ligands, or data base searches for novel peptide or peptidomimetic ligands. Peptide analogs that are resistant to proteolytic degradation have been synthesized by sequence truncation, incorporation of unnatural amino acids and cyclic amino acid derivatives, preparation of cyclic peptides, incorporation of partial nonpeptidal structures,[134] and C-terminal truncation, such as synthetic analogs of somatostatin.[104] The prediction of the tertiary structure of the reaction site can also be inferred from structures of homologous proteins, as in Case D, and this structure can be considered in the ligand template selection; this procedure is especially useful if the group of ligands are not conformationally constrained.

Case D — If the receptor active site conformation is unknown and if the bioactive ligands do not indicate the electrostatic and structural requirements of the active site, then the protein can be modeled from the known experimental structures of homologous family members[13,135] and this model can then be used in drug design, as in Case A or Case C. Distance geometry is one procedure to build protein structures from sequence alignments with homologues of known structure; multiple sequence alignments are translated into distance and chirality constraints, which are then used to obtain an ensemble of conformations for the unknown structure that are compatible with the rules employed.[9] Other protein secondary prediction approaches compare the sequence to a pattern of environments found in a fold,[92,136] to a series of model folds,[95] or to a template generated from structurally conserved residues. The basic steps in comparative model building using structurally conserved regions *(SCRs)* are as follows:

1. Superimpose structures of homologous proteins in three dimensions and identify the SCRs where the corresponding α-carbons overlap within 0.5 to 1 Å.[68,69,137,138] Usually spatial superpositions of homologous structures indicate deletions and insertions occur mainly in surface loops or on surface residues. The core secondary structural elements are usually in similar relative positions, if they exist, but may differ in length.

2. Identify sequence homology patterns from the alignment of the tertiary position of the α-carbons in the known structures and use these to define a consensus template.[139] The search template is more effective when physicochemical profiles, such as averaged state propensities, hydropathy, and amphipathic moment information are used.[76,80,98,140–142] If there are many homologous sequences, a very successful approach is to use insertions and deletions as parsing or beginning points, and to use the pattern of sequence variability at each residue position and the optimal lengths of secondary structures to distinguish between strand and helix.[136,143]

3. Align the unknown sequence to the template, allowing no deletions or insertions in the SCRs. If there is no clear characteristic sequence pattern for a SCR, such as might occur on external SCRs, then alternative alignments must be considered and propagated in the model construction. The variable regions *(VR)* are loosely aligned based on residue length and type.

4. Construct the three-dimensional model coordinates. The main chain coordinates in the SCRs are taken from the template and the side chains are positioned in one of several ways,[70] such as by using the template for up to the β-carbon, by sequential addition of atoms with intermediate molecular dynamics refinement, or by using side chain rotamer libraries.[144] The tentative starting structures for the VRs and possibly the N- and C-termini can be constructed in several ways, such as by using coordinates from a homologous protein if the sequences match in residue length and character, by conformational searches[68,69,137,138] by energy evaluations, or by data base searches for structural fragments of the correct residue length having a conformation at the terminal amino acids that fit the α-carbon positions at the ends of the loop of the template.[13,62]

5. The model structure is refined to remove bad contacts, bound water seen in the crystal structures is added (unless it overlaps a new atom or is now buried in a hydrophobic pocket), and an energy minimization is performed. The different confidence levels associated with SCRs vs. VRs can be considered in defining constraints used in the minimization or in molecular dynamics.

Comparative modeling procedures can also be used to predict structures for unknown sequences that have one, or no structurally characterized homologous sequences. If only one homologous structure is known, a model can be generated using an alignment of the unknown sequence with (1) a series of model folds[95] or (2) the known structure alone, or with structurally uncharacterized homologous proteins using sequence homology methods.[143,145,146] When aligning the sequence to a series of model folds,[95] molecular field theory is used for the long-range interactions and one-dimensional statistical mechanics for the short-range interactions; since there are only a small discrete set of folding patterns, it is possible to examine all "potentially stable" structures.

If there are no known homologous sequences but there are a large number of related sequences, the protein secondary structure can still be accurately predicted.[143,145,146] Much of the core structure of the catalytic domain of the protein kinases was predicted from the alignment of 79 sequences using (1) insertions and deletions as parsing points and (2) the pattern of sequence variability at each residue position and the optimal lengths of secondary structures to distinguish between strand and helix.[143] A comparison of this approach with the Garnier-Osguthorpe-Robson[147] statistical method indicated that although the residue comparison was similar, the tertiary fold prediction success was much greater.

Case E — Although there have been several successful *de novo* proteins designed to have specific structures,[15,148] protein drug design efforts have focused on changing the function of characterized proteins by either modifying binding sites or by introducing new ligand binding sites.[149,150] If a protein of known structure is used as a scaffold onto which a new binding site is to be attached, the rules governing the selection of the side chain and its conformation can be formulated as semiempirical force fields,[38] or as empirical rules derived from proteins of known structures.[3] The latter approach involves factors such as the packing of interior residues, hydrogen bonding to interior residues and the solvent, and the correct distribution of hydrophobic and hydrophilic groups. These approaches are typified by work on insulin[151] and erythropoietin (EPO),[52] where the functional changes incorporated in the protein were directed by using information derived from the analysis of superfamily members and molecular biology strategies.

A limitation in modeling by homology is the contents of the data base used. These methods cannot predict any structural or functional element that is unique, infrequently observed, or not represented proportionately in the data bases searched. In addition, a data base search may find a strong sequence similarity between two fragments in functionally or evolutionarily unrelated proteins.[72–73] It appears that structure, and especially topology, is more conserved than sequence; there are now many examples where sequence similarity is less than 20% in functionally similar or dissimilar proteins and yet the same topology is observed.[4,92,95]

III. PROTEIN SEQUENCE ALIGNMENT

Sequence alignment is the central technique in homology modeling both in determining the SCR regions and in aligning the structurally uncharacterized sequence. The most widely used method for sequence comparison is *dynamic programming,* first described in protein applications by Needleman and Wunsch;[153] since then, many variations of the original algorithm and other dynamic programming algorithms have been described.[154,155] Dynamic programming methods find the minimum distance or maximum similarity between sequences, using penalties for mismatches, insertions, and deletions. These approaches are sensitive not only to the degree of similarity but also to the choice of gap penalty and scoring matrix, both of which have a weak theoretical foundation. Secondary structural elements can be used to distribute gap penalties, increasing penalties for insertions and deletions within highly conserved regions to emphasize the parsing or beginning points of secondary structural elements.[77] Another approach is to use variable gap penalties in distantly related family members.[81]

Generalizing the Needleman-Wunsch dynamic programming algorithm to multiple sequence alignment leads to a "dimensional effect". The algorithm operating on two sequences compares and scores them using a square array formed from the sequences themselves. Three sequences require a cube, and higher numbers of sequences require higher dimensional hypercubes. Beyond four or five sequences, dynamic programming is infeasibly slow. Currently, multiple sequence alignments are done either by improved algorithms[74,78,79,156,157] or adequate approximations. The theory of scoring multiple sequence alignments, however, remains problematic.[56,75,78] In practice, a comparison of results produced using variable gap weights and an assessment of statistical significance in alignments are required.[91,158]

The amount of similarity between sequences is defined using a score matrix that rates the alignment of each pair of amino acid residues. These matrices indicate the most likely amino acid substitutions based on statistical, chemical, or physical criteria. Several different matrices have been proposed which are adapted to specific purposes.[2,56,82,98,153,154,157,159] The PAM-250[160] mutation matrix measures a certain rate of evolutionary substitutability in a certain collection of proteins; although the PAM-250 matrix is most commonly used, the PAM-120 matrix generally is more appropriate for data base searches, while the PAM-200 matrix appears to be more suitable for comparing two specific proteins with suspected homology.[161] Other matrices consider hydrophobicity,[162,163] or different selection pressures of protein cores and protein exteriors.[164] A more radical choice of metric is to align by DNA base pairs using a codon substitution matrix.[165] Alignment improvements are achieved when the scoring matrices are corrected to remove statistical errors caused by over-counting near duplicates in sequence alignments against families.[156]

Sequence alignment algorithms that consider only sequence motifs, however, do not accurately predict protein structures for proteins with low homology or with many partial alignments of varying quality and relevance.[166] Despite the limited number of known protein structures, clearly proteins fall into distinct structural families, built from a limited number of sequence and structural motifs. Structural motifs, such as the α-helix, β-strand, or composite secondary structure patterns, such as the αα or βαβ, dominate all unrelated protein structures. To aid the identification and use of these motifs, several consistent data bases have been established.[23,45,136,167–173] These data bases include coordinates as well as derived data, such as secondary structure locations and solvent accessibility. For example, PROSITE contains 337 known motifs encoded as a list of allowed residue types at specific positions along the sequence.[174,175] The HSSP data base contains aligned sequences, secondary structure, sequence variability, and sequence profile; it effectively increases the number of known protein structures by a factor of five.[176]

IV. SECONDARY AND TERTIARY STRUCTURE PREDICTION

Sequence analysis techniques, especially applied to proteins where the number of family members is large, can successfully predict protein structure. To predict the secondary structure from the amino acid sequence artificial intelligence (AI), statistical, and joint prediction[177–179] methods have been developed. For the problem of predicting alpha and beta helices these methods all have success rates of 60 to 65%.[89,180–182] The results of these comparisons indicate that the accuracy of these methods depends on the nature of the protein and that successful prediction is not attributable to differences in folding type or the distinction between mono- and oligomeric proteins.[181] The low accuracy of secondary prediction methods has been attributed to the neglect of stabilizing long-range interactions between residues far apart along the primary structure.[73] There have been several attempts to take global effects into account,[48,183,184] but the local-sequence prediction methods are currently more popular. Tertiary structure prediction methods, including extensions of some of the above methods, have given promising results.[5,93,164,182,185–189] In this section different secondary prediction methods will be introduced, including AI, statistical, and Markov methods.

Several AI methods have been developed. Classical AI methods, such as PROMIS,[87] are complicated to outsiders but they permit the user to study the rules as they are applied to the sequence and allow for manual intervention and subsequent reevaluation. A common AI technique is the neural network approach, a learning algorithm developed for pattern recognition.[50,90,187,190–193] One implementation[194] uses a second network to fine-tune a prediction network; the first network runs through a sequence of predictions, and a second network is trained to recognize when the first one is making its best predictions. A current drawback to using neural networks is the small number of known protein structures which comprise the training set. One expectation is that the solution of more protein structures will lead to superior predictions.

Statistical modeling[195] applied to the prediction of secondary structures begins with the observation of Chou and Fasman[196] that certain residues have a propensity towards or against helices. Patterns of structure are built up by following these probabilities. Statistical methods are based on the derivation of structural preference values for single residues, pairs of residues, short oligopeptides, or short sequence patterns. Several common statistical methods include Gibrat-Garnier-Robson,[197] Ptitsyn-Finkelstein,[184] Nagano,[198] Qian-Sejnowski,[90] Garnier-Osguthorpe-Robson,[147] and Chou-Fasman.[196] One statistical modeling approach by Lambert and Scheraga[199–201] uses a small subset of properties (factors) to define the protein conformation; for amino acids, 10 relevant residue factors suffice.[202] Their optimization scheme built on probabilistic predictions of tripeptide conformations is only partly based on theoretical understanding of probabilistic optimization, so room for improvement may exist here. Hidden Markov models (see below) are one such direction.

Another category of statistical modeling programs employs concepts in differential geometry to identify amino acid propensities for occupying particular conformational states. Differential geometric parameters have been used to give a taxonomy to protein structure[203] and to make three-dimensional alignments.[204] In differential geometry, a helix can be described mathematically as a space curve. Just as the derivative identifies the best linear approximations to a plane curve, two differential geometric parameters, the curvature and torsion, identify the best helicoidal approximations to a space curve. Rackovsky and Scheraga[205–208] developed this approach, showing how geometric parameters provide a quantitative approach to secondary structure, conceptually simpler than the more direct conformation parameter approaches. Louie and Somorjai[209] imaginatively proposed the development of a quantitative helicoid dynamics.

Another geometric method is the tetrahedral lattice model of Hinds and Levitt.[84] A vertex on the lattice corresponds to one, two, or three residues. A protein is modeled as a self-avoiding path on this lattice, constrained within a likely volume. In exchange for coarseness, the model gives an accessible conformation space of less than a billion candidate structures for 60-residue long proteins. They use a pseudoenergy derived from a data base of known structures, and can use it to narrow the search down to less than 500 structures, which can then be analyzed using more sophisticated methods.

We are developing statistical methods based on the differential geometry structure of proteins using hidden Markov models (HMM).[210,211] The premise of a classical Markov model is that transitions from state to state are given by fixed probabilities. In HMMs, these states are not seen directly (hidden), but instead there is a fixed probabilistic function on the states that produce what is actually observed. Given the observed states, one seeks to recover the most likely sequence of hidden states. With HMMs there is an algorithm for hill climbing to the hidden Markov chain that is the most likely explanation for the observed data. (In a non-Markovian context, this form of hill climbing is known as the EM algorithm.[212,213]) In protein modeling, the hidden states are the differential geometric parameters. Since these are continuous parameters, we form discrete states by clumping parameter values together. The coarsest clumping leads to the common helix/strand/coil classification. More fine-grained clumping can lead to a more detailed prediction. Then, just like integration of a derivative gives the original function, one can integrate curvature and torsion to recover the original space curve. In this context the space curve defines the general features of calculated models much like that for the lattice models.[84]

V. CONFORMATIONAL ANALYSIS USING MOLECULAR SIMULATIONS

Molecular simulation calculations have become a standard approach to describe the conformational properties of macromolecules and to test structural hypotheses of designed molecules. The two methods used in molecular simulations are (1) *molecular mechanics* and (2) *molecular dynamics* or *Monte Carlo simulations*.[214–216] Molecular mechanics calculations can generate a tertiary molecular conformation or the relative conformational energies of the multiple conformational states available to the molecule. Molecular dynamics or Monte Carlo studies can indicate the atomic motions in the molecule; thus, dynamics and thermodynamic properties which indicate entropy, enthalpy, and free energy differences can be calculated. In this section, approaches to approximate and bias the potential energy surface, minimize using molecular mechanics, and perform molecular calculations are outlined.

At the heart of modeling techniques is a series of potential energy functions, the force field, which replaces the quantum mechanical time-independent Schrödinger equation as a function of geometry.[217] The potential energy of the system as a function of the coordinates describes the multidimensional energy hyperspace of the system.[41,218] The force fields were formulated based upon physicochemical descriptions of molecular interactions.[219] The first force fields developed include classical energy functions for electrostatic interactions, van der Waals interactions defined in terms of dispersion and repulsion components, and torsional terms about rotatable dihedral angles; now terms are included to account for the Hookian behavior of the internal coordinates of the molecule, defined as bond lengths, bond angles, and dihedral angles. The potential energy of the system is expressed using the internal coordinates of the molecule, the distances between atoms, and an analytical, or target, function that is minimized. By calculating the energy of the system at a particular set of coordinate values, one explores the multidimensional energy surface.

Each term in the target function has several constants that are used to parameterize the force field so that theory and experimental data maximally agree. The parameterized force field is then used to search simultaneously each of the atomic positions for the most stable, lowest-energy conformation. The development of potential energy functions is a discipline in its own right, emphasizing the theoretical basis of force fields, their parameters, and the parameterization process.[220] An empirical force field developed for peptides, however, cannot be applied with confidence to organic molecules, and vice versa. The accuracy required in a force field depends on the properties of interest in the system. A force field may reproduce structural information very well compared to crystal structures but the calculated energies may not be accurate enough for quantitative energy descriptions of enthalpies.

Several force fields are routinely used in molecular energy calculations of peptides. Implementations of these include ECEPP/2,[221] AMBER,[222,223] GROMOS,[224] CHARMM,[225] CHARMm,[226] CVFF,[227] and MM2/3.[228] AMBER, CHARMM, and CVFF force fields have become widely used to the commercialization of software packages that contain these and modified versions of these force fields. In addition, there is a wealth of software and methods available for computer-assisted drug design which was recently reviewed[108,114] that is either the same or complementary to protein-based drug design.

The essential difficulty in using potential energy approaches to study conformational properties is the multiple-minimum problem on the global energy surface.[218,229–231] Multiple-conformation generation approaches that search the conformational hyperspace include (1) systematic search procedures that explore the complete conformational hyperspace, (2) modified minimization algorithms and/or target functions that allow the system to overcome local barriers, and (3) distance geometry methods.

Systematic conformations can be generated by the dihedral angle scanning method. This method is usually limited to about ten dihedral angles, since the number of initial starting structures increases exponentially with the number of rotatable bonds.[232] A less computationally expensive method is a buildup procedure, where dipeptides are built from the average geometry of amino acids, are energetically compared, then are used to build larger peptides.[233] This buildup procedure has been extended to the class of proteins that can be characterized solely by local interactions.[234]

The second approach is to *modify the minimization algorithm* and/or the *target function* through a reduction of dimensionality or by dynamic searching. In the reduction of dimensionality technique, the conformation is minimized to the global minimum in a space equal to the number of atoms squared.[235] In one molecular dynamics technique, the available thermal energy is used to cross energetic conformational barriers. Local minima are found during the simulation with periodic minimizations. To sample large molecules, simulations for milliseconds to minutes at 300° may be needed. Current molecular dynamics simulations, however, are limited to tens of nanoseconds where thousands of conformations can be sampled.[25,236–238] Another dynamic searching approach is to use the electrostatically driven Monte Carlo method that assumes that a polypeptide or protein molecule is driven toward the native structure by the combined action of electrostatic interactions and stochastic conformational changes associated with thermal movements.[231] Other successful dynamics approaches are to use simulated annealing[121,229,232,238–241] or the variable target function algorithms.[242]

The third approach is to use *distance geometry* methods to convert a set of distance constraints into an ensemble of conformers.[243,244] The starting structures are generated by a combination of model-built and random-start approaches. The trial distances are chosen randomly between upper and lower boundary conditions and the algorithm guarantees that the initial coordinates are the best multidimensional fit to the trial distances. The starting conformations are then optimized using Newton-Ralphson or conjugate-gradient minimization or the ellipsoid method.[129] In distance geometry, the time required is independent of the number of rotatable bonds and has approximately a quadratic time dependence on the number of atoms; thus, it is practical for large molecule explorations that are beyond the reach of systematic search procedures.

The local minima in many molecules, especially linear peptides such as some hormones and neurotransmitters, are separated by energy barriers that can be rapidly crossed at room temperature. Thus, in many cases, it would be misleading to suggest that the global minimum conformation for a protein in solution is the sole conformation, but rather represents only one conformer existing in a rapidly interconverting population of conformers. In addition, the time-average conformation often differs in solution and as a crystal and the exact conformation seen in the solution state may experience substantial fluctuations about the average. While the solution conformation(s) of peptides is accessible for examination by physical techniques, interpretation of such observations is difficult without knowledge of the number and relative population of the various conformers. In principle, a solution to this dilemma would involve calculation of the lowest energy conformers of the peptide in solution using theoretical methods, with a subsequent averaging of the experimental parameters over these states. In practice, with existing potential functions, one cannot accurately calculate the Boltzmann distribution over the lowest energy conformational states of a peptide.

Another consideration when selecting the simulation strategy is the effect of the solvent on the conformation. If the results of the simulation depend on charged interactions that may be modified by the presence of water or if hydrophobic effects are important then the solvent should be explicitly included using a solvent layer of at least two to three molecules in width around the protein.[245] A less rigorous and computationally less expensive approach is to define the dielectric constant to be greater than one or to use a distant-dependent function, such as a sigmoidal distance-dependent dielectric function.[246,247] These approximations, however, neglect packing effects, hydrogen bonding between solvent and protein, viscosity, the dynamic aspects of protein-solvent interactions, and dissimilar dielectric properties in different regions of the protein.[245] Another approach is to use a finite difference Poisson-Boltzmann method for calculating electrostatic interactions as a basis for a pairwise energy term that accounts for charge-solvent interactions.[248]

Known structural characteristics of a macromolecule often are explicitly included in the simulation using conformational constraints or restraints. There are also many applications in which biasing the target function is appropriate to test whether a particular conformation can be populated or to impose constraints or restraints on the molecular system to influence the computed energy pathway. A constraint is a fixed degree of freedom. Common types of constraints, which are especially used in the early stages of minimization, include rigid body treatments of molecules, torsion angle space minimizations, and Cartesian space atom fixing. For example, the structure of flexible loops in an immunoglobin[137,138] and trypsin[69] were predicted with the remainder of the protein fixed.

In contrast, a restraint is an energetic bias that tends to force the calculation toward certain restrictions. Four commonly used restraints are distance, torsional, template forcing, and tethering restraints. Applications of distance restraints include incorporating nuclear Overhauser effect (NOE) data or cyclizing linear molecules. Applications of torsional restraints include incorporating NMR coupling constant information and biasing the harmonic torque about a bond to study energy pathways across a barrier.[236] Template forcing is used to force one molecule to adopt the conformation of another[25,157,236,249,250] and the energy required to force the molecule into the required conformation can be used to evaluate how easily an analog can adopt the conformation of a given template. Tethering is a type of template forcing in which the atoms are restrained to their original positions rather than to positions in the template. Tethering is used in preliminary minimizations to remove large initial forces seen in X-ray crystal structures or when hydrogens are added.

One common technique used for incorporating the constraints and structural consequences of diverse forms of information about protein conformation is distance geometry.[185,243,244,251-255] The inputs to the algorithm can come from distances between specific pairs of atoms derived from chemical cross-linking, fluorescence transfer experiments, neutron diffraction, NMR, electron microscopy, Chou-Fasman secondary structures, etc.

A. MOLECULAR MECHANICS

Minimization of a molecule requires two steps. First an equation describing the energy of the system as a function of its coordinates is defined and evaluated for a given conformation. Target functions can be constructed that include external restraining terms to bias the minimization, in addition to the energy terms. Second, the conformation is adjusted to lower the value of the target function. The choice of the minimization algorithm depends on the size of the system and the level of optimization of the initial structure. Until the derivatives are much below 100 kcal mol^{-1} Å^{-1}, algorithms that assume the energy surface is quadratic, such as Newton-Ralphson, quasi-Newton-Ralphson, or conjugate gradients, are unstable. Steepest descents should be used until the derivative falls below this level, after which Newton-Ralphson or conjugate gradients can be used. Another minimization approach is to use the integration algorithm in molecular dynamics to adjust the average kinetic energy of the atoms. By gradually decreasing the temperature or setting it to a low value the total energy of the system can be reduced, yet the fluctuations in kinetic and potential energy can allow the system to overcome small barriers to settle to a lower minimum.[216,229,256] This procedure is used in dynamic annealing to search conformational space in conjunction with high-temperature dynamics.[257,258]

The choice of force field used in the minimization depends on the amount of distortion in the molecule. For highly distorted structures, cross-terms and Morse bond potentials in the force field can cause convergence problems due to the small restoring forces and the possibility of minima at nonphysical points on the potential energy surface. Until the large distortions are removed, force fields that have a simple quadratic functional form should be used. If hydrogen atoms are added to heavy atoms in crystal structures, large initial nonbonded repulsive forces may exist due to overlapping hydrogens. To remove these forces, 10 to 50 minimization steps should be performed with a high template forcing constant[236] to restrain the atoms to their starting positions while the starting energies are reduced to values requiring root mean square (rms) movements of less than about 0.3 Å for all the atoms in the molecule.

B. MOLECULAR DYNAMICS

The second component of molecular modeling is molecular dynamics.[38,237,259] This method is similar to molecular mechanics in that atomic interactions are modeled by a classical force field approximation, but it differs in that it also simulates thermal atomic motions by solving Newton's equations of motion for the system. Thus, molecular dynamics can be used to study molecular motions and their time scales. In addition, the average atomic kinetic energy enables the molecule to cross multidimensional energy barriers which leads to a sampling of the population distribution. Also calculations can be made of the approximate free energy differences for chemical process, such as ligand binding, solubility, and partition coefficients. Molecular dynamics studies have been used to elucidate structures from NMR experiments, to refine X-ray crystallographic structures or molecular models from poor starting structures, and to calculate the free energy change resulting from mutations in proteins.[38]

Molecular dynamics simulations provide information about the accessibility of conformational states available to a molecule. There are several different approaches used in molecular dynamics for controlling the temperature and pressure. A constant volume and energy ensemble is usually not used in equilibrium steps if a specific temperature is desired since there is no energy flow facilitated by temperature coupling and scaling. It is used to explore the constant energy surface of the conformational space, or to avoid the perturbation introduced by setting the temperature or pressure. A constant volume and temperature ensemble is used when conformational searching is carried out in a vacuum and if no periodic boundary conditions are used. Both low-temperature and high-temperature (900 K) dynamics are typically employed. High-temperature dynamics greatly increase the efficiency of producing conformational transitions, although structural distortions, such as the in peptide bond, must be considered and geometrical consistency must be attained.

A typical dynamics study includes the following steps. An initial molecular conformation is chosen based on NMR or X-ray crystallography data, comparative model building, or preliminary molecular modeling. Each atom is assigned an initial velocity, usually from a Maxwellian distribution, at the temperature of interest. The Newtonian equations of motion, which define the acceleration of each mass at its coordinates due to the potential, are integrated. The whole system is allowed to evolve in a space defined by the coordinates and the momenta of the atoms. The system is then equilibrated. After the equilibration step, the system is simulated further and any relevant properties or averages are computed as time and ensemble averages.

A major limitation to molecular dynamics is the amount of computer time required for the simulation. Typically several hundred picoseconds of simulation are performed for 10,000 atoms using time increments of a femtosecond. A femtosecond time increment, however, samples only high-frequency motions. For small systems, dynamics runs on the order of 30 psec are the minimum required for meaningful results; shorter runs, however, can be useful for sampling conformation space in minimization strategies. To increase the accuracy of the predicted equilibrium properties without increasing the simulation time, the number of degrees of freedom that are explicitly included in the calculation can be reduced, such as by using a dielectric constant in the Coulombic calculation rather than including solvent molecules.[222,225]

Molecular simulations, particularly in conjunction with chemical modifications of peptides, can provide information about the complex phenomena of ligand-receptor interactions. This information can then be further translated into synthetic compounds which are evaluated for bioactivity. An example of peptide drugs which were designed using molecular simulations is the constrained analogs of gonadotropin releasing hormone. The design strategy incorporated NMR data, energy minimization, template forcing, and molecular dynamics.[25,249]

VI. MOLECULAR MODELING USING NMR DATA

Tertiary conformations of proteins can be derived using X-ray crystallography, but the technique is limited by the availability of good crystalline samples from proteins. In addition, the X-ray-derived conformation is dependent on the crystal packing conditions, yet the conformations of proteins are often solvent-, temperature-, and pH-dependent. Developments in NMR spectroscopy make it possible to study the solution conformations in various solvents, at different pH values, at different salt concentrations, and at different temperatures. NMR spectroscopy, coupled with lattice and comparative modeling building procedures, has increased our ability to determine tertiary structures.

The general use of NMR to determine the three-dimensional structure of nonaggregating proteins is a well-established procedure;[260–263] with isotopic labeling the molecular weight limit may reach up to 40,000.[264] Computer programs are available and are being developed to automatically acquire and process the NMR data, assign and simulate the resonances, and calculate the three-dimensional structure of proteins. Since the uncertainties in the protein structure determined from the NMR data depend on the uncertainties of the individual constraints used in the calculations and on the computational methods, it is important to understand the available structural information in the

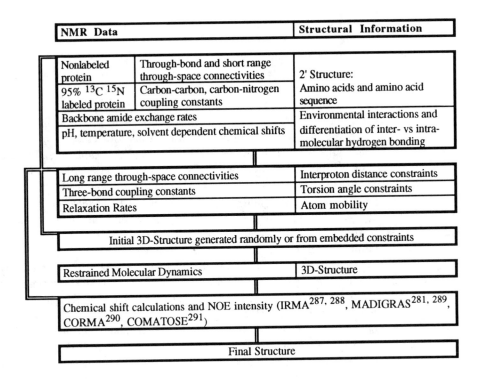

FIGURE 1. Structural characterization of proteins by NMR.

NMR data, the minimum amount of spectral data required, the precision of the information that is available from the NMR data, and the capabilities of the available software.

The structural information in NMR data are distance constraints based on the nuclear Overhauser effect (NOE), stereospecific assignments based on NOEs and coupling constants, and angular constraints based on coupling constraints (Figure 1). The amount of NMR data required for sequential characterization of the protein depends on whether or not, and by how much, it is isotopically enriched with ^{15}N or ^{13}C. If the protein is unlabeled, the sequential assignments are derived from the through-bond and short-range through-space connectivities and from the amide exchange rates. With the complete backbone assignments and possibly some of the protons on the β-carbons, in combination with the intensity of the proton short-range NOEs and the α-carbon (C_α) chemical shifts, a preliminary secondary structure can be proposed.[265–268] In cases where the secondary structure is well defined, backbone amide proton exchange data can be used to assign hydrogen-bonding constraints. The appropriateness of the assignments of the regular secondary structural elements can be estimated from the number and uniqueness of the set of NOE distance parameters.[261] Preliminary secondary structures can enhance the subsequent analysis of NOE data used in the tertiary characterization of the protein, or the structures can be used directly in generating a tertiary structure through homology searching,[9,12] interactive model building,[267–269] or molecular modeling.

To characterize the tertiary structure from NMR data, the maximum number of conformational constraints is sought through a complete assignment of all peaks, including the amino acid side chains. Potential caveats in this approach include incomplete information about the number of spin systems, missing sequential NOE data, the erroneous inclusion of nonsequential NOE information, relatively poor performance in regions of irregular structure such as loop regions, and unassignable proton resonances in the backbone amide and C_α regions. Ambiguities can often be eliminated by increasing the number of dimensions in the NMR experiment[270–272] and by isotopically labeling the protein. If a protein can be 95% ^{13}C and ^{15}N labeled, then both the amide and C_α proton resolutions are increased and the secondary structure can be assigned without relying on short-range NOEs,

TABLE 1
Several Computational Model Building Approaches

Approach	Program
Monte Carlo simulations	IMPACT[393,394]
Double iterative Kalman filter	PROTEAN,[395,396] FILMAN[397]
Minimization of torsion angle space with a variable target function, or series of ellipsoids	GEOM,[410,411] ELLIPSE,[380] DISMAN,[242] (CALIBA, HABAS,[295] GLOMSA, DIANA[298,412])
Metric matrix distance geometry	DISGEO,[381,413] (EMBED, VEMBED[414]), DSPACE,[378,415] DGEOM,[401] NMRGraph[416]
Energy minimization and molecular dynamics	CHARMM,[225] AMBER,[223,417,418] ECEPP/2,[221] FANTOM (ECEPP/2),[296] PEACS,[392] GROMOS,[224] DISCOVER,[419] (XPLOR;[385,402] XPLOR/dg[255])

which are a complicated function of the time-averaged motions and distances of an ensemble of molecules;[273] with no NOE data, a preliminary assignment of β-strands and α-helices can be proposed based on conformational shifts of the assigned C_α proton resonances.[274-278]

NMR data, however, are generally not sufficient to define a unique tertiary structure due to the number and distribution of the constraints. The regions with insufficient NMR constraints are most often surface loops, internal, and especially, external side chains, most probably due to solution mobility. Consequently, the experimental data are combined with empirical energy constraints to both deduce and refine a group of about 20 conformers which characterize the solution conformation of the protein (Figure 1, Table 1). The conformers are further refined in three-dimensional space using molecular modeling. Next systematic comparisons of the spectra predicted from the conformers with the actual data can be used to modify the NMR constraints and new conformers can be generated.

The extent of the deviations between conformers is usually expressed from the average of the pairwise root mean square deviations (rmsd) among the conformers,[279] or as the average of the rmsd between the individual conformers and the mean structure. The latter approach, however, will not distinguish between an accurately determined structure with one ill-defined region, and a poorly reproduced structure on the average. An acceptable structure determination is characterized by a small deviation between the conformers,[261] consequently the efficiency of the algorithm at sampling the possible conformational space must be considered. This presentation of the conformation is qualitatively related to the characterization of a crystal structure using experimental, apparent temperature factors.[280] Assuming identical crystal and solution conformations, the accuracy of the conformation can be measured by the rmsd of the average NMR-derived structure from the X-ray crystallography structure. It is important to note that NMR constraints indicate the behavior of a set of possibly conformationally diverse conformers; if on the NMR time scale the interconversion between conformations is fast, the NOE constraints do not necessarily indicate the average structure and there may be no single conformation that is consistent with the NOE data. The chemical shift data, however, are the average chemical shift in all the conformations present and the contribution from each species is weighted by its population.

The amount of NMR data required to define a tertiary structure is determined by how highly correlated the distances are and if the distances between residues far apart are restricted conformationally. The accuracy of the structures is dependent on the number and accuracy of the structural constraints.[281] An estimate of the dependence of the NMR-derived protein structures on the quality and quantity of the NMR data was published from the NMR-derived average conformations of crambin.[282] The study indicated that

1. The global rmsd between the reference and the NMR-derived structures decrease exponentially as the number of constraints increase and, after about 30% of all the potential constraints were included in the calculation, the errors asymptotically approach a limiting value

2. Increasing the assumed precision of the constraints has the same qualitative effect as increasing the number of constraints

3. For a comparable number of constraints per residue, the precision of the NMR-derived structures is less for a larger than for a smaller protein[282]

In general, however, the quantity of the constraints is more important than the quality.[283] The best determined tertiary structures were generated from data sets that contain on the average ten interresidue NOEs per amino acid residue with over 30 interresidue NOEs for internal residues. The best defined stereospecific assignments have up to 50% of the β-methylene protons, most of the nondegenerate valine γ-methyl protons, leucine δ-methyl, glycine α-methylene, proline δ-methylene protons, and some glutamine γ-methylene protons.[284] Typical results for the displacement of the conformers determined from NMR data from globular proteins with 5 to 10 NOE assignments per residue are 1 Å rmsd for all backbone hydrogens (excluding residues from the C- and N-termini) and 2 Å for side chain hydrogens independent of the computational method selected.[285,286]

Computer programs are available that can perform one or more of the steps required to extract the structural information from the raw, time domain NMR data, some of which are available with the commercial spectrometers. Two computer packages that can perform all of the required steps to produce tertiary structures from the raw data are from New Methods Research[292] and from Wüthrich and coworkers.[129,268,293–300] In addition, programs exist which simulate the response of spin systems to the experimental and processing parameters used to generate the NMR data;[293,294,301–305] these can be useful for verifying resonance assignments. The first step in processing the NMR data is to extract chemical shifts, coupling constants, and relaxation rates from the data. Software is available which can accomplish this task using one of several methods.[306–311] Most commonly, a Fourier transformation is applied,[260,297,311,312] followed by phasing[313] and correcting the baseline,[300,314] extracting chemical shift and coupling values,[45,315–323] and measuring the relative intensity of each signal.[324,325] There are cases, however, when the Fourier transform does not retrieve the required information from the signal, such as when the NMR signal is acquired with less than one-fourth of the number of points required to sufficiently digitize the fine structure in the data. Alternative methods,[326,327] such as linear prediction,[321,328–332] maximum entropy,[333–335] or Bayesian analysis,[336] have been developed that extract the information from the NMR data and which do not suffer from the same limitations seen in the Fourier transform.

The second step is to identify the spin systems from the chemical shift and scalar coupling information. The list of spin systems indicates the amino acid residue types which are compatible with the data. Several algorithms which automate this task[297,308,318,319,337,338] have been published since a recent review of the subject.[339]

The third step is to quantify or classify the NOE intensity data.[298,309–311] This is a nontrivial exercise since contributions to the NOE intensities can be from noise, zero- and double-quantum coherence, leakage due to alternative relaxation mechanisms, and molecular motions.[284,291,340–342] It is estimated that the assumption of isotropic motion in biopolymers leads to errors of about 10% in the NOE intensity estimate in regions of rapid molecular motion.[290,343] There are several procedures for defining the distance constraints from the NOE intensities: (1) to assign short ($1.8 \leq 2.7$ Å), medium ($1.8 \leq 3.3$ Å), and long ($1.8 \leq 5$ Å) ranges to the distances obtained from an initial slope[261] or model-independent[344,345] method; (2) to use data from a full relaxation matrix calculation to refine distance estimates.[287–291,340–343,346–354]

A promising model-independent procedure for measuring relaxation rates for each proton solely from a single NOE build-up curve is to use LPSVD to compute the relaxation matrix elements, where no assumptions about the relaxation of unmeasured spins or the motion of the molecule are required.[345] Although the full relaxation matrix analysis could remove errors caused by molecular motions and spin system interactions, the usual approach has been to classify the NOE signals into ranges, thereby avoiding any problems associated with variable internal mobility and r^{-6} averaging. This approach can produce errors in the data, such as a systematic compression of the structures and underestimating of NOE intensities from regions undergoing reorientations that are faster than the rigid tumbling of the protein, but the proximity constraints and the large number of constraints enable protein structures to be defined with reasonably high resolution. For globular proteins, the quality of the constraints was found to contribute less to the final structure than the quantity of the constraints.[283] Another promising approach is to characterize the internal motions of bonds through a mapping of the spectral density functions using heteronuclear NMR relaxation rates.[355]

The fourth step is to deduce the sequence and regular secondary structural elements of the spin systems in the peptide based on knowledge of the peptide sequence, the spin system assignments, and the NOE distance measurements. Several algorithms are available for semiautomated sequence-specific peak assignments[297,315,319,321,356-365] or recognition of secondary structural elements.[366] An empirical pattern recognition scheme is used for secondary structure identification.[367,368] Helices, peripheral β-strands, and interior strands of β-sheets can be more confidently assigned by measuring the amide proton exchange rates and by detecting similar $^3J_{NH\alpha}$ coupling constants in those regions.[266]

The fifth step is to discern stereospecific assignments of the diastereotopic protons in CH_2 groups and of methyl protons in valine and leucine residues from NOE and coupling data.[298,369-371] The inclusion of these stereospecific assignments significantly improves the precision of distance geometry structures.[295,298,370] If stereospecific assignments are not available in the NMR data, then pseudoatoms,[372,373] computer-generated assignment of the prochiral protons,[374,375] or the average of the distances to the relevant protons can be used.[376]

The final step is to determine the tertiary structure of the protein from the secondary assignments and the long-range NOE data. The model building methods[238,241,285,286,373] all attempt to locate the global minimum region of a target function that contains both stereochemical and experimental NMR terms. There are two approaches to modeling: (1) in real space where minimization or molecular dynamics techniques are applied in either Cartesian or torsion angle space; (2) in distance space using metric matrix distance geometry to project $N(N-1)/2$ dimensional distance constraints from N atoms to three-dimensional Cartesian space through a process known as embedding[244,254,255,285,373,377-380] (Table 1).

The first algorithms implemented in the structural analysis of NMR data used metric matrix distance geometry.[381,382] Subsequent procedures included a restrained least-squares minimization in torsion angle space of an ellipsoid algorithm[129,380] or a variable target function[242,383,384] that assigns pairs of diastereotopic substituents[298] in reduced computational time;[299] restrained molecular dynamics,[241] alone[385] or in combination with model building[269] or with distance geometry calculations;[238,386-388] hybrid metric matrix distance geometry with dynamical simulated annealing;[239,257,375,389-392] Monte Carlo simulations;[393,394] double iterated Kalman filter;[282,395-398] as well as substantial improvements to the first metric matrix distance geometry algorithms.[243,244,253-255] Several evaluations of these methods have been published.[244,254,282,283,379,398-400] An early comparison[282] of DGEOM,[401] XPLOR,[385,402] and PROTEAN[395,396] indicates that (1) all three methods reproduce the general features of the protein structure, (2) XPLOR produces structures that are most accurate and precise, (3) PROTEAN is as accurate as DGEOM but is less precise. Since these evaluations, further improvements in the algorithms have been made so the potential applicability of each algorithm is still unclear.

Recent publications indicate that the most popular approaches are (1) metric matrix distance geometry followed by simulated annealing using molecular dynamics, (2) structure determination using a variable target function that is optionally followed by a molecular mechanics energy minimization, and (3) double iterated Kalman filter. Even with the best NMR-derived structures, molecular dynamics simulations can usually improve the structures, especially by including solvent in the final stages of the simulations.[388,403]

Whether or not a single low-energy structure is found which is consistent with the NOE data, it is possible that there may be several dynamically interchanging conformers in the system. Conformationally heterogeneous populations can be studied in several ways. Structural improvements have been achieved in a restrained molecular dynamics approach using time-averaged restraints, which represent the time-averaged, nonlinear character of the NOE data and reduce the weight of the experimental nonphysical term in the force field;[392,404] the generated structures and the

NMR data can be in better agreement by 70 to 80%, the mobility of the structures can be decreased, and a wide range of conformers can be generated.[388,405] A second approach, which emphasizes feasible exchange systems rather than an average structure, interprets the experimental constraints in terms of the dynamic structure of a multiconformational protein with exchange rates between 10^3 to 10^7 s^{-1}.

A typical protocol for studying a protein with a heterogeneous population of conformers is as follows:

1. Check if the sample is aggregated by varying the concentration from 0.5 to 12 mM (or up to the concentration required for the ^{13}C studies) and check for changes in chemical shift ($\Delta\delta$) or linewidths (LW).
2. Assign the resonances and measure the torsional angle constraints.
3. Assess the solvent accessibility of the amides by measuring $\Delta\delta/\Delta T$ over a temperature range of 25 to 55°.
4. Measure the spatial proximity of the protons by assigning the resonances in the NOESY (Nuclear Overhauser Enhancement and Exchange Spectroscopy) data sets. One way to eliminate the contribution of spin diffusion is to use the linear portion of the buildup curves from data sets acquired at t_m values of 50, 150, 250, 500, and 750 msec. *Trans*-peptides will have no NOE peaks at αH_i–αH_{i+1} or NH_i–αH_{i+1}.
5. Assess the nsec motions of the atoms by measuring the ^{13}C relaxation times. Variations in relaxation times at different atoms indicate variations in the correlation times.
6. Measure the 10 to 10^3 s^{-1} motions of the atoms. After taking into account solvent viscosity, measure the $\Delta LW/\Delta T$ over a temperature range of –80 to 25°. If the LW increases or splits into two peaks, there exists more than one conformation.
7. Generate a group of conformers that satisfy the NMR constraints and empirical constraints (VDW radii, etc.). If there exists more than one conformation at room temperature, first use only the strong NOEs and the $^3J_{NH-H\alpha}$ values to generate potential conformations. Back-calculate the NMR spectra from the minimized structures to check the structures.
8. Cluster energy minimized structures according to conformational similarities between peptide linkages.
9. Run molecular dynamics simulations. If the LW indicates there is an equilibrium mixture of conformations, run in 1-fsec steps over 100 psec with snapshots every pico second and analyze the molecular dynamics trajectories to delineate peptide backbone motions.
10. Compare the NMR data with the results from molecular dynamics and select the best potential structures for the equilibrium mixture.

The success of the structures derived from the NMR data depends on the number and accuracy of the experimental data, the quality of the preliminary structures, the accuracy of the force field used in the molecular dynamics simulation, and the simulation time used to search the conformational space. Acceptable conformers have small deviations from the idealized covalent geometry, an internal energy of about –5 kcal/residue, and nonglycine residue backbone torsion angles in allowed regions of Ramachandran plots. Other criteria which should be considered are realistic apolar/polar surface ratios, charge distributions, and hydrogen bond patterns.

With the NMR-derived structures, other questions can be addressed, such as unfolding kinetics, the dynamics of changes in conformation after ligand binding, and conformational equilibria between conformational states, and nanosecond or millisecond to second internal dynamics.[406–409] In addition more elaborate approaches to drug design can be considered.

VII. STRATEGIES FOR ANTIBODY-BASED PEPTIDE DESIGN

Biologically active peptides have been developed based on the analysis of the active site of biomolecules. To develop the bioactive peptides, the active site of the molecule under consideration must be known. The highly homologous antibodies are well suited to the process of comparative model building.[69,137,138] Information derived from anti-receptor antibody sequences, or related biologically significant proteins, can lead to the development of peptides and pharmacophores that bind the active sites of receptors.[135]

Monoclonal antibodies have been useful in the identification of reactive functional groups, including the RGDS binding sequence in integrins which promote cell attachment to fibronectin, vitronectin, laminin, collagen, fibrinogen, and von Willebrand factor. The structural domains of integrins have been correlated with ligand binding by cross-linking to peptides containing RGD, a ligand recognition motif for several but not all integrins. Peptides containing the RGD tract exhibit a variety of bioactivities in studies from inhibition of platelet binding to inhibition of tumor metastasis. While the platelet and endothelial cytoadhesins may interact with similar peptidic sequences, they recognize different steric and electronic features.[420]

Taub et al.[421] analyzed the hypervariable loop structures of PAC1, which is an IgM-kappa murine monoclonal antibody that binds to the platelet fibrinogen receptor and is inhibited by both fibrinogen and RGD-containing peptides. They identified a conserved RYDT binding sequence in the third complementary determining region (CDR3) of the PAC1 heavy chain (H3) that, if present in the proper conformation, might behave like the RGD sequence in fibrinogen. A 21-residue synthetic peptide encompassing the H3 region inhibited fibrinogen-dependent platelet aggregation as well as the binding of PAC1 (K_i = 10 mM) and fibrinogen (K_i = 5 mM) to activated platelets. The RYD region of H3 seems central to its function, since substitution of the tyrosine with glysine increased the inhibitory potency of the peptide by 10-fold, while replacing the tyrosine with D-alanine or inverting the RYD sequence greatly reduced the inhibitory potency. Thus, the RYD sequence within H3 of PAC1 appears to mimic the RGD receptor recognition sequence in fibrinogen.[421]

PAC1 was then compared with other murine monoclonal antibodies, three of which contain the RYD sequence in H3.[422] The OP-G2 and LJ-CP3 antibodies bound to the $\alpha_{IIb}\beta_3$ receptor and, unlike with PAC1, platelet activation was not required for binding. To determine the structural features in H3 which impose the differential competition with fibrinogen for binding to $\alpha_{IIb}\beta_3$, we utilized a comparative modeling procedure previously used to model regions of antibodies and related molecules.[105] Superimposing the C_α coordinates indicated the consensus regions between the H3 structures. With the location of the framework region, a model was built for the various loops by searching the crystallographic data base for loops of the same size as the putative loop to be modeled.

A characterization of the loop regions indicates the RYD sites in H3 regions of PAC1, LJ-CP3, and OP-G2 were positioned roughly in a β-turn.[422] Molecular modeling of the loop also reveals that in OP-G2 and LJ-CP3, the angles at which the RYD tract is presented relative to the plane of the antigen binding pocket are nearly identical, and differ from that seen in PAC1. This suggests that the similar biological activity seen in OP-G2 and LJ-CP3 might be directly related to the conformational properties of the H3 region of these two antibodies. Studies of RGD peptides of varying lengths, however, indicate that platelet interactions are sensitive to the length of the RGD presenting peptides. Thus, it is also possible that the differences in activity between the antibodies is a function of the depth of the RGD binding pocket on $\alpha_{IIb}\beta_3$ which is accessible to the antigen binding surface of the antibodies.

The availability of structure-function data of the bioactive ligands, in addition to the native sequence of adhesion extracellular matrix proteins, indicates the steric and electronic requirements for selective activation of different integrins. Recent NMR and theoretical assessments of RGD-containing proteins and peptides, such as viper venoms and conformationally constrained, cyclic RGD peptides, have confirmed earlier speculations that RGD cell adhesion sites adopt β-turn conformations.[423–425] As was seen in the antibodies, the precise location of the RGD tract in the turn region of both linear and cyclic peptides is influential in establishing fine specificity. In addition the flanking residues can affect the conformation and environment of the RGD tract.

To understand the structural features that may lead to the differential competition between RGD peptides, PAC1, LJ-CP3, OP-G2, and fibrinogen for $\alpha_{IIb}\beta_3$, a model for the $\gamma^{400-411}$ region of fibrinogen was generated.[426] Molecular dynamics calculations indicate a structural similarity between the H3 region of PAC1 and the $\gamma^{400-411}$ fragment. This is consistent with the experimental data showing cross-reactivity of $\gamma^{400-411}$ with the anti-idiotypic PAC1 antibodies.[427] The dynamics calculations implicated the KQAGD tract of $\gamma^{400-411}$ as a turn region, with the HHLGGA region as an extended β-strand. This is similar to the PAC1 H3 structure in that the amino-terminal region has an extended conformation with the RYD tract centered within a β-turn. Recent NMR studies also indicate a turn property for this region.[428]

It appears that the Lys[406] or Gln[407] within the β-turn of $\gamma^{400-411}$ might be spatially disposed to participate in hydrogen bonding or salt bridges with residues within the $\alpha_{IIb}\beta_3$ RGD-binding site. The KQAGDV fragment has been shown to inhibit fibrinogen binding to platelets with a K_i similar to that of the full $\gamma^{400-411}$ peptide.[429] Moreover, a substitution of Arg for Lys[406] decreased the inhibitory potency of the $\gamma^{400-411}$ peptide by 15-fold, while substitution of Arg for Ala[408] increased the inhibitory potency by 6-fold. Further studies are underway to precisely define the effects of these substitutions on the conformation of $\gamma^{400-411}$.

Finally, conformational studies of the PAC1 H3 region indicates a similarity with RGD constrained cyclic RGD hexa- and pentapeptides. For example, superimposing the cyclic cRGDFV with the RYD tract of PAC1 suggests that Tyr in PAC1 is contained within a β-turn formed by the backbone NH group of Asp and the backbone carbonyl group of Arg. Both the cyclic peptide and the H3 of PAC1 can be stabilized by a hydrogen bond between the backbone NH group of Arg and the backbone carbonyl group of Asp. This hydrogen bonding arrangement is not observed with OP-G2 and LJ-CP3. Distance measurements between the guanidinium head group of Arg and the carboxylate group of Asp in the H3 regions of the three antibodies show distance spreads of 8.5 to 11 Å. Thus, the differential selectivity of the anti-$\alpha_{IIb}\beta_3$ antibodies may be related to the spatial relationships between the Arg and Asp residues. These same spatial relationships have also been recently noted in conformational analysis of bioactive RGD peptides[430] and would be expected to influence the biological properties of mimetic compounds that position amide and carboxylate groups in relative proximity.

Further studies are underway to correlate activities with distinct conformational features observed among the other RYD-containing antibodies, cyclic peptides, and peptidomimetics that display selective inhibition. Subsequently, optimizing the conformation of cyclic peptides and peptidomimetics may provide agents to treat vascular disease and perhaps as inhibitors for the metastasis of cancer cells.

VIII. SUMMARY

Current biocomputational approaches to peptide-based drug design combine knowledge-based systems involving crystallographic, NMR, and sequence data bases with experimentally determined constraints, conformational energy analysis, and molecular modeling. The general approaches to the design of bioactive proteins, peptides, and other pharmacophores are similar, depending on the type of structure-function information that is available about ligand-receptor interactions. With knowledge of the electrostatic and structural requirements to activate or inhibit a receptor, pharmacophores or mimetics can be designed or identified in data bases containing three-dimensional coordinates of small bioorganic compounds. The challenge is to see how much information can be extracted from the protein and pharmacophore data bases and how this information can be applied to design a compound that will interact optimally with a unique endogenous molecule and which may also have enhanced metabolic stability, oral bioactivity, and decreased side effects.

ACKNOWLEDGMENT

This work was partially supported by grants from the American Cancer Society, the American Heart Association, and the National Institutes of Health.

REFERENCES

1. **Williams, W.V. and Weiner, D.B.,** *Biologically Active Peptides: Design, Synthesis, and Utilization,* Technomic Publications, Lancaster, PA, 1993.
2. **Dayhoff, M.O., Schwartz, R.M., and Orcutt, B.C.,** in *Atlas of Protein Structure and Sequence,* Dayhoff, M.O., Ed., National Biomedical Research Foundation, Washington, D.C., 1978.
3. **Chothia, C.,** Principles that determine the structure of proteins, *Annu. Rev. Biochem.,* 53, 537, 1984.
4. **Thornton, J.M., Flores, T.P., Jones, D.T., and Swindells, M.B.,** Protein structure. Prediction of progress at last [news], *Nature,* 354, 105, 1991.
5. **Blundell, T.L., Sibanda, B.L., Sternberg, M.J., and Thornton, J.M.,** Knowledge-based prediction of protein structures and the design of novel molecules, *Nature,* 326, 347, 1987.
6. **Thornton, J.M. and Gardner, S.P.,** Protein motifs and data-base searching, *Trends Biochem. Sci.,* 14, 300, 1989.
7. **Crippen, G.M.,** Prediction of protein folding from amino acid sequence over discrete conformation spaces, *Biochemistry,* 30, 4232, 1991.
8. **Clark, D.A., Barton, G.J., and Rawlings, C.J.,** A knowledge-based architecture for protein sequence analysis and structure prediction, *J. Mol. Graph.,* 8, 94, 1990.
9. **Havel, T.F. and Snow, M.E.,** A new method for building protein conformations from sequence alignments with homologues of known structure, *J. Mol. Biol.,* 217, 1, 1991.
10. **Blundell, T., Carney, D., Gardner, S., Hayes, F., Howlin, B., Hubbard, T., Overington, J., Singh, D.A., Sibanda, B.L., and Sutcliffe, M.,** 18th Sir Hans Krebs lecture. Knowledge-based protein modelling and design, *Eur. J. Biochem.,* 172, 513, 1988.
11. **Greer, J.,** Comparative model-building of the mammalian serine proteases, *J. Mol. Biol.,* 153, 1027, 1981.
12. **Schiffer, C.A., Caldwell, J.W., Kollman, P.A., and Stroud, R.M.,** Prediction of homologous protein structures based on conformational searches and energetics, *Proteins,* 8, 30, 1990.
13. **Greer, J.,** Comparative modeling of homologous proteins, *Methods Enzymol.,* 202, 239, 1991.
14. **Evans, B.E., Rittle, K.E., Bock, M.G., DiPardo, R.M., Freidinger, R.M., Whitter, W.L., Lundell, G.F., Veber, D.F., Anderson, P.S., and Chang, R.S.,** Methods for drug discovery: development of potent, selective, orally effective cholecystokinin antagonists, *J. Med. Chem.,* 31, 2235, 1988.
15. **Sander, C., Vriend, G., Bazan, F., Horovitz, A., Nakamura, H., Ribas, L., Finkelstein, A.V., Lockhart, A., Merkl, R., and Perry, L.J.,** Protein design on computers. Five new proteins: Shpilka, Grendel, Fingerclasp, Leather, and Aida, *Proteins,* 12, 105, 1992.
16. **Milner-White, E.J.,** Predicting the biological conformations of short polypeptides, *Trends Pharmacol. Sci.,* 10, 70, 1989.
17. **Breckenridge, R.J.,** Molecular recognition: models for drug design, *Experientia,* 47, 1148, 1991.
18. **Kuntz, I.D.,** Structure-based strategies for drug design and discovery [see comments], *Science,* 257, 1078, 1992.
19. **Lewis, R.A., Roe, D.C., Huang, C., Ferrin, T.E., Langridge, R., and Kuntz, I.D.,** Automated site-directed drug design using molecular lattices, *J. Mol. Graph.,* 10, 66, 1992.
20. **Lybrand, T.P.,** Molecular simulation and drug design, *J. Pharm. Belg.,* 46, 49, 1991.
21. **Mager, P.P.,** Computer-assisted series design in chemical synthesis of bioactive compounds, *Med. Res. Rev.,* 11, 375, 1991.
22. **Nilakantan, R., Bauman, N., and Venkataraghavan, R.,** A method for automatic generation of novel chemical structures and its potential applications to drug discovery, *J. Chem. Inf. Comput. Sci.,* 31, 527, 1991.
23. **Martin, Y.C.,** 3D database searching in drug design, *J. Med. Chem.,* 35, 2145, 1992.
24. **Pearlman, R.S.,** 3D-searching: an overview of a new technique for computer-assisted molecular design, *Nida. Res. Monogr.,* 112, 62, 1991.
25. **Struthers, R.S., Tanaka, G., Koerber, S.C., Solmajer, T., Baniak, E.L., Gierasch, L.M., Vale, W., Rivier, J., and Hagler, A.T.,** Design of biologically active, conformationally constrained GnRH antagonists, *Proteins,* 8, 295, 1990.
26. **Hutchins, C. and Greer, J.,** Comparative modeling of proteins in the design of novel renin inhibitors, *Crit. Rev. Biochem. Mol. Biol.,* 26, 77, 1991.
27. **Lowman, H., Cunningham, B., and Wells, J.,** Mutational analysis and protein engineering of receptor-binding determinants in human placental lactogen, *J. Biol. Chem.,* 266, 10982, 1991.
28. **Cunningham, B. and Wells, J.,** High-resolution epitope mapping of hGH-receptor interactions by alanine-scanning mutagenesis, *Science,* 244, 1081, 1989.
29. **Cunningham, B., Henner, D., and Wells, J.,** Engineering human prolactin to bind to the human growth hormone receptor, *Science,* 247, 1461, 1990.
30. **de Vos, A., Ultsch, M., and Kossiakoff, A.,** Human growth hormone and extracellular domain of its receptor: crystal structure of the complex, *Science,* 255, 306, 1992.
31. **Sharif, M. and Hanley, M.R.,** Stepping up the pressure, *Nature,* 357, 279, 1992.
32. **Anfinsen, C.B., Haber, E., Sela, M., and White, F.H., Jr.,** The kinetics of formation of native ribonuclease during oxidation of the reduced polypeptide chain, *Proc. Natl. Acad. Sci. U.S.A.,* 47, 1309, 1961.

33. **Shakhnovich, E.I. and Gutin, A.M.,** Implications of thermodynamics of protein folding for evolution of primary sequences, *Nature,* 346, 773, 1990.

34. **Baldwin, R.L.,** How does protein folding get started?, *Trends Biochem. Sci.,* 14, 291, 1989.

35. **Baldwin, R.L.,** Pieces of the folding puzzle, *Nature,* 346, 409, 1990.

36. **Creighton, T.E.,** Molecular chaperones. Unfolding protein folding [news; comment], *Nature,* 352, 17, 1991.

37. **Behe, M.J., Lattman, E.E., and Rose, G.D.,** The protein-folding problem: the native fold determines packing, but does packing determine the native fold?, *Proc. Natl. Acad. Sci. U.S.A.,* 88, 4195, 1991.

38. **Karplus, M. and Petsko, G.A.,** Molecular dynamics simulations in biology, *Nature,* 347, 631, 1990.

39. **Yapa, K., Weaver, D.L., and Karplus, M.,** Beta-sheet coil transitions in a simple polypeptide model, *Proteins,* 12, 237, 1992.

40. **Staley, J.P. and Kim, P.S.,** Role of a subdomain in the folding of bovine pancreatic trypsin inhibitor, *Nature,* 344, 685, 1990.

41. **Head-Gordon, T., Stillinger, F.H., and Arrecis, J.,** A strategy for finding classes of minima on a hypersurface: implications for approaches to the protein folding problem, *Proc. Natl. Acad. Sci. U.S.A.,* 88, 11076, 1991.

42. **Ngo, J.T. and Marks, J.,** Computational complexity of a problem in molecular structure prediction, *Protein Eng.,* 5, 313, 1992.

43. **Taylor, W.R. and Orengo, C.A.,** Protein structure alignment, *J. Mol. Biol.,* 208, 1, 1989.

44. **Nussinov, R. and Wolfson, H.J.,** Efficient detection of three-dimensional structural motifs in biological macromolecules by computer vision techniques, *Proc. Natl. Acad. Sci. U.S.A.,* 88, 10495, 1991.

45. **Vriend, G. and Sander, C.,** Detection of common three-dimensional substructures in proteins, *Proteins,* 11, 52, 1991.

46. **Kubota, Y., Takahashi, H., Yoshino, T., and Tsuchiya, T.,** Chaos-theoretical analysis of possible structural quantities for globular proteins, *Biochim. Biophys. Acta,* 1079, 73, 1991.

47. **Sklenar, H., Etchebest, C., and Lavery, R.,** Describing protein structure: a general algorithm yielding complete helicoidal parameters and a unique overall axis, *Proteins,* 6, 46, 1989.

48. **Cohen, F.E., Abarbanel, R.M., Kuntz, I.D., and Fletterick, R.J.,** Turn prediction in proteins using a pattern-matching approach, *Biochemistry,* 25, 266, 1986.

49. **Rooman, M.J., Rodriguez, J., and Wodak, S.J.,** Relations between protein sequence and structure and their significance, *J. Mol. Biol.,* 213, 337, 1990.

50. **Holley, L.H. and Karplus, M.,** Protein secondary structure prediction with a neural network, *Proc. Natl. Acad. Sci. U.S.A.,* 86, 152, 1989.

51. **Sheridan, R.P. and Venkataraghavan, R.,** A systematic search for protein signature sequences, *Proteins Struct. Funct. Genet.,* 14, 16, 1992.

52. **Richards, F.M. and Kundrot, C.E.,** Identification of structural motifs from protein coordinate data: secondary structure and first-level supersecondary structure, *Proteins,* 3, 71, 1988.

53. **Mitchell, E.M., Artymiuk, P.J., Rice, D.W., and Willett, P.,** Use of techniques derived from graph theory to compare secondary structure motifs in proteins, *J. Mol. Biol.,* 212, 151, 1990.

54. **Rooman, M.J., Rodriguez, J., and Wodak, S.J.,** Automatic definition of recurrent local structure motifs in proteins, *J. Mol. Biol.,* 213, 327, 1990.

55. **Rackovsky, S.,** Quantitative organization of the known protein X-ray structures. I. Methods and short-length-scale results, *Proteins,* 7, 378, 1990.

56. **Vingron, M. and Argos, P.,** Motif recognition and alignment for many sequences by comparison of dot-matrices, *J. Mol. Biol.,* 218, 33, 1991.

57. **Edelman, J. and White, S.H.,** Linear optimization of predictors for secondary structure. Application to transbilayer segments of membrane proteins, *J. Mol. Biol.,* 210, 195, 1989.

58. **Singh, J. and Thornton, J.M.,** SIRIUS. An automated method for the analysis of the preferred packing arrangements between protein groups, *J. Mol. Biol.,* 211, 595, 1990.

59. **Leszczynski, J.F. and Rose, G.D.,** Loops in globular proteins: a novel category of secondary structure, *Science,* 234, 849, 1986.

60. **Chakrabartty, A., Schellman, J.A., and Baldwin, R.L.,** Large differences in the helix propensities of alanine and glycine, *Nature,* 351, 586, 1991.

61. **MacArthur, M.W. and Thornton, J.M.,** Influence of proline residues on protein conformation, *J. Mol. Biol.,* 218, 397, 1991.

62. **Summers, N.L. and Karplus, M.,** Modeling of globular proteins. A distance-based data search procedure for the construction of insertion/deletion regions and Pro----non-Pro mutations, *J. Mol. Biol.,* 216, 991, 1990.

63. **Lee, C. and Subbiah, S.,** Prediction of protein side-chain conformation by packing optimization, *J. Mol. Biol.,* 217, 373, 1991.

64. **Heringa, J. and Argos, P.,** Side-chain clusters in protein structures and their role in protein folding, *J. Mol. Biol.,* 220, 151, 1991.

65. **Kabsch, W. and Sander, C.,** Dictionary of protein secondary structure: pattern recognition of hydrogen-bonded and geometrical features, *Biopolymers,* 22, 2577, 1983.

66. **Factor, A.D. and Mehler, E.L.,** Graphical representation of hydrogen bonding patterns in proteins, *Protein Eng.,* 4, 421, 1991.

67. **Gregoret, L.M. and Cohen, F.E.,** Novel method for the rapid evaluation of packing in protein structures, *J. Mol. Biol.,* 211, 959, 1990.

68. **Bruccoleri, R.E. and Karplus, M.,** Prediction of the folding of short polypeptide segments by uniform conformational sampling, *Biopolymers,* 26, 137, 1987.

69. **Moult, J. and James, M.N.,** An algorithm for determining the conformation of polypeptide segments in proteins by systematic search, *Proteins,* 1, 146, 1986.

70. **Summers, N.L. and Karplus, M.,** Modeling of side-chains, loops, and insertions in proteins, *Methods Enzymol.,* 202, 156, 1991.

71. **Bordo, D. and Argos, P.,** Evolution of protein cores. Constraints in point mutations as observed in globin tertiary structures, *J. Mol. Biol.,* 211, 975, 1990.

72. **Argos, P.,** Analysis of sequence-similar pentapeptides in unrelated protein tertiary structures. Strategies for protein folding, *J. Mol. Biol.,* 197, 331, 1987.

73. **Kabsch, W. and Sander, C.,** On the use of sequence homologies to predict protein structure: identical pentapeptides can have completely different conformations, *Proc. Natl. Acad. Sci. U.S.A.,* 81, 1075, 1984.

74. **Lipman, D.J., Altschul, S.F., and Kececioglu, J.D.,** A tool for multiple sequence alignment, *Proc. Natl. Acad. Sci. U.S.A.,* 86, 4412, 1989.

75. **Altschul, S.F., Carroll, R.J., and Lipman, D.J.,** Weights for data related by a tree, *J. Mol. Biol.,* 207, 647, 1989.

76. **Gribskov, M., Luthy, R., and Eisenberg, D.,** Profile analysis, *Methods Enzymol.,* 183, 146, 1990.

77. **Subbiah, S. and Harrison, S.C.,** A method for multiple sequence alignment with gaps, *J. Mol. Biol.,* 209, 539, 1989.

78. **Carrillo, H. and Lipman, D.,** The multiple sequence alignment problem in biology, *Siam. J. Appl. Math.,* 48, 1073, 1988.

79. **Barton, G.J. and Sternberg, M.J.,** Flexible protein sequence patterns. A sensitive method to detect weak structural similarities, *J. Mol. Biol.,* 212, 389, 1990.

80. **Depiereux, E. and Feytmans, E.,** Simultaneous and multivariate alignment of protein sequences: correspondence between physicochemical profiles and structurally conserved regions (SCR), *Protein Eng.,* 4, 603, 1991.

81. **Sibbald, P.R. and Argos, P.,** Weighting aligned protein or nucleic acid sequences to correct for unequal representation, *J. Mol. Biol.,* 216, 813, 1990.

82. **Saqi, M.A. and Sternberg, M.J.,** A simple method to generate non-trivial alternate alignments of protein sequences, *J. Mol. Biol.,* 219, 727, 1991.

83. **Fujiyoshi-Yoneda, T., Yoneda, S., Kitamura, K., Amisaki, T., Ikeda, K., Inoue, M., and Ishida, T.,** Adaptability of restrained molecular dynamics for tertiary structure prediction: application to Crotalus atrox venom phospholipase A2, *Protein Eng.,* 4, 443, 1991.

84. **Hinds, D.A. and Levitt, M.,** A lattice model for protein structure prediction at low resolution, *Proc. Natl. Acad. Sci. U.S.A.,* 89, 2536, 1992.

85. **Lau, K.F. and Dill, K.A.,** A lattice statistical mechanics model of the conformational and sequence spaces of proteins, *Macromolecules,* 22, 3986, 1989.

86. **Skolnick, J. and Kolinski, A.,** Simulations of the folding of a globular protein, *Science,* 250, 1121, 1990.

87. **King, R.D. and Sternberg, M.J.,** Machine learning approach for the prediction of protein secondary structure, *J. Mol. Biol.,* 216, 441, 1990.

88. **Stolorz, P., Lapedes, A., and Xia, Y.,** Predicting protein secondary structure using neural net and statistical methods, *J. Mol. Biol.,* 225, 363, 1992.

89. **Hirst, J.D. and Sternberg, M.J.,** Prediction of ATP-binding motifs: a comparison of a perceptron-type neural network and a consensus sequence method, *Protein Eng.,* 4, 615, 1991.

90. **Qian, N. and Sejnowski, T.J.,** Predicting the secondary structure of globular proteins using neural network models, *J. Mol. Biol.,* 202, 865, 1988.

91. **van Heel, M.,** A new family of powerful multivariate statistical sequence analysis techniques, *J. Mol. Biol.,* 220, 877, 1991.

92. **Bowie, J.U., Luthy, R., and Eisenberg, D.,** A method to identify protein sequences that fold into a known three-dimensional structure, *Science,* 253, 164, 1991.

93. **Lüthy, R., Bowie, J.U., and Eisenberg, D.,** Assessment of protein models with three-dimensional profiles, *Nature,* 356, 83, 1992.

94. **Hendlich, M., Lackner, P., Weitckus, S., Floeckner, H., Froschauer, R., Gottsbacher, K., Casari, G., and Sippl, M.J.,** Identification of native protein folds amongst a large number of incorrect models. The calculation of low energy conformations from potentials of mean force, *J. Mol. Biol.,* 216, 167, 1990.

95. **Finkelstein, A.V. and Reva, B.A.,** A search for the most stable folds of protein chains, *Nature,* 351, 497, 1991.

96. **Sippl, M.J.,** Calculation of conformational ensembles from potentials of mean force. An approach to the knowledge-based prediction of local structures in globular proteins, *J. Mol. Biol.,* 213, 859, 1990.

97. **Holm, L. and Sander, C.,** Database algorithm for generating protein backbone and side-chain co-ordinates from a C alpha trace application to model building and detection of co-ordinate errors, *J. Mol. Biol.,* 218, 183, 1991.

98. **Sali, A. and Blundell, T.L.,** Definition of general topological equivalence in protein structures. A procedure involving comparison of properties and relationships through simulated annealing and dynamic programming, *J. Mol. Biol.,* 212, 403, 1990.

99. **Okamoto, Y., Fukugita, M., Nakazawa, T., and Kawai, H.,** Alpha-helix folding by Monte Carlo simulated annealing in isolated C-peptide of ribonuclease A, *Protein Eng.,* 4, 639, 1991.

100. **Balaji, V.N. and Ramnarayan, K.,** Computer assisted design of peptidomimetic drugs, in *Biologically Active Peptides: Design, Synthesis, and Utilization,* Williams, W.V. and Weiner, D.B., Eds., Technomic Publications, Lancaster, PA, 1993, 35.

101. **Paul, P.K., Burney, P.A., Campbell, M.M., and Osguthorpe, D.J.,** The conformational preferences of γ-lactam and its role in constraining peptide structures, *J. Comput. Aided Mol. Des.,* 4, 239, 1990.

102. **Bhandary, K.K. and Kopple, K.D.,** Conformation of cyclic octapeptides. VI. Structure of cyclo-bis-(-L-abanyl-glycyl-L-prolyl-L-phenylanyl-) tetrahydrate, *Acta Crystallogr.,* C47, 1280, 1991.

103. **Cerrini, S., Gavuzzo, E., Lucente, G., Luisi, G., Pinnen, F., and Radics, L.,** Ten-membered cyclotripeptides. III. Synthesis and conformation of cyclo(-Me beta Ala-Phe-Pro-) and cyclo(-Me beta Ala-Phe-DPro-), *Int. J. Pept. Protein Res.,* 38, 289, 1991.

104. **Wynants, C., Tourwe, D., Kazmierski, W., Hruby, V.J., and Van, B.G.,** Conformation of two somatostatin analogues in aqueous solution. Study by NMR methods and circular dichroism, *Eur. J. Biochem.,* 185, 371, 1989.

105. **Williams, W.V., Kieber-Emmons, T., Von, F.J., Greene, M.I., and Weiner, D.B.,** Design of bioactive peptides based on antibody hypervariable region structures. Development of conformationally constrained and dimeric peptides with enhanced affinity, *J. Biol. Chem.,* 266, 5182, 1991.

106. **Gilon, C., Halle, D., Chorev, M., Selinger, Z., and Byk, G.,** Backbone cyclization: a new method for conferring conformational constraint on peptides, *Biopolymers,* 31, 745, 1991.

107. **Sherman, D.B., Spatola, A.F., Wire, W.S., Burks, T.F., Nguyen, T.M., and Schiller, P.W.,** Biological activities of cyclic enkephalin pseudopeptides containing thioamides as amide bond replacements, *Biochem. Biophys. Res. Commun.,* 162, 1126, 1989.

108. **Hermann, R.B. and Herron, D.K.,** OVID and SUPER: two overlap programs for drug design, *J. Comput. Aided Mol. Design,* 5, 511, 1991.

109. **DesJarlais, R.L., Sheridan, R.P., Seibel, G.L., Dixon, J.S., Kuntz, I.K., and Venkataraghavan, R.,** Using shape complementarity as an initial screen in designing ligands for a receptor binding site of known three dimensional structure, *J. Med. Chem.,* 31, 722, 1988.

110. **DesJarlais, R.L., Seibel, G.L., Kuntz, I.D., Furth, P.S., Alvarez, J.C., DeMontellano-Ortiz, P.R., DeCamp, D.L., Babe, L.M., and Craik, C.S.,** Structure-based design of nonpeptide inhibitors specific for the human immunodeficiency virus 1 protease, *Proc. Natl. Acad. Sci. U.S.A.,* 87, 6644, 1990.

111. **Brint, A.T. and Willett, P.,** Pharmacoric pattern matching in files of 3-D chemical structures: comparison of geometric searching algorithms, *J. Mol. Graphics,* 5, 49, 1987.

112. **Sheridan, R.P., Rusinko, A., Nilakantan, R., and Venkataraghavan, R.,** Searching for pharmacores in large coordinate data bases and its use in drug design, *Proc. Natl. Acad. Sci. U.S.A.,* 86, 8165, 1989.

113. **Van Drie, H.H., Weininger, D., and Martin, Y.C.,** ALADDIN: an integrated tool for computer-assisted molecular design and pharmacore recognition from geometric, steric, and substructure searching of three-dimensional molecular structures, *J. Comput. Aided Mol. Design,* 3, 225, 1989.

114. **Cohen, N.C., Blaney, J.M., Humblet, C., Gund, P., and Barry, D.C.,** Molecular modeling software and methods for medicinal chemistry, *J. Med. Chem.,* 33, 883, 1990.

115. **Bartlett, P.A., T., S.G., Telfer, S.J., and Waterman, S.,** CAVEAT: a program to facilitate the structure-derived design of biologically active molecules, in *Molecular Recognition in Chemical and Biological Problems,* Roberts, S.M., Ed., Royal Society of Chemistry, Cambridge, 78, 182, 1989.

116. **Goodford, P.J.A.,** A computational procedure for determining energetically favorable binding sites on biologically important macromolecules, *J. Med. Chem.,* 28, 849, 1985.

117. **DesJarlais, R.L., Sheridan, R.P., Dixon, J.S., Kuntz, I.K., and Venkataraghavan, R.,** Docking flexible ligands to macromolecular receptors by molecular shape, *J. Med. Chem.,* 29, 2149, 1986.

118. **Hirono, S. and Kollman, P.A.,** Relative binding free energy calculations of inhibitors to two mutants (Glu46----Ala/Gln) of ribonuclease T1 using molecular dynamics/free energy perturbation approaches, *Protein Eng.,* 4, 233, 1991.

119. **Moon, J.B and Howe, W.J.,** Computer design of bioactive molecules: a method for receptor-based de novo ligand design, *Proteins,* 11, 314, 1991.

120. **Ooi, T., Oobatake, M., Nemethy, G., and Scherago, H.A.,** Accessible surface areas as a measure of the thermodynamic parameters of hydration of peptides, *Proc. Natl. Acad. Sci. U.S.A.,* 84, 3086, 1987.

121. **Kirkpatrick, S., Gelatt, C.D., and Vecchi, M.P.,** Optimization by simulated annealing, *Science,* 220, 671, 1983.

122. **Smellie, A.S., Crippen, G.M., and Richards, W.G.,** Fast drug-receptor mapping by site-directed distances: a novel method of predicting new pharmacological leads, *J. Chem. Inf. Comput. Sci.,* 31, 386, 1991.

123. **Bacon, D.J. and Moult, J.,** Docking by least-squares fitting of molecular surface patterns, *J. Mol. Biol.,* 225, 849, 1992.

124. **Caflisch, A., Niederer, P., and Anliker, M.,** Monte Carlo docking of oligopeptides to proteins, *Proteins,* 13, 223, 1992.

125. **Goodsell, D.S. and Olson, A.J.,** Automated docking of substrates to proteins by simulated annealing, *Proteins,* 8, 195, 1990.

126. **Hart, T.N. and Read, R.J.,** A multiple-start Monte Carlo docking method, *Proteins,* 13, 206, 1992.

127. **Kasinos, N., Lilley, G.A., Subbarao, N., and Haneef, I.,** A robust and efficient automated docking algorithm for molecular recognition, *Protein Eng.,* 5, 69, 1992.

128. **Shoichet, B.K. and Kuntz, I.D.,** Protein docking and complementarity, *J. Mol. Biol.,* 221, 327, 1991.

129. **Billeter, M., Engeli, M., and Wüthrich, K.,** The ellipsoid algorithm as a method for the determination of polypeptide conformations from experimental distance constraints and energy minimization, *J. Comput. Chem.,* 8, 132, 1987.

130. **Cramer, R.D., Patterson, D.E., and Bunce, J.D.,** Comparative molecular field analysis (CoFMA), *J. Am. Chem. Soc.,* 110, 5959, 1988.

131. **Doweyko, A.M.,** The hypothetical active site lattice. An approach to modeling active sites from data on inhibitor molecules, *J. Med. Chem.,* 31, 1396, 1988.

132. **Dean, P.M.,** *Molecular Foundations of Drug-Receptor Interaction,* Cambridge University Press, Cambridge, 1987.

133. **Sheridan, R.P. and Venkataraghavan, R.,** New methods of computer aided drug design, *Acc. Chem. Res.,* 20, 322, 1987.

134. **Freidinger, R.M.,** Non-peptide ligands for peptide receptors, *Trends Pharmacol. Sci.,* 10, 270, 1989.

135. **Kieber-Emmons, T., Krowka, J. F., Boyer, J., Ugen, K. E., Williams, W. V., Morrow, W. J. W., and Weiner, D. B.,** Immunological characteristics of the putative CD4-binding site of the HIV-1 envelope protein, *Pathobiology,* 60, 187, 1992.

136. **Lüthy, R., McLachlan, A.D., and Eisenberg, D.,** Secondary structure-based profiles: use of structure-conserving scoring tables in searching protein sequence databases for structural similarities, *Proteins,* 10, 229, 1991.

137. **Fine, R.M., Wang, H., Shenkin, P.S., Yarmush, D.L., and Levinthal, C.,** Predicting antibody hypervariable loop conformations. II. Minimization and molecular dynamics studies of MCPC603 from many randomly generated loop conformations, *Proteins,* 1, 342, 1986.

138. **Shenkin, P. S., Yarmush, D. L., Fine, R. M., Wang, H. J., and Levinthal, C.,** Predicting antibody hypervariable loop conformation. I. Ensembles of random conformations for ringlike structures, *Biopolymers,* 26, 2053, 1987.

139. **Bashford, D., Chothia, C., and Lesk, A.M.,** Determinants of a protein fold. Unique features of the globin amino acid sequences, *J. Mol. Biol.,* 196, 199, 1987.

140. **Webster, T.A., Lathrop, R.H., and Smith, T.F.,** Prediction of the common structural domain in acid t-RNA synthetases through use of a new pattern-directed inference system, *Biochemistry,* 26, 6950, 1987.

141. **Overington, J., Johnson, M.S., Sali, A., and Blundell, T.L.,** Tertiary structural constraints on protein evolutionary diversity: templates, key residues and structure prediction, *Proc. R. Soc. Lond. (Biol.),* 241, 132, 1990.

142. **Geourjon, C., Deleage, G., and Roux, B.,** ANTHEPROT: an interactive graphics software for analyzing protein structures from sequences, *J. Mol. Graph.,* 9, 188, 1991.

143. **Benner, S.A. and Gerloff, D.,** Patterns of divergence in homologous proteins as indicators of secondary and tertiary structure: a prediction of the structure of the catalytic domain of protein kinases, *Adv. Enzyme Regul.,* 31, 121, 1991.

144. **Ponder, J.W. and Richards, F.M.,** Tertiary templates for proteins. Use of packing criteria in the enumeration of allowed sequences for different structural classes, *J. Mol. Biol.,* 193, 775, 1987.

145. **Crawford, I.P., Niermann, T., and Kirshner, K.,** Predictions of secondary structure by evolutionary comparison: application to the α-subunit of tryptophan synthase, *Proteins,* 1, 118, 1987.

146. **Taylor, W.R. and Green, N.M.,** The predicted secondary structures of the nucleotide-binding sites of six cation-transporting ATPases lead to a probable tertiary fold, *Eur. J. Biochem.,* 179, 241, 1989.

147. **Garnier, J., Osguthorpe, D.J., and Robson, B.,** Analysis of the accuracy and implications of simple methods for predicting the secondary structure of globular proteins, *J. Mol. Biol.,* 120, 97, 1978.

148. **Hecht, M.H., Richardson, J.S., Richardson, D.C., and Ogden, R.C.,** De novo design, expression, and characterization of felix: a four-helix bundle protein of native-like sequence, *Science,* 249, 884, 1990.

149. **Hellinga, H.W. and Richards, F.M.,** Construction of new ligand binding sites in proteins of known structure. I. Computer-aided modeling of sites with pre-defined geometry, *J. Mol. Biol.,* 222, 763, 1992.

150. **Hellinga, H.W. and Richards, F.M.,** Construction of new ligand binding sites in proteins of known structure. II. Grafting of a buried transition metal binding site into *Escherichia coli* Thioredoxin, *J. Mol. Biol.,* 222, 787, 1992.

151. **Shoelson, S.E., Lu, Z.X., Parlautan, L., Lynch, C.S., and Weiss, M.A.,** Mutations at the dimer, hexamer, and receptor-binding surfaces of insulin independently affect insulin-insulin and insulin-receptor interactions, *Biochemistry,* 31, 1757, 1992.

152. **Fuh, G., Cunningham, B. C., Fukunaga, R., Nagata, S., Goeddel, D.V., and Wells, J.A.,** Rational design of potent antagonists to the human growth hormone receptor, *Science,* 256, 1677, 1992.

153. **Needleman, S.B. and Wunsch, C.D.,** A general method applicable to the search for similarities in the amino acid sequence of two proteins, *J. Mol. Biol.,* 48, 443, 1970.

154. **Pearson, W.R. and Miller, W.,** Dynamic programming algorithms for biological sequence comparison, *Methods Enzymol.,* 210, 575, 1992.

155. **Barton, G.J.,** Protein multiple sequence alignment and flexible pattern matching, *Methods Enzymol.,* 183, 403, 1990.

156. **Huang, X. and Miller, W.,** A time-efficient, linear-space local similarity algorithm, *Adv. Appl. Math.,* 12, 337, 1991.

157. **Taylor, W.R.,** Hierarchical method to align large numbers of biological sequences, *Methods Enzymol.,* 183, 456, 1990.

158. **Mott, R.F., Kirkwood, T.B., and Curnow, R.N.,** Tests for the statistical significance of protein sequence similarities in data-bank searches, *Protein Eng.,* 4, 149, 1990.

159. **Tyson, H.,** Relationships between amino acid sequences determined through optimum alignments, clustering, and specific distance patterns: application to a group of scorpion toxins, *Genome,* 35, 360, 1992.

160. **Dayhoff, M.O., Barker, W.C., and Hunt, L.T.,** Establishing homologies in proteins, *Methods Enzymol.,* 91, 524, 1983.

161. **Altschul, S.F.,** Amino acid substitution matrices from an information theoretic perspective, *J. Mol. Biol.,* 219, 555, 1991.

162. **Engelman, D.M., Steitz, T.A., and Goldman, A.,** Identifying non-nonpolar transbilayer helices in amino acid sequences of membrane proteins, *Annu. Rev. Biophys. Biophys. Chem.,* 15, 321, 1986.

163. **Bowie, J.U., Clarke, N.D., Pabo, C.O., and Sauer, R.T.,** Identification of protein folds: matching hydrophobicity patterns of sequence sets with solvent accessibility patterns of known structures, *Proteins,* 7, 257, 1990.

164. **Garratt, R.C., Thornton, J.M., and Taylor, W.R.,** An extension of secondary structure prediction towards the prediction of tertiary structure, *Febs Lett.,* 280, 141, 1991. [Published erratum appears in *FEBS Lett.,* 280(2), 401, 1991.]

165. **Dorit, R.L., Schoenbach, L., and Gilbert, W.,** How big is the universe of exons?, *Science,* 250, 1377, 1990.

166. **Rooman, M.J. and Wodak, S.J.,** Weak correlation between predictive power of individual sequence patterns and overall prediction accuracy in proteins, *Proteins,* 9, 69, 1991.

167. **Pascarella, S. and Argos, P.,** A data bank merging related protein structures and sequences, *Protein Eng.,* 5, 121, 1992.

168. **Gonnet, G.H., Cohen, M.A., and Benner, S.A.,** Exhaustive matching of the entire protein sequence database, *Science,* 256, 1443, 1992.

169. **Akrigg, D., Attwood, T.K., Bleasby, A.J., Findlay, J.B., North, A.C., Maughan, N.A., Parry, S.D., Perkins, D.N., and Wootton, J.C.,** SERPENT — an information storage and analysis resource for protein sequences, *Comput. Appl. Biosci.,* 8, 295, 1992.

170. **Islam, S.A. and Sternberg, M.J.,** A relational database of protein structures designed for flexible enquiries about conformation, *Protein Eng.,* 2, 431, 1989.

171. **Huysmans, M., Richelle, J., and Wodak, S.J.,** SESAM: a relational database for structure and sequence of macromolecules, *Proteins,* 11, 59, 1991.

172. SERATUS, IDITUS, distributed by Oxford Molecular, Oxford, England.

173. Oracle Management System, distributed by Oracle Corporation, Bristol, U.K.

174. **Bairoch, A.,** PROSITE Dictionary of Protein Sites and Patterns, University of Geneva, Geneva, Switzerland, 1992.

175. **Sternberg, M.J.,** PROMOT: a FORTRAN program to scan protein sequences against a library of known motifs, *Comput. Appl. Biosci.,* 7, 257, 1991.

176. **Sander, C. and Schneider, R.,** Database of homology-derived protein structures and the structural meaning of sequence alignment, *Proteins,* 9, 56, 1991.

177. **Viswanadhan, V.N., Denckla, B., and Weinstein, J.N.,** New joint prediction algorithm (Q7-JASEP) improves the prediction of protein secondary structure, *Biochemistry,* 30, 11164, 1991.

178. **Zhang, X., Mesirov, J.P., and Waltz, D.L.,** Hybrid system for protein secondary structure prediction, *J. Mol. Biol.,* 225, 1049, 1992.

179. **Parker, J.M. and Hodges, R.S.,** Prediction of surface and interior regions in proteins. II. Predicting secondary structure in regions bound by surface exposed regions, *Pept. Res.,* 4, 355, 1991.

180. **Garnier, J.,** Protein structure prediction, *Biochimie,* 72, 513, 1990.

181. **Nishikawa, K. and Noguchi, T.,** Predicting protein secondary structure based on the amino acid sequence, *Methods Enzymol.,* 202, 31, 1991.

182. **Fasman, G.D.,** The development of the prediction of protein structure, in *Prediction of Protein Structure and Principles of Protein Conformation,* Fasman, G.D., Ed., Plenum Press, New York, 1989, 193.

183. **Taylor, W.R. and Thornton, J.M.,** Recognition of super-secondary structure in proteins, *J. Mol. Biol.,* 173, 487, 1984.

184. **Ptitsyn, O.B. and Finkelstein, A.V.,** Theory of protein secondary structure and algorithm of its prediction, *Biopolymers,* 22, 15, 1983.

185. **Goel, N.S. and Ycas, M.,** On the computation of the tertiary structure of globular proteins II, *J. Theor. Biol.,* 77, 253, 1979.

186. **Jones, D.T., Taylor, W.R., and Thornton, J.M.,** A new approach to protein fold recognition, *Nature,* 358, 86, 1992.

187. **Kneller, D.G., Cohen, F.E., and Langridge, R.,** Improvements in protein secondary structure prediction by an enhanced neural network, *J. Mol. Biol.,* 214, 171, 1990.

188. **Garnier, J. and Levin, J.M.,** The protein structure code: what is its present status?, *Comput. Appl. Biosci.,* 7, 133, 1991.

189. **Sternberg, M.J. and Zvelebil, M.J.,** Prediction of protein structure from sequence, *Eur. J. Cancer,* 26, 1163, 1990.

190. **Vieth, M. and Kolinski, A.,** Prediction of protein secondary structure by an enhanced neural network, *Acta Biochim. Pol.,* 38, 335, 1991.

191. **McGregor, M.J., Flores, T.P., and Sternberg, M.J.,** Prediction of beta-turns in proteins using neural networks, *Protein Eng.,* 2, 521, 1989. [Published erratum appears in *Protein Eng.,* 3(5), 459, 1990.]

192. **Hirst, J.D. and Sternberg, M.J.,** Prediction of structural and functional features of protein and nucleic acid sequences by artificial neural networks, *Biochemistry,* 31, 7211, 1992.

193. **Ferran, E.A. and Ferrara, P.,** Clustering proteins into families using artificial neural networks, *Comput. Appl. Biosci.,* 8, 39, 1992.

194. **Muskal, S.M. and Kim, S.H.,** Predicting protein secondary structure content. A tandem neural network approach, *J. Mol. Biol.,* 225, 713, 1992.

195. **Karlin, S., Bucher, P., Brendel, V., and Altschul, S.F.,** Statistical methods and insights for protein and DNA sequences, *Annu. Rev. Biophys. Chem.,* 20, 175, 1991.

196. **Chou, P.Y. and Fasman, G.D.,** Prediction of the secondary structure of proteins from their amino acid sequence, *Adv. Enzymol.,* 47, 45, 1978.

197. **Gibrat, J., Garnier, J., and Robson, B.,** Further developments of protein secondary structure prediction using information theory: new parameters and consideration of residue pairs, *J. Mol. Biol.,* 198, 425, 1987.

198. **Nagano, K.,** Triplet formation in helix prediction applied to the analysis of super-secondary structures, *J. Mol. Biol.,* 109, 251, 1977.

199. **Lambert, M.H. and Scheraga, H.A.,** Pattern recognition in the prediction of protein structure. I. Tripeptide conformational probabilities calculated from the amino acid sequence, *J. Comp. Chem.,* 10, 770, 1989.

200. **Lambert, M.H. and Scheraga, H.A.,** Pattern recognition in the prediction of protein structure. II. Chain conformation from a probability-directed search procedure, *J. Comp. Chem.,* 10, 798, 1989.

201. **Lambert, M.H. and Scheraga, H.A.,** Pattern recognition in the prediction of protein structure. III. An importance-sampling minimization procedure, *J. Comp. Chem.,* 10, 817, 1989.

202. **Kidera, A., Konishi, Y., Oka, M., Ooi, T., and Scheraga, H.A.,** Statistical analysis of the physical properties of the 20 naturally occurring amino acids, *J. Protein Chem.,* 4, 23, 1985.

203. **Rackovsky, S. and Goldstein, D.A.,** Protein comparison and classification: a differential geometric approach, *Proc. Natl. Acad. Sci. U.S.A.,* 85, 777, 1988.

204. **Zuker, M. and Somorjai, R.L.,** The alignment of protein structures in three dimensions, *Bull. Math. Biol.,* 51, 55, 1989.

205. **Rackovsky, S. and Scheraga, H.A.,** Differential geometry and polymer conformation. I. Comparison of protein conformations, *Macromolecules,* 11, 1168, 1978.

206. **Rackovsky, S. and Scheraga, H.A.,** Differential geometry and polymer conformation. II. Development of a conformational distance function, *Macromolecules,* 13, 1440, 1980.

207. **Rackovsky, S. and Scheraga, H.A.,** Differential geometry and polymer conformation. III. Single-site and nearest-neighbor distributions, and nucleation of protein folding, *Macromolecules,* 15, 1340, 1981.

208. **Rackovsky, S. and Scheraga, H.A.,** Differential geometry and polymer conformation. IV. Conformational and nucleation properties of individual amino acids, *Macromolecules,* 15, 1340, 1982.

209. **Louie, A.H. and Somorjai, R.L.,** Differential geometry of proteins. Helical approximations, *J. Mol. Biol.,* 168, 143, 1983.

210. **Levinson, S.E., Rabiner, L.R., and Sondhi, M.M.,** An introduction to the application of the theory of probabilistic functions of a Markov process to automatic speech recognition, *The Bell Syst. Tech. J.,* 62, 1035, 1983.

211. **Rabiner, L.R.,** A tutorial on hidden Markov models and selected applications in speech recognition, *Proc. IEEE,* 77, 257, 1989.

212. **Dempster, A.P., Laird, N.M., and Rubin, D.B.,** Maximum likelihood from incomplete data via the EM algorithm, *J. R. Stat. Soc.,* 108, 1, 1977.

213. **Lawrence, C.E. and Reilly, A.A.,** An expectation maximization (EM) algorithm for the identification and characterization of common sites in unaligned biopolymer sequences, *Proteins,* 7, 41, 1990.

214. **Skolnick, J. and Kolinski, A.,** Dynamic Monte Carlo simulations of a new lattice model of globular protein folding, structure and dynamics, *J. Mol. Biol.,* 221, 499, 1991.

215. **Ripoll, D.R., Vasquez, M.J., and Scheraga, H.A.,** The electrostatically driven Monte Carlo method: application to conformational analysis of decaglycine, *Biopolymers,* 31, 319, 1991.

216. **Chou, K.C. and Carlacci, L.,** Simulated annealing approach to the study of protein structures, *Protein Eng.,* 4, 661, 1991.

217. **Hagler, A.T., Huler, E., and Lifson, S.,** Energy functions for peptides and proteins. I. Derivation of a consistent force field including the hydrogen bond from amide crystals, *J. Am. Chem. Soc.,* 96, 5319, 1974.

218. **Meirovitch, H. and Scheraga, H.,** An approach to the multiple-minimum problem in protein folding, involving a long-range geometrical restriction and short-, medium-, and long-range interactions, *Macromolecules,* 14, 1250, 1981.

219. **Burkert, U. and Allinger, N.L.,** *Molecular Mechanics,* American Chemical Society, Washington, D.C., 1982.

220. **Momany, F.A., Mcguire, R.F., Burgess, A.W., and Scheraga, H.A.,** Energy parameters in polypeptides. VII. Geometric parameters, partial atomic charges, nonbonded interactions, hydrogen bond interactions, and intrinsic torsional potentials for the naturally occurring amino acids, *J. Phys. Chem.,* 79, 2361, 1975.

221. **Nemethy, G., Pottle, M.S., and Scherago, H.A.,** Energy parameters in polypeptides. IX. Updating of geometrical parameters, nonbonded interactions, and hydrogen bond interactions for the naturally occurring amino acids [ECEPP/2], *J. Phys. Chem.,* 87, 1883, 1983.

222. **Weiner, S.J., Kollman, P.A., Nguyen, D.T., and Case, D.A.,** An all-atom force field for simulations of proteins and nucleic acids., *J. Comput. Chem.,* 7, 230, 1986.

223. **Singh, U.C., Weiner, P.K., Case, D.A., Caldwell, J., and Kollman, P.A.,** AMBER 3.0, distributed by Department of Pharmaceutical Chem., University of California, San Francisco, CA.

224. **van Gunsteren, W.F. and Berendsen, H.C.J.,** GROMOS (Groningen Molecular Simulation), distributed by BIOMOS B.V., Nijenborgh, 16, 9747 AG Groningen, the Netherlands.

225. **Brooks, B.R., Bruccoleri, R.E., Olafson, B.D., States, D.J., Swaminathan, S., and Karplus, M.,** CHARMM: a program for macromolecular energy minimization and dynamics calculations, *J. Comput. Chem.,* 4, 187, 1983.

226. **CHARMm,** distributed by Polygen Corporation, Waltham, MA.

227. **Hagler, A.T., Lifson, S., and Dauber, P.,** Consistent force field studies of intermolecular forces in hydrogen bonded crystals. II. A benchmark for the objective comparisons of alternative force fields, *J. Am. Chem. Soc.,* 101, 5122, 1979.

228. **Allinger, N.L., Yuh, Y.H., and Lii, J.H.,** Molecular mechanics. The MM3 forcefield for hydrocarbons, *J. Am. Chem. Soc.,* 111, 8551, 1989.

229. **Morales, L.B., Garduno, J.R., and Romero, D.,** Applications of simulated annealing to the multiple-minima problem in small peptides, *J. Biomol. Struct. Dyn.,* 8, 721, 1991.

230. **Ripoll, D.R. and Scheraga, H.A.,** The multiple-minima problem in the conformational analysis of polypeptides. III. An electrostatically driven Monte Carlo method: tests on enkephalin, *J. Protein Chem.,* 8, 263, 1989.

231. **Ripoll, D.R., Piela, L., Vasquez, M., and Scheraga, H.A.,** On the multiple-minima problem in the conformational analysis of polypeptides. V. Application of the self-consistent electrostatic field and the electrostatically driven Monte Carlo methods to bovine pancreatic trypsin inhibitor, *Proteins,* 10, 188, 1991.

232. **Smith, G.M. and Veber, D.F.,** Computer-aided, systematic search of peptide conformations constrained by NMR data, *Biochem. Biophys. Res. Commun.,* 134, 907, 1986.

233. **Vasquez, M. and Scheraga, H.A.,** Use of the buildup and energy-minimization procedures to compute low-energy structures of the backbone of enkephalin, *Biopolymers,* 24, 1437, 1985.

234. **Simon, I., Glasser, L., and Scheraga, H.A.,** Calculation of protein conformation as an assembly of stable overlapping segments: application to bovine pancreatic trypsin inhibitor, *Proc. Natl. Acad. Sci. U.S.A.,* 88, 3661, 1991.

235. **Purisima, E.O. and Scheraga, H.A.,** An approach to the multiple-minima problem by relaxing dimensionality, *Proc. Natl. Acad. Sci. U.S.A.,* 83, 2782, 1986.

236. **Struthers, R.S., Rivier, J., and Hagler, A.T.,** Design of peptide analogs: theoretical simulation of conformation, energetics and dynamics, in *Conformationally Directed Drug Design: Peptides and Nucleic Acids as Templates or Targets,* Vida, J.A. and Gordon, M., Eds., American Chemical Society, Washington, D.C., 1984.

237. **Hagler, A.T.,** Theoretical simulation of conformation, energetics and dynamics of peptides, in *The Peptides,* Hruby, V.J. and Meienhofer, J., Eds., Academic Press, New York, 7, 213, 1985.

238. **Clore, G.M., Brünger, A.T., Karplus, M., and Gronenborn, A.M.,** Application of molecular dynamics with inter-proton distance restraints to three-dimensional nuclear Overhauser effect spectra, *J. Mol. Biol.,* 191, 523, 1986.

239. **Nilges, M., Gronenborn, A.M., Brünger, A.T., and Clore, G.M.,** Determination of three-dimensional structures of proteins by simulated annealing with interproton distance restraints. Application to crambin, potato carboxypeptidase inhibitor and barley serine proteinase inhibitor 2, *Protein Eng.,* 2, 27, 1988.

240. **Bruccoleri, R.E. and Karplus, M.,** Conformational sampling using high-temperature molecular dynamics, *Biopolymers,* 29, 1847, 1990.

241. **Scheek, R.M., van Gunsteren, W.F., and Kaptein, R.,** Molecular dynamics simulation techniques for determination of molecular structures from nuclear magnetic resonance data, *Methods Enzymol.,* 177, 204, 1989.

242. **Braun, W. and Go, N.,** Calculation of protein conformations by proton-proton distance constraints. A new efficient algorithm, *J. Mol. Biol.,* 186, 611, 1985.

243. **Crippen, G.M. and Havel, T.F.,** *Distance Geometry and Molecular Conformation,* Research Studies Press, Taunton, U.K., 1988.

244. **Havel, T.F.,** An evaluation of computational strategies for use in the determination of protein structure from distance constraints obtained by nuclear magnetic resonance, *Prog. Biophys. Mol. Biol.,* 56, 43, 1991.

245. **Brooks 3d, C.L. and Karplus, M.,** Solvent effects on protein motion and protein effects on solvent motion. Dynamics of the active site region of lysozyme, *J. Mol. Biol.,* 208, 159, 1989.

246. **Hingerty, B.E., Ritchie, R.H., Ferrell, T.L., and Turner, J.E.,** Dielectric effects in biopolymers: the theory of ionic saturation revisited, *Biopolymers,* 24, 427, 1985.

247. **Daggett, V., Kollman, P.A., and Kuntz, I.D.,** Molecular dynamics simulations of small peptides: dependence on dielectric model and pH, *Biopolymers,* 31, 285, 1991.

248. **Gilson, M.K. and Honig, B.,** The inclusion of electrostatic hydration energies in molecular mechanics calculations, *J. Comput. Aided Mol. Design,* 5, 5, 1991.

249. **Baniak, E.2., Rivier, J.E., Struthers, R.S., Hagler, A.T., and Gierasch, L.M.,** Nuclear magnetic resonance analysis and conformational characterization of a cyclic decapeptide antagonist of a cyclic decapeptide antagonist of GnRH, *Biochemistry,* 26, 2642, 1987.

250. **Taylor, W.R.,** A template based method of pattern matching in protein sequences, *Prog. Biophys. Mol. Biol.,* 54, 159, 1989.

251. **Havel, T.F., Kuntz, I.D., and Crippen, G.M.,** The theory and practice of distance geometry, *Bull. Math. Biol.,* 45, 665, 1983.

252. **Kuntz, I.D., Crippen, G.M., Kollman, P.A., and Kimelman, D.,** Calculation of protein tertiary structure, *J. Mol. Biol.,* 106, 983, 1976.

253. **Easthope, P.L. and Havel, T.F.,** Computational experience with an algorithm for tetrangle inequality bound smoothing, *Bull. Math. Biol.,* 51, 173, 1989.

254. **Crippen, G.M. and Havel, T.F.,** Global energy minimization by rotational energy embedding, *J. Chem. Inf. Comput. Sci.,* 30, 222, 1990.

255. **Kuszewski, J., Nilges, M., and Brünger, A.T.,** Sampling and efficiency of metric matrix distance geometry: a novel partial metrization algorithm, *J. Biomol. NMR,* 2, 33, 1992.

256. **Berendsen, H.J., Postma, J.P.M., van Gunsteren, W.F., DiNola, A., and Haak, J.R.,** Molecular dynamics with coupling to an external bath, *J. Chem. Phys.,* 81, 3684, 1984.

257. **Nilges, M., Clore, G.M., and Gronenborn, A.M.,** Determination of three-dimensional structures of proteins from interproton distance data by hybrid distance geometry-dynamical simulated annealing calculations, *FEBS Lett.,* 229, 317, 1988.

258. **van Schaik, R.C., F., v. G.W., and Berendsen, H.J.,** Conformational search by potential energy annealing: algorithm and application to cyclosporin A, *J. Comput. Aided Mol. Design,* 6, 97, 1992.

259. **Levitt, M. and Sharon, R.,** Accurate simulation of protein dynamics in solution, *Proc. Natl. Acad. Sci. U.S.A.,* 85, 7557, 1988.

260. **Ernst, R.R., Bodenhausen, G., and Wokaun, A.,** *Principles of Nuclear Magnetic Resonance in One and Two Dimensions,* Clarendon Press, Oxford, 1986.

261. **Wüthrich, K.,** *NMR of Proteins and Nucleic Acids,* John Wiley & Sons, New York, 1986.

262. **Wüthrich, K.,** Six years of protein structure determination by NMR spectroscopy: what have we learned?, *Ciba Found. Symp.,* 161, 136, 1991.

263. **Clore, G.M. and Gronenborn, A.M.,** Structures of larger proteins in solution: three- and four-dimensional heteronuclear NMR spectroscopy, *Science,* 252, 1390, 1991.

264. **Wüthrich, K.,** Protein structure determination in solution by nuclear magnetic resonance spectroscopy, *Science,* 243, 45, 1989.

265. **Basus, V.J.,** Proton nuclear magnetic resonance assignments, *Methods Enzymol.,* 177, 132, 1989.

266. **Wüthrich, K., Spitzfaden, C., Memmert, K., Widmer, H., and Wider, G.,** Protein secondary structure determination by NMR. Application with recombinant human cyclophilin, *FEBS Lett.,* 285, 237, 1991.

267. **Zuiderweg, E.R.P., Billeter, M., Boelens, R., Sheek, R.M., Wüthrich, K., and Kaptein, R.,** Spatial arrangement of the three α helices in the solution conformation of *E. coli lac* repressor DNA-binding domain, *FEBS Lett.,* 174, 243, 1984.

268. **Billeter, M., Engeli, M., and Wüthrich, K.,** Interactive program for the investigation of protein structures based on 1H NMR experiments [CONFOR], *J. Mol. Graphics,* 3, 79, 1985.

269. **Kaptein, R., Zuiderweg, E.R.P., Scheek, R.M., Boelens, R., and van Gunsteren, W.F.,** A protein structure from nuclear magnetic resonance data: lac repressor headpiece, *J. Mol. Biol.,* 182, 179, 1985.

270. **Vuister, G.W., Boelens, R., Padilla, A., Kleywegt, G.J., and Kaptein, R.,** Assignment strategies in homonuclear three-dimensional 1H NMR spectra of proteins, *Biochemistry,* 29, 1829, 1990.

271. **Clore, G.M. and Gronenborn, A.M.,** Two-, three-, and four-dimensional NMR methods for obtaining larger and more precise three-dimensional structures of proteins in solution, *Annu. Rev. Biophys. Chem.,* 20, 29, 1991.

272. **Bax, A., Ikura, M., Kay, L.E., Barbato, G., and Spera, S.,** Multidimensional triple resonance NMR spectroscopy of isotopically uniformly enriched proteins: a powerful new strategy for structure determination, *Ciba Found. Symp.,* 161, 108, 1991.

273. **Ikura, M., Kay, L.E., and Bax, A.,** A novel approach for sequential assignment of 1H, 13C, and 15N spectra of proteins: heteronuclear triple-resonance three-dimensional NMR spectroscopy. Application to calmodulin, *Biochemistry,* 29, 4659, 1990.

274. **Szilagyi, L. and Jardetzky, O.,** α-Proton chemical shifts and secondary structure in proteins, *J. Magn. Reson.,* 83, 441, 1989.

275. **Pastore, A. and Saudek, V.,** The relationship between chemical shift and secondary structure in proteins, *J. Magn. Reson.,* 90, 165, 1990.

276. **Wishart, D.S., Sykes, B.D., and Richards, F.M.,** The chemical shift index: a fast and simple method for the assignment of protein secondary structure through NMR spectroscopy, *Biochemistry,* 31, 1647, 1992.

277. **Williamson, M.P.,** Secondary-structure dependent chemical shifts in proteins, *Biopolymers,* 29, 1423, 1990.

278. **Gao, Y., Veitch, N.C., and Williams, R.J.P.,** The value of chemical shift parameters in the description of protein solution structures, *J. Biomol. NMR,* 1, 457, 1991.

279. **McLachlan, A.D.,** Gene duplications in the structural evolution of chymotrypsin, *J. Mol. Biol.,* 128, 49, 1979.

280. **Billeter, M., Kline, A.D., Huber, R., and Wüthrich, K.,** Comparison of the high-resolution structures of the alpha-amylase inhibitor Tendamistat determined by nuclear magnetic resonance in solution and by X-ray diffraction in single crystals, *J. Mol. Biol.,* 206, 677, 1989.

281. **Thomas, P.D., Basus, V.J., and James, T.L.,** Protein solution structure determination using distances from two-dimensional nuclear Overhauser effect experiments: effect of approximations on the accuracy of derived structures, *Proc. Nat. Acad. Sci. U.S.A.,* 88, 1237, 1991.

282. **Liu, Y., Zhao, D., Altman, R., and Jardetzky, O.,** A systematic comparison of three structure determination methods from NMR data: dependence upon quality and quantity, *J. Biomol. NMR,* 2, 373, 1992.

283. **Wagner, G., Braun, W., Havel, T.F., Schaumann, T., Go, N., and Wüthrich, K.,** Protein structures in solution by nuclear magnetic resonance and distance geometry. The polypeptide fold of the basic pancreatic trypsin inhibitor determined using two different algorithms, DISGEO and DISMAN, *J. Mol. Biol.,* 196, 611, 1987.

284. **Wagner, G.,** NMR investigations of protein structure, *Prog. NMR Spectrosc.,* 22, 101, 1990.

285. **Kuntz, I.D., Thomason, J.F., and Oshiro, C.M.,** Distance geometry, *Methods Enzymol.,* 177, 159, 1989.

286. **Clore, G.M. and Gronenborn, A.M.,** Determination of three-dimensional structures of proteins and nucleic acids in solution by nuclear magnetic resonance spectroscopy, *Crit. Rev. Biochem. Mol. Biol.,* 24, 479, 1989.

287. **Boelens, R., Koning, T.M.G., van der Marel, G.A., van Boom, J.H., and Kaptein, R.,** Iterative procedure of structure determination from proton-proton NOE's using a full relaxation matrix approach. Applications to a DNA octamer, *J. Magn. Reson.,* 82, 290, 1989.

288. **Boelens, R., Koning, T.M.G., and Kaptein, R.,** Determination of biomolecular structures from proton-proton NOE's using a relaxation matrix approach, *J. Mol. Struct.,* 173, 299, 1988.

289. **James, T.L. and Borgias, B.A.,** Determination of DNA and protein structures in solution via complete relaxation matrix analysis of 2D NOE spectra, in *Frontiers of NMR in Molecular Biology, UCLA Symp. Mol. Cell. Biol.,* Live, D., Armitage, I., and Patel, D., Eds., Alan R. Liss, New York, 1989, 109.

290. **Keepers, J.W. and James, T.L.,** A theoretical study of distance determination from NMR. Two-dimensional nuclear Overhauser effect spectra [CORMA], *J. Magn. Reson.,* 57, 404, 1984.

291. **Borgias, B.A. and James, T.L.,** COMATOSE, a method for constrained refinement of macromolecular structure based on two-dimensional nuclear Overhauser effect spectra, *J. Magn. Reson.,* 79, 493, 1988.

292. NMR1, NMR2, distributed by New Methods Research, Syracuse, NY.

293. **Widmer, H. and Wüthrich, K.,** Simulation of two-dimensional NMR experiments using numerical density matrix calculations [SPHINX, LINSHA], *J. Magn. Reson.,* 70, 270, 1986.

294. **Widmer, H. and Wüthrich, K.,** Simulated two-dimensional NMR crosspeak fine structure for ^{1}H spin systems in polypeptides and polydeoxynucleotides, *J. Magn. Reson.,* 74, 316, 1987.

295. **Güntert, P., Braun, W., Billeter, M., and Wüthrich, K.,** Automated stereospecific proton NMR assignments and their impact on the precision of protein structure determinations in solution [HABAS], *J. Am. Chem. Soc.,* 111, 3997, 1989.

296. **Schaumann, T., Braun, W., and Wüthrich, K.,** The program FANTOM for energy refinement of polypeptides using a Newon-Raphson minimizer in torsion angle space, *Biopolymers,* 29, 679, 1990.

297. **Eccles, C., Güntert, P., Billeter, M., and Wüthrich, K.,** Efficient analysis of protein 2D NMR spectra using the software package EASY, *J. Biomol. NMR,* 1, 111, 1991.

298. **Güntert, P., Braun, W., and Wüthrich, K.,** Efficient computation of three-dimensional protein structures in solution from nuclear magnetic resonance data using the program DIANA and the supporting programs CALIBA, HABAS and GLOMSA, *J. Mol. Biol.,* 217, 517, 1991.

299. **Güntert, P. and Wüthrich, K.,** Improved efficiency of protein structure calculations from NMR data using the program DIANA with redundant dihedral angle constraints [REDAC], *J. Biomol. NMR,* 1, 447, 1991.

300. **Güntert, P. and Wüthrich, K.,** FLATT — A procedure for high-quality baseline correction of multidimensional NMR spectra, *J. Magn. Reson.,* 96, 403, 1992.

301. **Bock, I. and Rösch, P.,** Simulation of the cross-peak fine structure in 2D NMR spectroscopy by analytical calculation of the density operator in a product operator basis [MARS], *J. Magn. Reson.,* 74, 177, 1987.

302. **Case, D.A.,** Dynamical simulation of rate constants in protein-ligand interactions, *Prog. Biophys. Mol. Biol.,* 52, 39, 1988.

303. **Berendsen, H.J.,** Dynamic simulation as an essential tool in molecular modeling, *J. Comput. Aided Mol. Design,* 2, 217, 1988.

304. **de Bouregas, F.S. and Waugh, J.S.,** ANTIOPE, a program for computer experiments on spin dynamics, *J. Magn. Reson.,* 96, 280, 1992.

305. **Andersen, N.H., Lai, X.N., Hammen, P.K., and Marschner, T.M.,** Computer-aided conformational analysis based on NOESY signal intensities, *Basic Life Sci.,* 56, 95, 1990.

306. **Hoffman, R.E. and Levy, G.C.,** Modern methods of NMR data processing and data evaluation, *Prog. NMR Spectrosc.,* 23, 211, 1991.

307. **Hoch, J.C., Burns, M.M., and Redfield, C.,** Computer assisted assignment of two-dimensional NMR spectra of proteins., *UCLA Symp. Mol. Cell. Biol. New Ser.,* 109, 167, 1990.

308. **Hoch, J.C., Redfield, C., and Stern, A.S.,** Computer-aided analysis of protein NMR spectra, *Curr. Opin. Struct. Biol.,* 1, 1036, 1991.

309. **Zolnai, Z.,** Bruker Aspect 1000/3000 processing, postprocessing, and storage of 2D NMR data, distributed by Operations Assistant, National Magnetic Resonance Facility at Madison, Biochemistry Dept., 420 Henry Mall, Madison, Wisconsin, 53706.

310. **Zolnai, Z., Westler, W.M., Ulrich, E.L., and Markley, J.L.,** Drafting table and light-box software for multidimensional NMR spectral analysis (pixi): the personal computer workstation, *J. Magn. Reson.,* 88, 511, 1990.

311. FELIX 2.05, DSPACE, distributed by Hare Research, Inc., Suite 104, 18943–120th Ave. NE, Bothell, WA 98011.

312. **Szalma, S., Pelczer, I., Borer, P.N., and Levy, G.C.,** Selective discrete Fourier transformation, an alternative approach for multidimensional NMR data processing, *J. Magn. Reson.,* 91, 194, 1991.

313. **Hoffman, R.E., Delaglio, F., and Levy, G.C.,** Phase correction of two-dimensional NMR spectra using DISPA, *J. Magn. Reson.,* 98, 231, 1992.

314. **Dietrich, W., Ründel, C.H., and Neumann, M.,** Fast and precise baseline correction of one- and two-dimensional NMR spectra, *J. Magn. Reson.,* 91, 1, 1991.

315. **Billeter, M.,** Computer-assisted resonance assignments, *Methods Enzymol.,* 177, 150, 1989.

316. **Meier, B.U. and Ernst, R.R.,** Cross-peak analysis in 2D NMR spectroscopy by recursive multiplet contraction, *J. Magn. Reson.,* 79, 540, 1988.

317. **Neidig, K.P. and Kalbitzer, H.R.,** Improvement of 2D NMR spectra by matching symmetry-related spectral features, *Magn. Reson. Chem.,* 26, 848, 1988.

318. **Pfändler, P. and Bodenhausen, G.,** Topological classification of fragments of coupling networks and multiplet patterns in two-dimensional NMR spectra, *J. Magn. Reson.,* 79, 99, 1988.

319. **Weber, P.L., Malikayil, J.A., and Mueller, L.,** Automated elucidation of J connectivities in proton NMR spectra, *J. Magn. Reson.,* 82, 419, 1989.

320. **Kleywegt, G.J., Boelens, R., and Kaptein, R.,** A versatile approach toward partially automatic recognition of crosspeaks in 2D ^{1}H NMR spectra [STELLA], *J. Magn. Reson.,* 88, 601, 1990.

321. **Cieslar, C., Holak, T.A., and Oschkinat, H.,** Three-dimensional tocsy-tocsy processing using linear prediction, as a potential technique for automated assignment, *J. Magn. Reson.,* 89, 184, 1990.

322. **Garrett, D.S., Powers, R., Gronenborn, A.M., and Clore, G.M.,** A common sense approach to peak picking in two-, three-, and four-dimensional spectra using automatic computer analysis of contour diagrams [CAPP], *J. Magn. Reson.,* 95, 214, 1991.

323. **Billeter, M., Neri, D., Gottfreid, O., Quain, Y.Q., and Wüthrich, K.,** Precise vicinal coupling constants ^{3}J$_{NH\alpha}$ in proteins from nonlinear fits of J-modulated [^{15}N^{1}H]-COSY experiments, *J. Biomol NMR,* 2, 257, 1992.

324. **Denk, W., Baumann, R., and Wagner, G.,** Quantitative evaluation of cross-peak intensities by projection of two-dimensional NOE spectra on a linear space spanned by a set of reference resonance lines, *J. Magn. Reson.,* 67, 386, 1986.

325. **Eccles, C., Billeter, M., Güntert, P., and Wüthrich, K.,** S50 Abstracts 10th Meeting of the International Society of Magnetic Resonance, Morzine, France, July 16–21, 1989.

326. **Hoch, J.C.,** The Rowland NMR Toolkit, distributed by Rowland Institute for Science, 100 Cambridge Parkway, Cambridge, MA 02142.

327. **Hoch, J.C.,** Modern spectral analysis in nuclear magnetic resonance: alternatives to the Fourier transform, *Methods Enzymol.,* 176, 216, 1989.

328. **de Beer, R.,** *Linear Prediction and Hankel Singular Value Decomposition,* University of Delft, Dept. of Applied Physics, P.O. Box 5046, 2600 GA Delft, the Netherlands,

329. **Led, J.J. and Gesmar, H.,** Linear prediction enhancement of 2D heteronuclear correlated spectra of proteins, *J. Biomol. NMR,* 1, 237, 1991.

330. **Zhu, G. and Bax, A.,** Two-dimensional linear prediction for signals truncated in both dimensions, *J. Magn. Reson.,* 98, 192, 1992.

331. **Uike, M., Uchiyama, T., and Minamitani, H.,** Comparison of linear prediction methods based on singular value decomposition, *J. Magn. Reson.,* 99, 363, 1992.

332. **Delsuc, M.A.,** GIFA, D.C.S.O., Ecole Polytechnique, 91128, Palaiseau, France.

333. Maximum Entropy Reconstruction, distributed by Maximum Entropy Data Consultants, Ltd., 33 North End, Meldreth, Royston SG8 6NR, England.

334. **Jones, J.A. and Hore, P.J.,** The maximum entropy method and Fourier transformation compared, *J. Magn. Reson.,* 91, 276, 1991.

335. **Stern, A.S. and Hoch, J.C.,** A new, storage efficient algorithm for maximum-entropy spectrum reconstruction, *J. Magn. Reson.,* 97, 255, 1992.

336. **Bretthorst, G.L.,** Baysian analysis, *J. Magn. Reson.,* 88, 533, 1990.

337. **Weber, P.L. and Mueller, L.,** Use of nitrogen-15 labeling for automated three-dimensional sorting of cross peaks in protein 2D nmr spectra, *J. Magn. Reson.,* 81, 430, 1989.

338. **Kleywegt, G.J., Boelens, R., Cox, M., Llinas, M., and Kaptein, R.,** Computer-assisted assignment of 2D proton NMR spectra of proteins: basic algorithms and application to phoratoxin b, *J. Biomol. NMR,* 1, 23, 1991.

339. **Markley, J.L.,** Two-dimensional nuclear magnetic resonance spectroscopy of proteins: an overview, *Methods Enzymol.,* 176, 12, 1989.

340. **Borgias, B.A. and James, T.L.,** Two-dimensional nuclear Overhauser effect: complete relaxation matrix analysis, *Methods Enzymol.,* 176, 169, 1989.

341. **Clore, G.M. and Gronenborn, A.M.,** Assessment of errors involved in determination of interproton distance ratios and distances by one- and two-dimensional NOE measurements, *J. Magn. Reson.,* 61, 158, 1985.

342. **Olejniczak, E.T., Gampe, R.T.J., and Fesik, S.W.,** Accounting for spin diffusion in the analysis of 2D NOE data, *J. Magn. Reson.,* 67, 28, 1986.

343. **LeMaster, D.M., Kay, L.E., Brunger, A.T., and Prestegard, J.H.,** Protein dynamics and distance determination by NOE measurements, *FEBS Lett.,* 236, 71, 1988.

344. **van de Ven, F.J.M., Blommers, M.J.J., Schonten, R.E., and Hilbers, C.W.,** Calculation of interproton distances from NOE intensities. A relaxation matrix approach without requirement of a molecular model [NO2DI], *J. Magn. Reson.,* 94, 140, 1991.

345. **Malliavin, T.E., Celsuc, M.A., and Lallemand, J.Y.,** Computation of relaxation elements from incomplete NOESY data sets, *J. Mol. Biol. NMR,* 2, 349, 1992.

346. **Lefevre, J., Lane, A.N., and Jardetzky, O.,** Solution structure of the Trp Operator of *Escherichia coli* determined by NMR, *Biochemistry,* 26, 5076, 1987.

347. **Yip, P.F. and Case, D.A.,** A new method for refinement of macromolecular structures based on nuclear Overhauser effect spectra, *J. Magn. Reson.,* 83, 643, 1989.

348. **Gippert, G.P., Yip, P.F., Wright, P.E., and Case, D.A.,** Computational methods for determining protein structures from NMR data, *Biochem. Pharmacol.,* 40, 15, 1990.

349. **James, T.L., Borgias, B.A., Bianucci, A.M., and Zhou, N.,** Determination of DNA and protein structures in solution via complete relaxation matrix analysis of 2D NOE spectra, *Basic Life Sci.,* 56, 135, 1990.

350. **South, T.L., Kim, B., R., H.D., and Summers, M.F.,** Zinc fingers and molecular recognition. Structure and nucleic acid binding studies of an HIV zinc finger-like domain, *Biochem. Pharmacol.,* 40, 123, 1990.

351. **Bonvin, A.M.J.J., Boelens, R., and Kaptein, R.,** Direct NOE refinement of biomolecular structures using 2D NMR data, *J. Biomol. NMR,* 1, 305, 1991.

352. **Fejzo, J., Krezel, A.M., Westler, W.M., Macura, S., and Markley, J.L.,** Refinement of the NMR solution structure of a protein to remove distortions arising from neglect of internal motion, *Biochemistry,* 30, 3807, 1991.

353. **Mertz, J.E., Güntert, P., Wüthrich, K., and Braun, W.,** Complete relaxation matrix refinement of NMR structures of proteins using analytically calculated dihedral angle derivatives of NOE intensities, *J. Biomol. NMR,* 1, 257, 1991.

354. **Liu, H., Thomas, P.D., and James, T.L.,** Averaging of cross-relaxation rates and distances for methyl, methylene, and aromatic rings due to motion or overlap. Extraction of accurate distances iteratively via relaxation matrix analysis of 2D NOE spectra [MADIGRAS], *J. Magn. Reson.,* 98, 163, 1992.

355. **Peng, J.W. and Wagner, G.,** Mapping the spectral density functions using heteronuclear NMR relaxation rates, *J. Magn. Reson.,* 98, 308, 1992.

356. **Eads, C.D. and Kuntz, I.D.,** Programs for computer-assisted sequential assignment of proteins, *J. Magn. Reson.,* 82, 467, 1989.

357. **Billeter, M., Basus, V.J., and Kuntz, I.D.,** A program for semi-automatic sequential resonance assignments in protein proton nuclear magnetic resonance spectra [SEQASSIGN], *J. Magn. Reson.,* 76, 400, 1988.

358. **Grahn, H., Delagio, F., Delsuc, M.A., and Levy, G.C.,** Multivariate data analysis for pattern recognition in two-dimensional NMR, *J. Magn. Reson.,* 77, 294, 1988.

359. **Cieslar, C., Clore, G.M., and Gronenborn, A.M.,** Computer-aided sequential assignment of protein proton NMR spectra, *J. Magn. Reson.,* 80, 119, 1988.

360. **Kraulis, P.J.,** Ansig: a program for the assignment of protein proton two-dimensional NMR spectra by interactive computer graphics, *J. Magn. Reson.,* 84, 627, 1989.

361. **Kleywegt, G.J., Lamerichs, R.M.J.N., Boelens, R., and Kaptein, R.,** Toward automated assignment of protein ^{1}H NMR spectra [CLAIRE], *J. Magn. Reson.,* 85, 189, 1989.

362. **Van De Ven, F.J.M.,** Prospect, a program for automated interpretation of 2D NMR spectra of proteins, *J. Magn. Reson.,* 86, 633, 1990.

363. **Nelson, S.J., Schneider, D.M., and Wand, A.J.,** Implementation of the main chain directed assignment strategy. Computer assisted approach, *Biophys. J.,* 59, 1113, 1991.

364. **Wand, A.J. and Nelson, S.J.,** Refinement of the main chain directed assignment strategy for the analysis of 1H NMR spectra of proteins, *Biophys. J.,* 59, 1101, 1991.

365. **Oschkinat, H., Holak, T.A., and Cieslar, C.,** Assignment of protein NMR spectra in the light of homonuclear 3D spectroscopy: an automatable procedure based on 3D TOCSY-TOCSY and 3D TOCSY-NOESY, *Biopolymers,* 31, 699, 1991.

366. **Brugge, J.A., Buchanan, B.G., and Jardetzky, O.,** Toward automating the process of determining polypeptide secondary structure from proton NMR data, *J. Comput. Chem.,* 9, 662, 1988.

367. **Wüthrich, K., Billeter, M., and Braun, W.,** Polypeptide secondary structure determination by nuclear magnetic resonance observation of short proton-proton distances, *J. Mol. Biol.,* 180, 715, 1984.

368. **Pardi, A., Billeter, M., and Wüthrich, K.,** Calibration of the angular dependence of the amide proton-alpha carbon proton coupling constants, 3J(NH), in a globular protein, *J. Mol. Biol.,* 180, 741, 1984.

369. **Neri, D., Szyperski, T., Otting, G., Senn, H., and Wüthrich, K.,** Stereospecific nuclear magnetic resonance assignments of the methyl groups of valine and leucine in the DNA-binding domain of the 434 repressor by biosynthetically directed fractional 13C labeling, *Biochemistry,* 28, 7510, 1989.

370. **Driscoll, P.C., Gronenborn, A.M., and Clore, G.M.,** The influence of stereospecific assignments on the determination of three-dimensional structures of proteins by nuclear magnetic resonance spectroscopy. Application to the sea anemone protein BDS-I, *Febs Lett.,* 243, 223, 1989.

371. **Nilges, M., Clore, G.M., and Gronenborn, A.M.,** Proton NMR stereospecific assignments by conformational database searches, *Biopolymers,* 29, 813, 1990.

372. **Wüthrich, K., Billeter, M., and Braun, W.,** Pseudo-structures for the 20 common amino acids for use in studies of protein conformation by measurements of intramolecular proton-proton distance constraints with nuclear magnetic resonance, *J. Mol. Biol.,* 169, 949, 1983.

373. **Braun, W.,** Distance geometry and related methods for protein structure determination from NMR data, *Q. Rev. Biophys.,* 19, 115, 1987.

374. **Pardi, A., Hare, D.R., Selsted, M.E., Morrison, R.D., Bassolino, D.A., and Bach, A.C.J.,** Solution structure of the rabbit neutrophil defensin NP-5, *Mol. Biol.,* 201, 625, 1988.

375. **Nilges, M., Clore, G.M., and Gronenborn, A.M.,** Determination of three-dimensional structures of proteins from interproton distance data by dynamical simulated annealing from a random array of atoms. Circumventing problems associated with folding, *FEBS Lett.,* 239, 129, 1988.

376. **Neuhaus, D. and Williamson, M.,** *The Nuclear Overhauser Effect in Conformational and Structural Analysis,* VCH Publishers, New York, 1989.

377. **de Vlieg, J., Scheek, R.M., van Gunsteren, W.F., Berendsen, H.J., Kaptein, R., and Thomason, J.,** Combined procedure of distance geometry and restrained molecular dynamics techniques for protein structure determination from nuclear magnetic resonance data: application to the DNA binding domain of lac repressor from Escherichia coli, *Proteins,* 3, 209, 1988.

378. **Nerdal, W., Hare, D.R., and Reid, B.R.,** Three-dimensional structure of the wild-type lac Pribnow promoter DNA in solution. Two-dimensional nuclear magnetic resonance studies and distance geometry calculations, *J. Mol. Biol.,* 201, 717, 1988.

379. **Metzler, W.J., Hare, D.R., and Pardi, A.,** Limited sampling of conformational space by the distance geometry algorithm: implications for structures generated from NMR data, *Biochemistry,* 28, 7045, 1989.

380. **Havel, T.F.,** The sampling properties of some distance geometry algorithms applied to unconstrained polypeptide chains: a study of 1830 independently computed conformations, *Biopolymers,* 29, 1565, 1990.

381. **Havel, T. and Wüthrich, K.,** A distance geometry program for determining the structures of small proteins and other macromolecules from nuclear magnetic resonance measurements of intramolecular proton-proton proximities in solution, *Bull. Math. Biol.,* 46, 673, 1984.

382. **Havel, T.F. and Wüthrich, K.,** An evaluation of the combined use of nuclear magnetic resonance and distance geometry for the determination of protein conformations in solution, *J. Mol. Biol.,* 182, 281, 1985.

383. **Vásquez, M. and Scherago, H.A.,** Variable target function and built up procedures for the calculation of protein conformation. Application to bovine pancreatic trypsin inhibitor using limited simulated nuclear magnetic resonance data, *J. Mol. Struct. Dyn.,* 5, 757, 1988.

384. **Rico, M., Santoro, J., Carlos, G., Bruix, M., Neira, J.L., Nieto, J.L., and Herranz, J.,** 3D structure of bovine pancreatic ribonuclease A in aqueous solution: an approach to tertiary structure determination from a small basis of ¹H NMR NOE correlations, *J. Biomol. NMR,* 1, 283, 1991.

385. **Brünger, A.T.,** Three dimensional structure of proteins determined by molecular dynamics with interproton distance restraints: application to cambrin, *Proc. Nat. Acad. Sci. U.S.A.,* 83, 3801, 1986.

386. **Clore, G.M., Gronenborn, A.M., Brünger, A.T., and Karplus, M.J.,** The solution conformation of a heptadecapeptide comprising the DNA binding helix F of the cyclic AMP receptor protein of *Escherichia coli:* combined use of H-nuclear magnetic resonance and restrained molecular dynamics, *J. Mol. Biol.,* 186, 435, 1985.

387. **Kaptein, R., Boelens, R., Scheek, R.M., and van Gunsteren, W.F.,** Protein structures from NMR, *Biochemistry,* 27, 5389, 1988.

388. **de Vlieg, J. and van Gunsteren, W.F.,** Combined procedures of distance geometry and molecular dynamics for determining protein structure from nuclear magnetic resonance data, *Methods Enzymol.,* 202, 268, 1991.

389. **Driscoll, P.C., Gronenborn, A.M., Beress, L., and Clore, G.M.,** Determination of the three-dimensional solution structure of the antihypertensive and antiviral protein BDS-I from the sea anemone Anemonia sulcata: a study using nuclear magnetic resonance and hybrid distance geometry-dynamical simulated annealing, *Biochemistry,* 28, 2188, 1989.

390. **Habazettl, J., Cieslar, C., Oschkinat, H., and Holak, T.A.,** 1H NMR assignments of sidechain conformations in proteins using a high-dimensional potential in the simulated annealing calculations, *FEBS Lett.,* 268, 141, 1990.

391. **Hoch, J.C., Ed.,** *Computational Aspects of the Study of Biological Macromolecules,* Plenum Press, New York, 1991.

392. **van Gunsteren, W.F., Gros, P., Torda, A.E., Berendsen, H.J., and van Schaik, R.C.,** On deriving spatial protein structure from NMR or X-ray diffraction data [PEACS], *Ciba Found. Symp.,* 161, 150, 1991.

393. **Bassolino, D.A., Hirata, F., Kitchen, D.B., Kominos, D., Pardi, A., and Levy, R.M.,** Determination of protein structures in solution using NMR data and IMPACT, *Int. J. Supercomput. Appl.,* 2, 41, 1988.

394. **Levy, R.M., Bassolino, D.A., Kitchen, D.B., and Pardi, A.,** Solution structures of proteins from NMR data and modeling: alternative folds for neutrophil peptide 5, *Biochemistry,* 28, 9361, 1989.

395. **Altman, R.B. and Jardetzky, O.,** Heuristic refinement method for determination of solution structure of proteins from nuclear magnetic resonance data, *Methods Enzymol.,* 177, 218, 1989.

396. **Altman, R.B., Pachter, R., Carrara, E., and Jardetzky, O.,** PROTEAN2, *Quantum Chemistry Program Exchange,* 10, 596, 1990.

397. **Koehl, P., Lefevre, J.F., and Jardetzky, O.,** Computing the geometry of a molecule in dihedral angle space using NMR-derived constraints. A new algorithm based on optimal filtering, *J. Mol. Biol.,* 223, 299, 1992.

398. **Pachter, R., Altman, R.B., Czaplicki, J., and Jardetzky, O.,** Comparison of the NMR solution structures of cyclosporin A determined by different techniques, *J. Magn. Reson.,* 92, 468, 1991.

399. **Clore, G.M., Nilges, M., Brunger, A.T., Karplus, M., and Gronenborn, A.M.,** A comparison of the restrained molecular dynamics and distance geometry methods for determining three-dimensional structures of proteins on the basis of interproton distances, *Febs Lett.,* 213, 269, 1987.

400. **Oshiro, C.M., Thomason, J., and Kuntz, I.D.,** Effects of limited input distance constraints upon the distance geometry algorithm, *Biopolymers,* 31, 1049, 1991.

401. **Blaney, J.,** DGEOM, distributed by DuPont Experimental Station, Building E328, Wilmington, DE, 19898.

402. **Brünger, A.T.,** XPLOR, distributed by Yale University, New Haven, CT.

403. **Holm, L. and Sander, C.,** Evaluation of protein models by atomic solvation preference, *J. Mol. Biol.,* 225, 93, 1992.

404. **Torda, A.E., Scheek, R.M., and van Gunsteren, W.F.,** Time-dependent distance restraints in molecular dynamics simulations, *Chem. Phys. Lett.,* 157, 289, 1989.

405. **Torda, A.E., Scheek, R.M., and van Gunsteren, W.F.,** Time-averaged nuclear Overhauser effect distance restraints applied to Tendamistat, *J. Mol. Biol.,* 214, 223, 1990.

406. **Gronenborn, A.M. and Clore, G.M.,** Protein structure determination in solution by two-dimensional and three-dimensional nuclear magnetic resonance spectroscopy, *Anal. Chem.,* 62, 2, 1990.

407. **Orrell, K.G., Sik, V., and Stephenson, D.,** Quantitative investigations of molecular stereodynamics by 1D and 2D NMR methods, *Prog. NMR Spectrosc.,* 22, 141, 1990.

408. **Brüschweiler, R., Blackledge, M., and Ernst, R.R.,** Multi-conformational peptide dynamics derived from NMR data: a new search algorithm and its application to antamanide [MEDUSA], *J. Biomol. NMR,* 1, 3, 1991.

409. **Neri, D., Billeter, M., Wider, G., and Wüthrich, K.,** NMR determination of residual structure in the urea-denatured protein, the 434-repressor, *Science,* 257, 1559, 1992.

410. **Sanner, M., Widmer, A., Senn, H., and Braun, W.,** GEOM: a new tool for molecular modelling based on distance geometry calculations with NMR data, *J. Comput. Aided Mol. Design,* 3, 195, 1989.

411. **Senn, H., Loosli, H.R., Sanner, M., and Braun, W.,** Conformational studies of cyclic peptide structures in solution from 1H-NMR data by distance geometry calculation and restrained energy minimization, *Biopolymers,* 29, 1387, 1990.

412. **Güntert, P., Qian, Y.Q., Otting, G., Muller, M., Gehring, W., and Wüthrich, K.,** Structure determination of the Antp (C39----S) homeodomain from nuclear magnetic resonance data in solution using a novel strategy for the structure calculation with the programs DIANA, CALIBA, HABAS and GLOMSA, *J. Mol. Biol.,* 217, 531, 1991.

413. **Havel, T.F.,** DISGEO, distributed by Div. of Biophysics, University of Michigan, Ann Arbor, MI 48109.

414. **Kuntz, I.D.,** EMBED, VEMBED, distributed by Dept. of Pharmaceutical Chem., University of California, San Francisco, CA 94143.

415. **Hare, D.R. and Reid, B.R.,** Three-dimensional structure of a DNA hairpin in solution: two-dimensional NMR studies and distance geometry calculations on d(CGCGTTTTCGCG), *Biochemistry,* 25, 5341, 1986.

416. NMRGraf, distributed by Biodesign, 199 S. Ros Robles Ave., Pasadena, CA 91101.

417. **Weiner, P.K. and Kollman, P.A.,** Amber: assisted model building with energy refinement. A general program for modeling molecules and their interactions, *J. Comput. Chem.,* 2, 287, 1981.

418. **Snow, M.E. and Crippen, G.M.,** Dimensional oscillation. A fast variation of energy embedding gives good results with the AMBER potential energy function, *Int. J. Pept. Protein Res.,* 38, 161, 1991.

419. INSIGHT2, DISCOVER, distributed by BIOSYM Technology, 10065 Barnes Canyon Rd., San Diego, CA 92121.

420. **Tranqui, L., Andrieux, A., Hudry, C.G., Ryckewaert, J.J., Soyez, S., Chapel, A., Ginsberg, M.H., Plow, E.F., and Marguerie, G.,** Differential structural requirements for fibrinogen binding to platelets and to endothelial cells, *J. Cell Biol.,* 108, 2519, 1989.

421. **Taub, R., Gould, R.J., Garsky, V.M., Ciccarone, T.M., Hoxie, J., Friedman, P.A., and Shattil, S.J.,** A monoclonal antibody against the platelet fibrinogen receptor contains a sequence that mimics a receptor recognition domain in fibrinogen, *J. Biol. Chem.,* 264, 259, 1989.

422. **Tomiyama, Y., Brojer, E., Ruggeri, Z.M., Shattil, S.J., Smiltneck, J., Gorski, J., Kumar, A., Kieber-Emmons, T., and Kunicki, T.J.,** A molecular model of RGD ligands. Antibody D gene segments that direct specificity for the integrin alpha IIb beta 3, *J. Biol. Chem.,* 267, 18085, 1992.

423. **Chen, Y., Pitzenberger, S.M., Garsky, V.M., Lumma, P.K., Sanyal, G., and Baum, J.,** Proton NMR assignments and secondary structure of the snake venom protein echistatin, *Biochemistry,* 30, 11625, 1991.

424. **Aumailley, M., Gurrath, M., Muller, G., Calvete, J., Timpl, R., and Kessler, H.,** Arg-Gly-Asp constrained within cyclic pentapeptides. Strong and selective inhibitors of cell adhesion to vitronectin and laminin fragment P1, *Febs Lett.,* 291, 50, 1991.

425. **Adler, M. and Wagner, G.,** Sequential 1H NMR assignments of kistrin, a potent platelet aggregation inhibitor and glycoprotein IIb-IIIa antagonist, *Biochemistry,* 31, 1031, 1992.

426. **Shattil, S.S., Weisel, J.W., and Kieber-Emmons, T.,** Use of monoclonal antibodies to study the interaction between an integrin adhesion receptor, GP IIb-IIa, and its physiological ligand, fibrinogen, *Immunomethods,* 1, 53, 1992.

427. **Abrams, C.S., Ruggeri, Z.M., Taub, R., Hoxie, J.A., Nagaswami, C., Weisel, J.W., and Shattil, S.J.,** Anti-idiotypic antibodies against an antibody to the platelet glycoprotein (GP) IIb-IIIa complex mimic GP IIb-IIIa by recognizing fibrinogen, *J. Biol. Chem.,* 267, 2775, 1992.

428. **Blumenstein, M., Matsueda, G.R., Timmons, S., and Hawiger, J.,** A β-turn is present in the 392–411 segment of the human fibrinogen γ-chain. Effects of structural changes in this segment on affinity to antibody 4A5, *Biochemistry,* 31, 10692, 1992.

429. **Chen, C.S. and Hawiger, J.,** Reactivity of synthetic peptide analogs of adhesive proteins in regard to the interaction of human endothelial cells with extracellular matrix, *Blood,* 77, 2200, 1991.

430. **Rao, S.N.,** Bioactive conformation of Arg-Gly-Asp by X-ray data analyses and molecular mechanics, *Peptide Res.,* 5, 148, 1992.

Chapter 8

ARCHITECTURE AND DESIGN OF ZINC PROTEIN-LIGAND COMPLEXES

David W. Christianson and Anastasia M. Khoury Christianson

CONTENTS

I. INTRODUCTION

The rational design of regulatory molecules which bind avidly to specific target proteins persists as a topic of intense interest in the fields of pharmaceutical chemistry and structural biology. Such compounds range from small, synthetic peptide mimics which inhibit enzymes to large protein hormones, engineered through molecular biological techniques, which regulate the function of a protein receptor. Regardless of the size of the regulatory molecule, there are structural principles that guide any molecular design strategy. These principles invariably derive from accurate stereochemical information derived from X-ray crystallographic investigations of proteins and their complexes with other molecules. It is particularly trendy to tout such an approach as "rational drug design"; however, it is not too often that an actual drug target provides a known three-dimensional structure to guide rational drug design. Nonetheless, a wealth of stereochemical information is available from X-ray crystallographic studies of proteins and also small molecules which can inspire molecular design efforts and, at the very least, provide a powerful starting point for the design of tight-binding protein ligands.

Inspired by the current pharmaceutical interest in zinc-requiring proteins, this chapter focuses on the rational design of ligands that recognize and bind to zinc metalloproteins. The coordination chemistry of zinc is versatile as the metal ion functions in biology, and this function is governed to a great degree by the protein residues, as well as the ligand functional groups, which coordinate to the metal ion. In fact, metalloprotein-ligand affinity is invariably dominated by zinc recognition and coordination phenomena. Zinc coordination polyhedra exhibit variation in number, charge, structure, and amino acid composition depending upon the functional nature of the metal ion. Although zinc hexahydrate, $Zn^{2+}(OH_2)_6$, predominates in aqueous solution, zinc is typically 4-coordinate with tetrahedral or distorted tetrahedral geometry in proteins. Additionally, it is possible for zinc in metalloproteins to become 5-coordinate upon the binding of ligand molecules with bidentate functional groups (e.g., see Matthews[92] and Christianson and Lipscomb[28]).

The coordination polyhedron of structural zinc is dominated by cysteine thiolates, and the metal ion is typically sequestered from solvent by its molecular environment; the coordination polyhedron of catalytic or regulatory zinc is dominated by histidine ligands, and the metal ion is exposed to bulk solvent and typically binds a solvent molecule.[126] Zinc-bound solvent molecules can often be displaced by the functional groups of rationally designed ligand molecules. Structural features of zinc coordination by ligand molecules, as well as the participation of ligand molecules in hydrogen bond networks with zinc ligands, govern the molecular recognition of rationally designed protein ligands.

In this review, a brief summary of the preferential coordination stereochemistry of zinc-ligand interactions precedes the discussion of several tight-binding protein-ligand complexes. The first of these examples relates an approach to rational drug design where the three-dimensional structure of the drug receptor, the zinc enzyme angiotensin-converting enzyme, is unknown — however, adequate information can be inferred from the known structures of two convergently related proteins, carboxypeptidase A and thermolysin. In contrast, the second example, human carbonic anhydrase II, presents a known three-dimensional structure to guide rational drug design efforts. Finally, the third example illustrates an approach to engineering protein-protein interactions where the three-dimensional structures of the drug, human growth hormone, and its receptor(s) were not initially available to guide structure-activity studies. Intriguingly, this hormone exhibits zinc-dependent dimerization properties that reflect a storage mechanism in secretory granules; moreover, the hormone displays zinc-dependent receptor recognition and discrimination.

II. STEREOCHEMICAL STATISTICS OF ZINC-LIGAND INTERACTIONS

Molecular structure data bases such as the Brookhaven Protein Data Bank[135] and the Cambridge Structural Database[134] are particularly useful in the analysis and design of protein ligands. Recurring stereochemical motifs identified among the independent three-dimensional structures in these data bases may be used to guide efforts in rational drug design. Moreover, this structural information is useful for the (re)design of protein-zinc binding sites.[137,140,141,143–146] Consequently, a brief summary of functional group-zinc coordination stereochemistry provides a useful structural reference for molecular design experiments.

A. CARBOXYLATE, CARBONYL

As shown in Figure 1, Lewis acids (i.e., metal ions or hydrogen bond donors) may exhibit *syn* or *anti* stereochemistry as they interact with the carboxylate anion. However, in a study of enzyme active sites, Gandour[50] first noticed that hydrogen bond donors to the carboxylates of aspartate and glutamate residues preferentially occur with *syn* stereochemistry. As a carboxylate-hydrogen bond donor interaction CO_2-H compares with a carboxylate-metal ion interaction, carboxylate-metal ion interactions are likewise expected to be most favorable with *syn* stereochemistry. Indeed, Carrell and colleagues[19] found that the *syn*-oriented lone electron pair of the carboxylate oxygen is preferred for cation binding in a survey of the Cambridge Structural Database, and bidentate carboxylate-metal

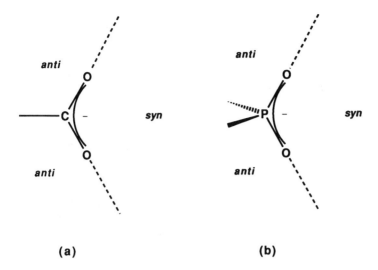

(a)   **(b)**

FIGURE 1. Stereochemical *syn/anti* designations for Lewis acids (i.e., metal ions or hydrogen bond donors) as they interact with carboxylates (a) and phosphates (b).

ion interactions typically occur when the metal-oxygen distance is in the range 2.3 to 2.6 Å. In particular, for 67 carboxylate-zinc interactions, 52 are *syn* (5 of these are bidentate) and 15 are *anti* (i.e., 78% *syn*/22% *anti*). Furthermore, zinc exhibits a strong preference to be in the plane of the carboxylate, which optimizes its interaction with the oxygen sp^2 lone electron pair. A scatterplot of carboxylate-zinc interactions retrieved from the Cambridge Structural Database, with superimposed "probability" contours[117] is reproduced in Figure 2.[19]

A survey of protein-metal ion interactions available in the Brookhaven Protein Data Bank reveals that *syn*-carboxylate-metal ion stereochemistry is similarly preferred.[21,27] It has been suggested that potent zinc enzyme inhibition arises from *syn*-oriented interactions between inhibitor carboxylates and active site zinc ions,[29,99] and the structures of such interactions may sample the reaction coordinate for enzymic catalysis in certain systems.[30]

The uncharged carbonyl, as found in the peptide backbone or in the carboxamide side chains of asparagine or glutamine, is commonly observed in protein structures as a calcium ligand, although some examples of carbonyl-zinc interactions are found in certain protein complexes.[23] Chakrabarti[22] outlines the stereochemistry of peptide carbonyl-metal ion interactions retrieved from refined protein structures. Among a total of 72 observations (62 calcium, 6 zinc, 2 copper, 1 magnesium, 1 sodium), 11 are defined as bidentate, i.e., another atom from one of the amino acids forming the peptide linkage also interacts with the metal ion. A scatterplot of unidentate and bidentate peptide carbonyl-metal ion interactions is reproduced in Figure 3. For unidentate interactions the C=O→M^{n+} angles tend to lie within the range 140 to 170°, but for bidentate examples this angle tends to lie within the range 110 to 130°; not unexpectedly, most metal ions tend to lie within 35° of the peptide plane.

The glycine chelate comprises a special example of bidentate carbonyl-metal ion coordination, where both the carbonyl oxygen and amino nitrogen of glycine coordinate to a metal ion to form a five-membered chelate (Figure 4). This binding mode characterizes important examples of biological zinc complexation. For example, avian pancreatic polypeptide hormone requires zinc for satisfactory crystallization.[11,112] Electron density maps of the protein reveal that the metal ion cross links three individual molecules and thereby stabilizes the crystal lattice; the amino-terminal glycine of one of these protein molecules engages zinc through a five-membered chelate. A five-membered glycine chelate is also observed in the binding of the pseudosubstrate glycyl-L-tyrosine to the zinc protease carboxypeptidase A,[31] and is reminiscent of the binding of a hydroxamate inhibitor to the zinc protease thermolysin through an analogous five-membered chelate structure.[62]

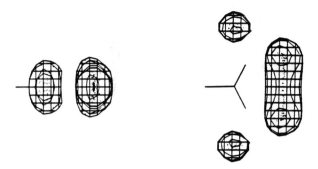

FIGURE 2. Two perpendicular orientations of 67 superimposed carboxylate-zinc interactions retrieved from the Cambridge Structural Database; the orientation on the right corresponds to the carboxylate group as represented in Figure 1. Contours highlight population clusters exhibiting *syn* coordination stereochemistry, with significant numbers of bidentate interactions. (Reprinted with permission from Carrell, C.J., Carrell, H.L., Erlebacher, J., and Glusker, J.P., *J. Am. Chem. Soc.,* 110, 8651, 1988. Copyright © 1988 American Chemical Society.)

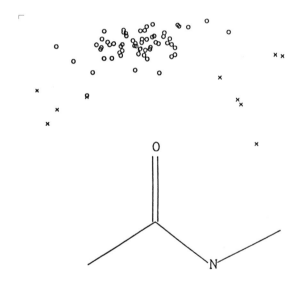

FIGURE 3. Peptide carbonyl-metal ion interactions retrieved from refined protein structures, where unidentate and bidentate examples are indicated by circles and crosses, respectively. (Reprinted with permission from Chakrabarti, P., *Biochemistry,* 29, 651, 1990. Copyright © 1990 American Chemical Society.)

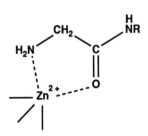

FIGURE 4. The peptide carbonyl and the free α-amino group of an amino acid may complex a metal ion to form a five-membered chelate. This binding mode characterizes several examples of biological zinc coordination.

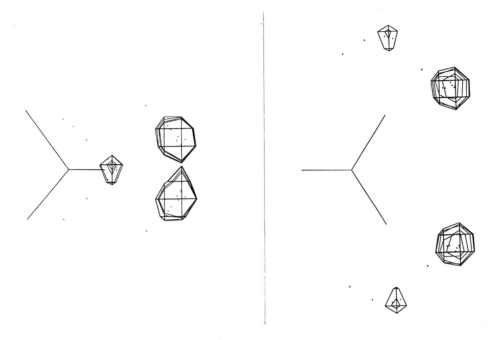

FIGURE 5. Perpendicular scatterplots of 12 unique and independent phosphinyl-zinc interactions retrieved from the Cambridge Structural Database; the orientation on the right corresponds to the phosphate group and its derivatives as represented in Figure 1. In this subset of the 108 phosphinyl-metal ion interactions reported by Alexander and colleagues[3] *syn*-unidentate coordination stereochemistry is preferred by 67%. Note the complete absence of bidentate phosphate-zinc interactions; this feature contrasts contrast with scatterplots of carboxylate-zinc interactions (Figure 2).

B. PHOSPHATE, PHOSPHINYL

The anionic phosphinyl portion ($-PO_2^-$) of the phosphate group may coordinate to metal ions stereochemistry with *syn* or *anti* stereochemistry, analogous to the carboxylate (Figure 1). In a study of the Cambridge Structural Database, Alexander and colleagues[3] reported that the phosphinyl group exhibits a 63% preference for *syn*-oriented metal interactions. A tendency toward out-of-plane geometry is observed, and unidentate $P=O \rightarrow M^{n+}$ coordination is preferred with an average angle of 141°. For the subset of 12 independent phosphinyl-zinc interactions, unidentate interactions are similarly preferred, and the *syn/anti* ratio is 8/4, or 67%/33%. The average $P=O \rightarrow Zn^{2+}$ angle is 136°, and the average zinc-oxygen distance is 1.94 Å. A scatterplot of phosphinyl-zinc interactions is found in Figure 5.

An intriguing contrast can be made between carboxylates and phosphates as each interacts with metal ions. In carboxylate-metal ion interactions, zinc and other metal ions typically prefer to be in the plane of the carboxylate group.[19] However, the metal ion engaged by phosphate prefers a location that is nearly 1 Å out of the plane of the phosphinyl group. Additionally, even though there are several examples of bidentate carboxylate-metal ion interactions (Figure 2), a symmetrically bidentate phosphinyl-metal ion interaction is not preferred (Figure 5).

C. IMIDAZOLE

The geometry of Zn^{2+} interactions with imidazole and its derivatives has been examined by Vedani and Huhta[129] in a study of the Cambridge Structural Database. Not surprisingly, metal ions prefer a head-on and in-plane approach to the sp^2 lone electron pair of the nitrogen atom (Figure 6).[129] This stereochemical result is in accord with the analysis of imidazole hydrogen bond geometry in small molecules,[127] histidine hydrogen bond geometry in protein structures,[66] and histidine-metal ion geometry in protein structures.[24]

The protonation state of histidine cannot be ascertained from X-ray crystallographic experiments, and this is of particular concern because of the near-physiological first pK_a of the imidazole side

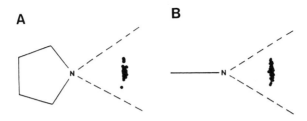

FIGURE 6. Superimposed Zn^{2+} interactions with sp^2 nitrogen-containing heterocycles retrieved from the Cambridge Structural Database demonstrate that the metal ion prefers a head-on and in-plane approach to the sp^2 lone electron pair of nitrogen; two perpendicular orientations (a) and (b) are shown. Dashed lines are at $\pm30°$ from the C-N-X (X = C, N) bisector. (Reprinted with permission from Vedani, A. and Huhta, D.W., *J. Am. Chem. Soc.*, 112, 4759, 1990. Copyright © 1990 American Chemical Society.)

chain (about 6.0 to 6.5). The second pK_a of histidine is about 14.0 to 14.5, and this value may be lowered about 2 units by imidazole coordination to a metal ion.[119] Hence, histidine is found in protein structures as the protonated imidazolium or as the neutral imidazole, and even sometimes as the deprotonated imidazolate.

Hydrogen bond interactions are important for the function of imidazole as a zinc ligand. For example, in a survey of zinc-binding motifs, Christianson and Alexander[26,27] reported that head-on and in-plane geometry is observed as histidine residues coordinate to zinc in certain metalloproteins, and head-on and in-plane geometry is likewise favored for carboxylate and carbonyl groups hydrogen bonded to these histidine metal ligands. It is intriguing that the imidazolate form of a histidine zinc ligand may be stabilized by hydrogen bonding. Molecular orbital calculations[102] suggest that a system in which a carboxylate hydrogen bonds to a histidine which, in turn, coordinates to a zinc ion is isoenergetic with a system in which the carboxylate is protonated and the histidine is negatively charged. That is to say, CO_2^----H-His$\rightarrow Zn^{2+}$ is isoenergetic with CO_2H---His$^-$ $\rightarrow Zn^{2+}$, and both forms could be in equilibrium at the zinc binding site. Resonance Raman spectroscopic experiments suggest the presence of both carboxylate-histidine-iron and carboxylic acid-histidinate-ion forms in the active site of cytochrome *c* peroxidase,[118] so it is reasonable to expect a comparable equilibrium in the zinc binding site.

D. THIOLATE

The sulfur atom is a favorable zinc ligand because of its size and polarizability. The thiol side chain of cysteine (with pK_a = about 8.5) is negatively charged as it complexes a metal ion in a protein; in addition to metal coordination, the cysteine thiol may simultaneously accept hydrogen bonds from other protein residues.[2,66] Hydrogen bond networks with cysteine metal ligands serve to preorient the ligand for metal complexation, thereby reducing the entropic cost of binding site organization.[25] In redox proteins such as cytochrome c_2, such networks modulate the high redox potential of the iron center.[17]

A detailed stereochemical analysis of cysteine-metal ion interactions is presented by Chakrabarti[20] in a survey of the Brookhaven Protein Data Bank. The average $S\rightarrow Zn^{2+}$ distance for 14 independent observations is 2.1 Å, the average C_β-$S\rightarrow Zn^{2+}$ angle is 112°, and the C_α-C_β-$S\rightarrow M^{n+}$ torsion angle distribution is trimodal, with peaks at $\pm90°$ and 180° (Figure 7). This coordination stereochemistry roughly corresponds to energetically favorable *trans/gauche* geometries for thiolate metal complex-ation, as shown in the Newman projection of Figure 8. Although the cysteine N-C_α-C_β-S torsion angle χ_1 tends toward *gauche*$^+$ (χ_1 = 300°; preferred) or *gauche*$^-$ (χ_1 = 60°) in the absence of metal coordination,[94] this angle preferentially tends toward a *trans* (χ_1 = 180°; preferred) or *gauche*$^-$ orientation where cysteine coordinates to a metal ion; and when χ_1 = 60° the C_α-C_β-$S\rightarrow M^{n+}$ torsion angles tend toward 90° and when χ_1 = 180° the C_α-C_β-$S\rightarrow M^{n+}$ torsion angles tend toward 180°.

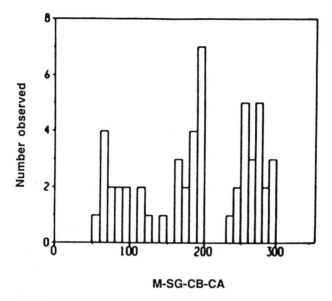

M-SG-CB-CA

FIGURE 7. The distribution of cysteine-metal ion torsion angles, C_α-C_β-S-M^{n+}, in protein structures is trimodal with peaks at ±90 and 180°; this distribution is valid for $M^{n+} = Zn^{2+}$. (Reprinted with permission from Chakrabarti, P., *Biochemistry,* 28, 6081, 1989. Copyright © 1989 American Chemical Society.)

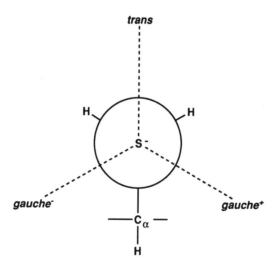

FIGURE 8. The stereochemistry of thiolate-metal ion coordination may be described by the *trans/gauche* convention. Note that the favorable coordination stereochemistry indicated here generally corresponds to the trimodal distribution of cysteine-metal ion torsion angles $C\alpha$-$C\beta$-S-M^{n+} found in Figure 7.

III. INHIBITORS OF THE ZINC PROTEASES

The zinc proteases comprise a class of enzymes which require a divalent zinc ion for the catalytic hydrolysis of a peptide substrate into carboxylate and amino products. Carboxypeptidase A (CPA) and thermolysin (TLN) comprise the best-characterized zinc proteases in terms of chemical, kinetic, and structural data (for recent reviews, see Matthews[92] and Christianson and Lipscomb[28]). Although these two proteases are not themselves the targets for drug therapy, they are related to metalloproteases

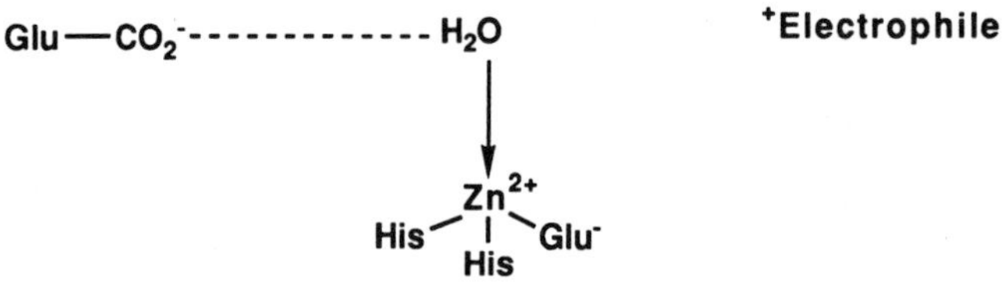

Glu—CO$_2^-$ - - - - - - - - - - - H$_2$O $^+$**Electrophile**

Zn^{2+}

His — | — Glu$^-$

His

FIGURE 9. Consensus structure of the zinc protease active site as found in carboxypeptidase A and thermolysin. The electrophile is the protonated, positively charged side chain of Arg or His. Although zinc is tetracoordinate in the resting enzyme, it may be pentacoordinate in catalysis and in certain enzyme-inhibitor complexes.

of unknown structure which are pharmaceutical targets. It is beneficial, then, to rely upon at least a basic similarity among the zinc proteases of known and unknown structure, that the relative constellations of catalytic residues within their active sites are related by a common catalytic requirement.

For reference, a consensus active site structure for the zinc proteases CPA and TLN is shown in Figure 9. A pair of catalytic residues in the zinc protease active site flank zinc-bound solvent in the active site groove — one of these residues is the carboxylate group of glutamate and the other residue, the positively charged side chain of arginine or histidine, functions as an electrophile. The carboxylate group hydrogen bonds to zinc-bound solvent in the resting enzyme, and this solvent molecule is displaced by the zinc-coordinating functional group of the proteases inhibitor. Further details of catalysis and inhibitor binding interactions in the active sites of CPA and TLN have been summarized elsewhere.[28,92] Analogous functional groups among enzyme active sites lead to parallels in the binding affinity of analogous inhibitors.

A pharmaceutically important zinc protease of unknown structure is angiotensin-converting enzyme (ACE), a target for the control of hypertension. ACE catalyzes the hydrolysis of the C-terminal His-Leu dipeptide from the decapeptide angiotensin I; the resultant octapeptide, angiotensin II, is a vasopressor which plays a role in blood pressure elevation. Given the similar catalytic and inhibitory properties among CPA, TLN, and ACE, it is likely that the consensus active site structure in Figure 9 also applies to ACE. Hence, CPA and TLN comprise important paradigms for rational drug design, insofar as their structural relationship with ACE allows for the rational design of inhibitors which are useful for the control of hypertension.

As a paradigm, CPA is particularly intriguing since large conformational changes of Tyr-248 and its associated backbone loop accompany the binding of ligands to the active site (reviewed by Christianson and Lipscomb[28]). The phenolic hydroxyl of this residue donates a hydrogen bond to the terminal carboxylate of a bound substrate, and it accepts an additional hydrogen bond from the amide proton of the penultimate peptide linkage of polypeptide substrates. These interactions, which require substantial reorganization of enzyme structure to accommodate ligand binding, provide a unique illustration of Koshland's "induced fit" hypothesis.[75] Conformational changes of comparable magnitude do not accompany ligand binding to TLN;[92] nevertheless, it is a reasonable generalization that to some degree, complementary structural changes on the part of the protein molecule, the ligand molecule, or a combination of both will accompany any protein-ligand association event.

With the consensus zinc protease active site of Figure 9 as a target, the design features of three different groups of inhibitors are now summarized. In these examples, it is important to realize that rational drug design may proceed without the three-dimensional structure of the actual drug receptor, ACE, but instead with the three-dimensional structures of the related proteins CPA and TLN.

A. THIOLATES

The discovery of potent thiolate-zinc interactions made by inhibitors of CPA provided a route toward the first rationally designed ACE inhibitor (Captopril) used in hypertension therapy.[38,39,106,107] The assumptions of the design strategy are summarized graphically in Figure 10. In short, just as

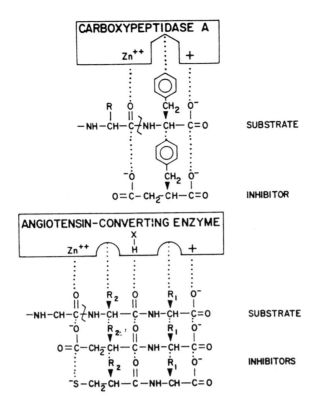

FIGURE 10. Presumed analogies among the substrate and inhibitor structures relevant to carboxypeptidase A and angiotensin-converting enzyme. (Reprinted with permission from Cushman, D.W., Cheung, H.S., Sabo, E.F., and Ondetti, M.A., *Biochemistry,* 16, 5484, 1977. Copyright © 1977 American Chemical Society.)

a thiolate zinc ligand can be successfully incorporated into a peptide analog inhibitor of CPA, the presumed analogy between the CPA and ACE active sites allows for the parallel construction of a peptide thiolate inhibitor of ACE. Dissociation constants for thiolate-bearing ACE inhibitors are in the nanomolar range.

Although the three-dimensional structure of the paradigm complex between CPA and 2-benzyl-3-mercaptopropanoic acid (K_i = 11 nM)[105] has not been studied by X-ray crystallographic methods, the structure of the complex between TLN and a related peptidylthiolate, (2-benzyl-3-mercaptopropanoyl)-L-alanylglycinamide (K_i = 0.75 μM)[104] reveals *gauche⁻* C-C-S→Zn^{2+} stereochemistry which characterizes energetically favorable[20] thiolate-zinc interactions (Figure 11).[98] It is reasonable to expect that similarly favorable stereochemistry, with either *gauche*- or *trans*-oriented C-C-S→Zn^{2+} coordination, predominates in CPA- and ACE-peptidylthiolate complexes. Favorable zinc-ligand coordination stereochemistry in these examples is certain to contribute substantially to effective protease inhibition.

B. CARBOXYLATES

Small molecule enzyme inhibitors designed to exploit zinc-carboxylate interactions with CPA and TLN have received considerable high-resolution X-ray crystallographic study.[12,30,71,86,99] In another example involving a protein-protein complex, the structure of the complex between CPA and the cleaved potato inhibitor has been determined by X-ray crystallographic methods to a resolution of 2.5 Å.[115] In all but one of the cited examples, inhibitor binding to the enzyme is characterized by favorable *syn* stereochemistry as the inhibitor carboxylate complexes zinc; in a few cases, the coordination to zinc is bidentate. The carboxylate group comprises one product of the proteolytic reaction catalyzed by the proteases, so it is not surprising that a carboxylate can be

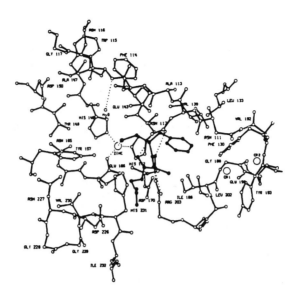

FIGURE 11. Structure of the complex between thermolysin and the thiolate-bearing inhibitor (2-benzyl-3-mercapto-propanoyl)-L-alanylglycinamide determined by X-ray crystallographic methods at 1.9 Å resolution. Enzyme atoms are shown as open circles, inhibitor atoms are shown as filled circles; note that the inhibitor thiolate coordinates to zinc with favorable *gauche* C-C-S→Zn^{2+} stereochemistry (see Figure 8). (Reprinted with permission from Monzingo, A.F. and Matthews, B.W., *Biochemistry*, 21, 3390, 1982. Copyright © 1982 American Chemical Society.)

readily incorporated within a zinc enzyme inhibitor such that optimal metal coordination stereochemistry is possible in the enzyme-inhibitor complex.

The carboxylate group may be incorporated into a so-called "bisubstrate" analog, an inhibitor which has structural features common to two substrates.[16] Interestingly, the bisubstrate analog 2-benzylsuccinate binds to TLN with a K_i of 3800 μ*M,* but it binds to CPA with a K_i of 0.45 μ*M.*[15,16,108] The three-dimensional crystal structure of the TLN-2-benzylsuccinate complex reveals an *anti*-oriented carboxylate-zinc interaction,[12] whereas that of the CPA-2-benzylsuccinate complex reveals the more favorable *syn*-oriented interaction (Figure 12).[86] It is possible that the different affinities of CPA and TLN for 2-benzylsuccinate may arise in part from the ability of the inhibitor to achieve optimal zinc coordination stereochemistry in CPA, and the inability of the inhibitor to do so in the TLN active site. Hence, optimal zinc-ligand coordination stereochemistry is a prerequisite for high-affinity enzyme-inhibitor complexes.

The carboxylate group comprises an effective carry-over design for the inhibition of zinc proteases such as ACE, for which no three-dimensional structural information is available. The successful application of the carry-over design of the TLN inhibitor *N*-(1-carboxy-3-phenylpropyl)-L-leucyl-L-tryptophan (CLT),[93] shown by Monzingo and Matthews[99] to bind with the *N*-carboxymethyl carboxylate coordinating to the active site zinc ion with *syn*-bidentate coordination stereochemistry (Figure 13), to an ACE inhibitor resulted in a drug now used in the control of hypertension (Enalaprilate; the orally administered prodrug is Enalapril).[57,109,110] The carboxylate group of Enalaprilate presumably binds to the Zn^{2+} ion of ACE in a manner similar to that observed in the TLN-CLT complex.[99] This success demonstrates that stereochemical conclusions regarding protein-ligand interactions in structurally characterized paradigms may be successfully applied to related systems of unknown structure.

C. PHOSPHATE DERIVATIVES

Tetrahedral phosphonamidates and phosphonates have been incorporated into inhibitors of CPA, TLN, and ACE in order to mimic the tetrahedral proteolytic transition state for substrate hydrolysis.[6-8,46-49,53,58-60,63,68,70,123] Such ligand design experiments are inspired by Pauling's hypothesis[111]

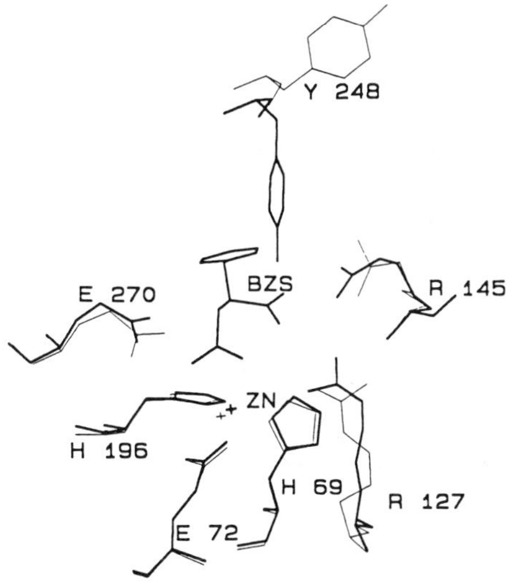

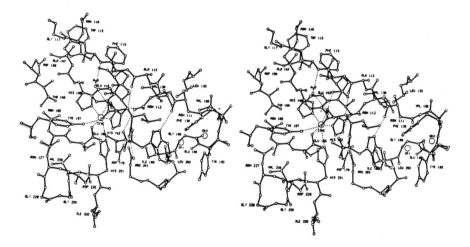

FIGURE 12. The binding of 2-benzylsuccinate to the active site of carboxypeptidase A determined in a recent X-ray crystallographic investigation. (Reprinted with permission from Mangani, S., Carloni, P., and Orioli, P., *J. Mol. Biol.*, 223, 573, 1992.) Note that the zinc ion is coordinated by the inhibitor carboxylate with favorable *syn*-bidentate stereochemistry (see Figure 2).

FIGURE 13. Stereoview of *N*-(1-carboxy-3-phenylpropyl)-L-leucyl-L-tryptophan (CLT) binding to the active site of thermolysin as determined by X-ray crystallographic methods.[99] Note that the zinc ion is coordinated by the inhibitor carboxylate with favorable *syn*-bindentate stereochemistry. (Reprinted with permission from Monzingo, A.F. and Matthews, B.W., *Biochemistry*, 23, 5724, 1984. Copyright © 1984 American Chemical Society.)

that the transition state of an enzyme-catalyzed reaction is bound several orders of magnitude more tightly to the enzyme than the substrate or product (Figure 14). Hence, if a tetrahedral phosphate derivative, such as a phosphonamidate or phosphonate, is an analog of the proteolytic transition state, it should bind much more tightly to the zinc protease than, for instance, a carboxylate-bearing product or product analog. Indeed, this strategy has recently led to the development of tetrahedral phosphonate inhibitors of CPA with dissociation constants in the femtomolar range.[70]

As discussed previously, the preferential stereochemistry of phosphate-zinc coordination differs from that of carboxylate-zinc coordination. The statistical analysis of structures in the Cambridge Structural Database suggests that symmetrically *syn*-bidentate zinc coordination is unfavorable for

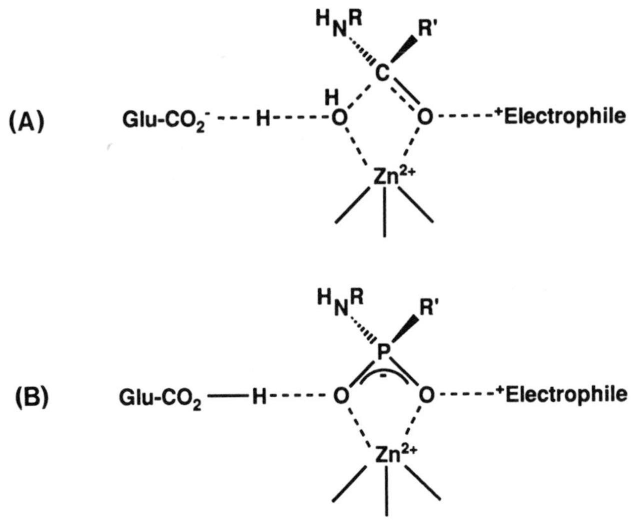

FIGURE 14. (a) The tetrahedral proteolytic transition state for peptide hydrolysis, as catalyzed in the consensus zinc protease active site of Figure 9; (b) a tetrahedral phosphonamidate is a stable, structural analog of the chemical transition state depicted in (a).

the tetrahedral phosphate derivative, e.g., as found in a phosphonate zinc protease inhibitor.[3,69] Instead, *syn*-unidentate coordination geometry is preferred, and this is consistent with molecular orbital considerations.[77,132] The structures of several enzyme-inhibitor complexes involving TLN[61,124] and CPA[29,72] reflect this stereochemical preference, where phosphinyl-containing inhibitors generally tend toward *syn*-unidentate zinc coordination. This avenue of effort has led to rationally designed phosphonate inhibitors of CPA with K_i values in the femtomolar range (Figure 15).[70,73]

Although suitably constructed phosphate derivatives exhibit appreciable inhibitory properties toward ACE, these derivatives have not yet yielded a rationally designed drug for hypertension therapy. For example, *N*-phosphoryl-L-alanine-L-proline is a competitive inhibitor of ACE with a K_i of 1.4 n*M*;[46] moreover, the phenethylphosphonyl derivative of this compound exhibits a threefold greater affinity for ACE than the parent inhibitor.[49] Certainly, the family of phosphate-derived ACE inhibitors holds great promise for yielding an antihypertensive agent with the desired pharmacological properties.

IV. INHIBITORS OF HUMAN CARBONIC ANHYDRASE II

There are several isozymes of carbonic anhydrase (CA) occurring throughout the animal, plant, and bacterial kingdoms. Isozyme II of this zinc-requiring hydrase is found in the erythrocyte and various other tissues of mammalian species and its chemical, kinetic, and structural properties have been reviewed.[25,81,114,136] Within the erythrocyte, CAII hydrates carbon dioxide to form bicarbonate ion plus a proton; most of the carbon dioxide generated during the process of respiration requires this CAII-catalyzed event for transport out of the cell.

X-ray crystallographic studies of native CAII have been reported at 2.0 Å resolution.[42,43] Important active site residues include Thr-199, Thr-200, Glu-106, His-64, Trp-209, Val-121, Val-143, the zinc ion (liganded by His-94, His-96, and His-119), and the zinc-bound hydroxide ion. A schematic view of the active site is found in Figure 16. The globular protein has a molecular weight of about 29 kDa and is comprised of a single polypeptide chain of 260 amino acids. The active site of the enzyme can be envisaged as a conical cavity extending 15 Å into the core of the protein; the zinc ion resides at the bottom of the conical cavity, and this cavity is roughly divisible into hydrophobic and hydrophilic halves.

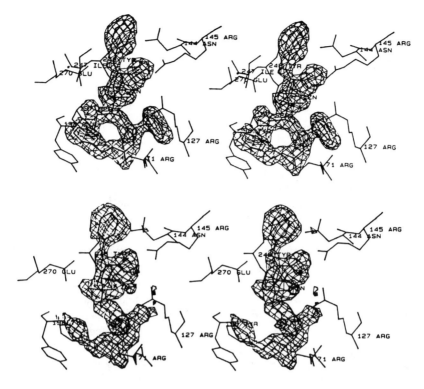

FIGURE 15. Stereoview of X-ray crystallographic electron density maps calculated for two phosphonate tripeptide analogs with K_i values measured in the femtomolar range.[70] Note that in each example the phosphinyl-zinc coordination stereochemistry is *syn*-unidentate. (Reprinted with permission from Kim, H. and Lipscomb, W.N., *Biochemistry*, 39, 8171, 1991. Copyright © 1991 American Chemical Society.)

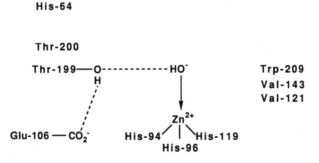

FIGURE 16. Schematic view of important amino acid residues in the carbonic anhydrase II active site. Note that there is a hydrophilic portion and a hydrophobic portion.

The molecular function of CAII plays an important role in the regulation of intraocular pressure. Since there is a significant excess concentration of bicarbonate ion in the posterior chamber of the aqueous humor,[74] and since bicarbonate ion concentration is linked to aqueous secretion,[45] there is a causal relationship between the ocular activity of CAII and intraocular pressure: the activity of CAII results in increased bicarbonate ion concentration, which in turn stimulates aqueous secretion — this results in elevated intraocular pressure. Hence, the inhibition of ocular CAII activity should inhibit the elevation of intraocular pressure. This physiological effect has been observed *in vivo* for the sulfonamide inhibitor Diamox (Figure 17).[9,14,56,87] Importantly, the regulation of intraocular pressure has proven to be a successful therapy for glaucoma, a disease which is accompanied by painful elevation of intraocular pressure (reviewed by Maren[88–90]). It is noteworthy that inhibitors

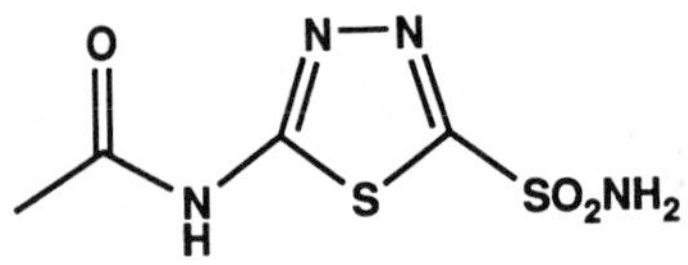

FIGURE 17. The sulfonamide Diamox (acetazolamide), a carbonic anhydrase II inhibitor used for the treatment of glaucoma.

of carbonic anhydrase also find application toward the treatment of epilepsy, acute mountain sickness, and space motion sickness.[89,91]

Persistent concerns regarding CAII-targeted drugs include the method of drug delivery, the stability of the drug molecule *in vivo*, and the ocular retention of the drug in concentrations sufficient for effective therapy. Hence, new leads in the design of CAII inhibitors may exhibit improved inhibitory properties as well as improved ocular retention. The rational design of sulfonamide CAII inhibitors is therefore a field of sustained pharmaceutical interest (e.g., see Baldwin et al.,[5] Graham et al.,[54] Graham et al.,[55] and references cited therein), and it is indeed fortunate that in this example the three-dimensional structure of the actual drug target is available to guide quantitative molecular design efforts. Recently, aspects of zinc coordination by ionized sulfonamides (which displace the zinc-bound hydroxide ion of the native enzyme) and the role of flanking hydrogen bond networks in CAII-inhibitor complexes have been studied by molecular mechanics and free energy perturbation methods,[95–97,128,129] and molecular orbital considerations of sulfonamide-metal coordination have been made in a recent *ab initio* study.[80] Importantly, it is possible to reproduce relative binding free energies for enzyme-inhibitor complexes that are consistent with experimental results, and theory therefore provides a unique route toward understanding the recognition and discrimination of tight-binding drug candidates in the CAII active site. The particularly avid binding of sulfonamide groups to zinc relies upon a hydrogen bond network between enzyme and inhibitor which involves the ionized, metal-bound sulfonamide nitrogen.[97,128,129] The structure of the CAII-Diamox complex has been reported at 1.9 Å resolution;[130] a scheme illustrating the hydrogen bond network involving the ionized sulfonamide group of this drug is found in Figure 18.

Structure-assisted drug design efforts have utilized the three-dimensional structure of CAII refined at 2.0 Å resolution,[42,43] as well as the refined three-dimensional structures of acetazolamide[130] and 3-acetoxymercury-4-aminobenzenesulfonamide[44] in their complexes with CAII. As in examples with carboxypeptidase A, conformational changes within the CAII active site also play an important role in ligand binding and rational ligand design. For instance, upon the binding of *R* or *S* stereoisomers of the thienothiopyran-2-sulfonamide shown in Figure 19, X-ray crystallographic analysis of each CAII-inhibitor complex reveals that His-64 undergoes a significant rotation about side chain torsion angle χ_1; this conformational change is sterically required to accommodate the binding of each inhibitor.[5] Intriguingly, the conformational change of His-64 is observed in other CAII-inhibitor complexes[147], and it is observed to occur in certain examples even when not sterically required by inhibitor binding.[138,142] Conformational changes in the inhibitor induced by its association with the enzyme are also apparent when the structure of the CAII-complexed *S* stereoisomer is compared with its uncomplexed structure in the crystal: the orientation of the isobutylamino group changes from pseudoequatorial to pseudoaxial with respect to the thiopyran ring. Additionally, a 14° difference in the N-S-C-S dihedral angle characterizes the binding of *R* and *S* stereoisomers to CAII; this difference, as well as the contrasting pseudoequatorial conformation of the isobutylamino group in the enzyme-bound *R* stereoisomer, must contribute to the better than 100-fold enhancement in binding affinity measured for the *S* stereoisomer relative to the *R* stereoisomer.

It should be noted that conformational changes have been observed for the side chain of His-64 in recent X-ray crystallographic investigations of uncomplexed CAII.[76,100,137] The imidazole side chain of this residue appears to be in dynamic equilibrium between two principle conformers,

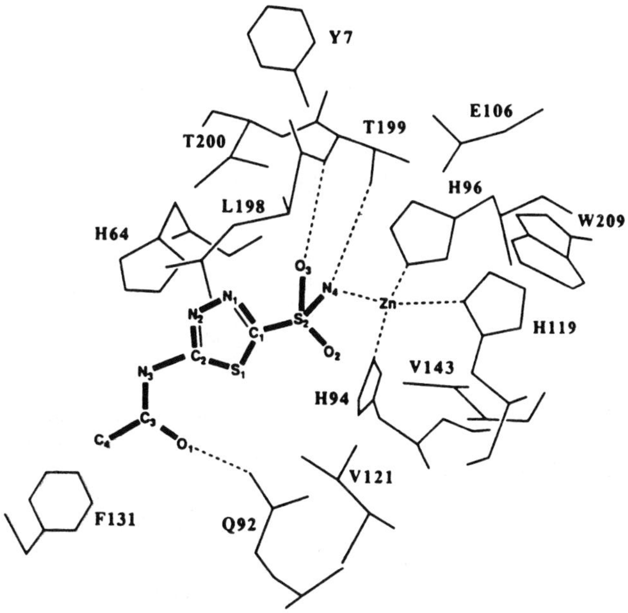

FIGURE 18. Hydrogen bond network and zinc coordination in the carbonic anhydrase II-Diamox complex. (Reprinted by permission of the publishers, Butterworth Heinemann Ltd. © 1990, from Vidgren, J., Liljas A., and Walker, N.P.C., *Int. J. Biol. Macromol.*, 12, 342, 1990.)

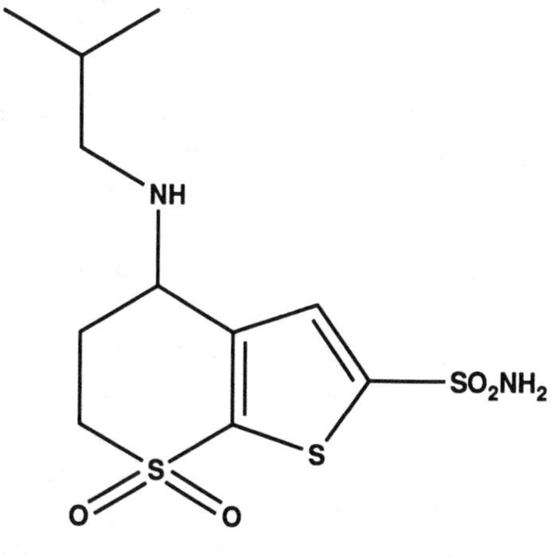

FIGURE 19. Thienothiopyran-2-sulfonamide inhibitor of carbonic anhydrase II. The *S* enantiomer binds more than 100-fold more tightly to the enzyme than the *R* enantiomer.

designated "in" (directed toward the active site) and "out" (directed away from the active site). The "in" conformation is the predominant conformer in the human blood enzyme at pH 8.5 and the recombinant wild-type enzyme at pH 8.0.[4,43] The "out" conformation places the imidazole group of His-64 closer to a more hydrophobic region largely defined by Trp-5, Gly-6, Tyr-7, and Phe-231. The "out" conformation is observed in the crystal structure of the Thr-200→Ser mutant of CAII at pH 8.0, where His-64 rotates about χ_1 by 105° (the mutant enzyme nonetheless exhibits normal CO_2 hydrase activity and proton transfer kinetics[76]). Additionally, a partially "out" conformation of His-64 is observed in the structure of the native blood enzyme at pH 5.7, where the side chain rotates by 64° about χ_1 relative to its conformation at pH 8.5.[100] A superposition of the various His-64 conformers observed in these X-ray crystallographic studies is found in Figure 20.

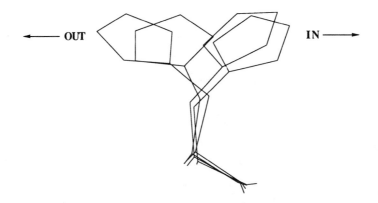

FIGURE 20. Superposition of residue His-64 from various crystallographic structure determinations of CAII summarized in the text, illustrating the reaction coordinate of conformational mobility. This reaction coordinate characterizes the movement of His-64 in certain CAII-drug complexes, and it may also accompany the catalytic role of His-64 as a proton shuttle. The "in" conformation points toward the active site; the "out" conformation points away from the active site.

At higher pH values, electron density corresponding to the "out" conformer of His-64 may sometimes be interpretable as solvent. Eriksson and colleagues[43] interpret and refine this density as a water molecule in pH 8.5 CAII at 2.2 Å resolution, and the relatively high thermal B factor of 29 Å^2 is indicative of disorder. Nair and Christianson[101] arrive at a similar conclusion in pH 9.5 CAII at 2.2 Å resolution, but these investigators note that the corresponding density is not inconsistent with an alternate conformer of His-64. In recombinant wild-type CAII at 2.1 Å resolution, the corresponding electron density is likewise ambiguous but is interpretable as a low-occupancy "out" conformer of His-64.[76]

It is difficult to establish the molecular factors responsible for shifting the in $\rightleftarrows$ out equilibrium of His-64. The imidazole group of this residue must be positively charged at pH 5.7, since its pKa is measured at 7.1.[18] Interestingly, the conformational change of His-64 is only 64° at pH 5.7, whereas it is measured at 105° in Thr-200→Ser CAII; perhaps the conformational change at low pH is smaller since the greater the χ_1 rotation to the "out" conformation, the closer the positively charged imidazolium group approaches the adjacent hydrophobic patch. However, the positively charged imidazolium group of His-64 might rotate away from the active site in order to minimize a repulsive interaction with the positively-charged zinc coordination polyhedron (zinc-bound water has a pK$_a$ of about 7,[40,113] so it would not be appreciably ionized at pH 5.7). Nevertheless, it is difficult to conclude that the conformational change at low pH results solely from the protonation of the side chain, since the "out" conformer predominates at pH 8.0 in Thr-200→Ser CAII.[76] A general conclusion may be that the mobility of His-64 is affected by changes in solvent structure which, in turn, arise from changes in protein structure, counterion concentration, or pH. Thus, it is not necessarily possible to design a tight-binding protein ligand from simply the structure of the uncomplexed protein — recognition and affinity may require dynamic changes in the protein structure to accommodate ligand binding.

V. ZINC-MEDIATED HORMONE-RECEPTOR COMPLEXES

Although hormone-receptor association does not exhibit nanomolar affinity in all biological examples, certain hormones are present at such low peripheral concentrations that potent hormone-receptor affinity and a significant excess of receptors per cell surface are required in order to elicit the biological response.[83] Human growth hormone (hGH), a 191-amino acid polypeptide, provides a timely example of these effects. Both hGH and human prolactin hormone (hPRL) are homologous proteins from a large family of hormones that regulate a wide variety of physiological effects, including linear bone growth, lactation, differentiation, and electrolyte balance.[23,67,103,131] These biological effects are initiated by the binding of the hormone to specific cellular receptors,[65] e.g., the

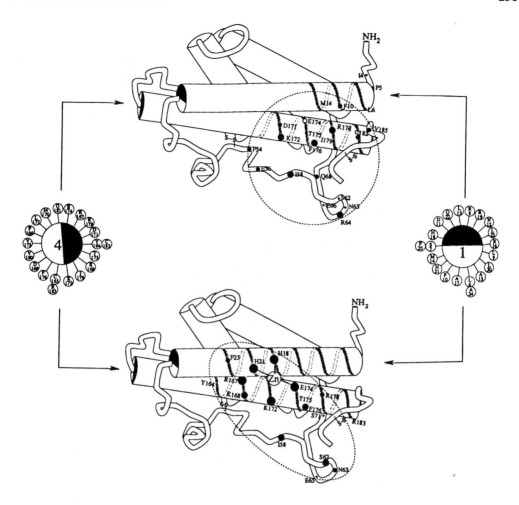

FIGURE 21. Functional maps of the hPRLbp (bottom) and hGHbp (top) binding domains on hGH (modeled by the three-dimensional structure of pGH) identified by scanning mutational analysis. Proposed zinc ligands are His-18, His-21, and Glu-174 when Zn^{2+} mediates the binding of hGH to hPRLbp but not to hGHbp; additionally, this binding site is also implicated in zinc-dependent hGH dimerization. (Reprinted with permission from Lowman, H.B., Bass, S.H., Simpson, N., and Wells, J.A., *Biochemistry*, 30, 10832, 1991. Copyright © 1991 American Chemical Society.)

binding of hGH to the lactose receptor stimulates lactation, and the binding of hGH to the somatogenic receptor initiates bone growth. Orally administered supplements of hGH promote growth in dwarf children; normal children release 50 to 100 n*M* hGH/day from the pituitary gland, whereas children suffering from dwarfism release only 10 n*M* hGH per day.[35]

The concentration of hGH at the receptor is about 2 n*M*.[64,120,121] However, Cunningham and colleagues[35] estimate the concentration of hGH in the millimolar range when released from the pituitary. This concentration gradient may be exploited in a storage mechanism for the hormone in secretory granules, where Zn^{2+} is present at high concentrations,[122] since hGH dimerizes in the presence of zinc with a K_d of 1 μ*M*.[35] Indeed, hGH and Zn^{2+} are present at about equimolar concentrations in the granules, so more than 99% of the hormone could be present in the form hGH_2-Zn^{2+}; additionally, the hGH_2-Zn^{2+} complex is more stable to denaturation than monomeric hGH.[35] These data are consistent with the observation that elevated zinc concentrations inhibit growth hormone release.[78,84,116] Intriguingly, the receptor specificity of hGH also exhibits zinc-dependence; although zinc does not affect hGH association with the hGH receptor, the binding of hGH to the prolactin receptor is extremely sensitive to the presence of Zn^{2+} (K_d = 0.03 n*M*).[33] Therefore, zinc is becoming increasingly recognized as a mediator of regulatory protein-protein complexes.

The only cloned human genes whose products are known to bind hGH are those corresponding to hGH receptor from liver[79] and the hPRL receptor from mammary gland.[13] However, it is not known whether the binding sites of these two receptors on hGH are identical. Additionally, it is not clear what pharmacological effect is associated with each receptor. Cunningham and Wells[37] and Cunningham and colleagues[35] address these issues by mapping sites on the hGH that interact with the extracellular domains of the hGH and hPRL receptors (hGHbp and hPRLbp, respectively). Intriguingly, the two receptor binding sites overlap, but are not identical (see Figure 21). Principally, the zinc-dependence of hormone specificity reflects these binding site differences: hGH requires zinc for tight binding to the hPRL receptor, but does not require zinc for binding to the hGH receptor.[35]

Scanning mutational analysis of hGH has identified specific side chains on hGH that mediate its binding to hPRLbp in the presence of zinc.[33] In particular, the imidazole side chains of His-18 and His-21, and the carboxylate side chain of Glu-174, are implicated for metal binding: when each residue is substituted by alanine, the affinity of the hGH variant toward hPRLbp is reduced by about 100-fold. When the amino acid sequence of hGH was mapped onto the known three-dimensional structure of porcine growth hormone (pGH; the only available structure from that family of hormones at the time[1]) it was found that His-18 and His-21 are on adjacent turns of an α-helix near the amino-terminus, and Glu-174 is proximally located on another α-helix near the carboxyl-terminus of the protein (see Figure 21). It is intriguing that the zinc coordination polyhedron of hGH is chemically identical to that found in the zinc proteases carboxypeptidase A and thermolysin.[28,92]

Using the hGH model constructed from the three-dimensional structure of pGH, functional maps of the entire hPRLbp and hGHbp binding domains on hGH have been constructed (Figure 21). This map highlights locations where alanine substitutions result in altered binding affinities with each receptor, and specific residues on hGH are identified that are important for the binding of *both* hGH and hPRL. For example, Glu-174$\rightarrow$Ala and His-21$\rightarrow$Ala mutations separately increase the binding affinity of hGH to hGHbp by four- and threefold, respectively.[37] However, both Glu-174 and His-21 are required residues for zinc-mediated binding of hGH to hPRLbp, as summarized above. Thus, hGH compromises binding to hGHbp for the sake of dual receptor specificity. More recent studies of hGH-hGHbp affinity have employed a phage display system which allows for the screening of nearly 10^6 hGH variants, and several variants have been isolated which exhibit enhanced affinity and specificity toward hGHbp.[85]

Recent biophysical and X-ray crystallographic studies indicate that hGH forms a 1:2 complex with hGHbp, and it is likely that the formation of the analogous complex on the cell surface is critical for signal transduction *in vivo*.[36,41,139] These studies reveal that there are two functionally distinct sites on hGH which interact with hGHbp, and hormone-receptor recognition proceeds in ordered fashion: the first hGHbp molecule binds to "site 1" on hGH (a region including the carboxyl-terminal α-helix), and the second hGHbp molecule binds to "site 2" on hGH (a region including the amino-terminus). Since hormone-induced receptor oligomerization is typically believed to occur with hormone$_2$-receptor$_2$ stoichiometry,[125,133] it is clear that the hormone-receptor$_2$ stoichiometry observed in the hGH-hGHbp$_2$ complex does not conform to such an oligomerization model for signal transduction. This stoichiometry does not necessarily characterize the interaction of hGH with hPRLbp, since deletions in the N-terminal portion of helix 1 reduce the binding affinity of hGH to somatogenic receptors to a greater extent than they reduce the binding affinity of hGH to lactogenic receptors.[10]

In addition to studying hGH-hGHbp interactions, Cunningham and coworkers have also engineered the binding of hPRL hormone to the hGH receptor.[34] The hPRL hormone is only 23% homologous to hGH. However, by modeling both hGH and hPRL to the known structure of pGH, Cunningham and colleagues[34] found that hGH and hPRL are 34% identical in the region that modulates hormone binding to hGHbp. This region is comprised of three discontinuous segments of the polypeptide sequence which is located on the same side of the folded protein: the first half of helix 1, the second half of loop 2, and the C-terminal portion of helix 4. Cunningham and coworkers[34] and Goffin and Martial[52] demonstrate by alanine substitutions that the binding specificities of hPRL for either hPRLbp or hGHbp can be altered by modifing residues on hPRL which have been identified by homology to pGH as residues that strongly modulate its binding to hGHbp. The rational redesign of binding specificity, as a result of only a few amino acid substitutions, confirms the structural relationships among hGH, hPRL, and perhaps other hormones from the same family.

VI. SUMMARY

Various structural factors which influence protein-ligand recognition and affinity have been addressed in this review, and these factors are relevant to the understanding and design of protein small molecule as well as hormone-receptor complexes. In the examples selected, a zinc ion plays a functional role in the formation and stability of the protein-drug complex. Given that a proven route toward engineering zinc protein ligands requires the incorporation of appropriate functional groups within the molecular framework of the ligand, then a detailed stereochemical intuition of functional group-zinc coordination architecture is a prerequisite to understanding successes (and failures) in ligand design. Additionally, consideration of the hydrogen bond networks that engage the zinc-bound atom of a coordinating functional group can facilitate design efforts. Nature designs metalloproteins in which the metal binding site is nested within an extensive array of hydrogen bonds and other weakly polar interactions, and successful design approaches will exploit potential interactions within these preexisting networks. In some examples, the three-dimensional structure of a suitably related protein will suffice for rational ligand design experiments *if* the structure of the actual drug receptor is unknown.

Finally, the degree of "induced fit" which accompanies any macromolecular association event should be appreciated. In no case is the native structure of an uncomplexed protein exactly equivalent to the structure of the protein in its ligand-complexed form; similarly, the bound conformation of a protein ligand (either a small molecule, or even another protein) is not equivalent to the conformation of the unbound ligand molecule. Therefore, the conformations of both protein and ligand molecules in the complex will not necessarily be minimum energy structures characteristic of the uncomplexed species.

Although the factors discussed in this review represent only a subset of those contributing to zinc protein-ligand affinity (other factors might include entropy effects related to desolvation and/or conformation, for example), they represent a first-order approach to the rational design of ligand molecules which target zinc-requiring proteins. There is no doubt that this group of proteins will continue to be thoroughly investigated by pharmaceutical chemists and structural biologists.

ACKNOWLEDGMENTS

D.W.C. thanks the National Institutes of Health for grants GM45614, GM49758, and HL51893 and A.M.K.C. thanks the NIH for a postdoctoral fellowship. D.W.C. is an Alfred P. Sloan research fellow and a Camille and Henry Drefus Teacher-Scholar.

REFERENCES

1. **Abdel-Meguid, S.S., Shieh, H.S., Smith, W.W., Dayringer, H.E., Violand, B.N., and Bentle, L.A.,,** *Proc. Natl. Acad. Sci. U.S.A.,* 84, 6434, 1987.
2. **Adman, E., Watenpaugh, K.D., and Jensen, L.H.,** *Proc. Natl. Acad. Sci. U.S.A.,* 72, 4854, 1975.
3. **Alexander, R.S., Kanyo, Z.F., Chirlian, L.E., and Christianson, D.W.,** *J. Am. Chem. Soc.,* 112, 933, 1990.
4. **Alexander, R.S., Nair, S.K., and Christianson, D.W.,** *Biochemistry* 30, 11064, 1991.
5. **Baldwin, J.J., Ponticello, G.S., Anderson, P.S., Christy, M.E., Murcko, M.A., Randall, W.C., Schwam, H., Sugrue, M.F., Springer, J.P., Gautheron, P., Grove, J., Mallorga, P., Viader, M.-P., McKeever, B.M., and Navia, M.A.,** *J. Med. Chem.,* 32, 2510, 1989.

6. **Bartlett, P.A. and Marlowe, C.K.,** *Biochemistry,* 22, 4618, 1983.
7. **Bartlett, P.A. and Marlowe, C.K.,** *Biochemistry,* 26, 8553, 1987.
8. **Bartlett, P.A. and Marlowe, C.K.,** *Science,* 235, 569, 1987.
9. **Becker, B.,** *Am. J. Ophthalmol.,* 37, 13, 1954.
10. **Binder, L., Gertler, A., Elberg, G., Guy, R., and Volgel, T.,** *Mol. Endocrinol.,* 4, 1060, 1990.
11. **Blundell, T.L., Pitts, J.E., Tickle, I.J., Wood, S.P., and Wu, C.-W.,** *Proc. Natl. Acad. Sci. U.S.A.,* 87, 4175, 1981.
12. **Bolognesi, M.C. and Matthews, B.W.,** *J. Biol. Chem.,* 254, 634, 1979.
13. **Boutin, J.M., Shirota, M., Jolicoeur, C., Lesueur, L., Ali, S., Guold, D., Djiane, J., and Kelly, P.A.,** *Mol. Endocrinol.,* 3, 1455, 1989.
14. **Breinin, G.M. and Gortz, H.,** *Arch. Ophthalmol.,* 52, 333, 1954.
15. **Byers, L.D. and Wolfenden, R.,** *J. Biol. Chem.,* 247, 606, 1972.
16. **Byers, L.D. and Wolfenden, R.,** *Biochemistry,* 12, 2070, 1973.
17. **Caffrey, M.S., Daldal, F., Holden H.M., and Cusanovich, M.A.,** *Biochemistry,* 30, 4119, 1991.
18. **Campbell, I.D., Lindskog, S., and White, A.I.,** *J. Mol. Biol.,* 98, 597, 1975.
19. **Carrell, C.J., Carrell, H.L., Erlebacher, J., and Glusker, J.P.,** *J. Am. Chem. Soc.,* 110, 8651, 1988.
20. **Chakrabarti, P.,** *Biochemistry,* 28, 6081, 1989.
21. **Chakrabarti, P.,** *Protein Eng.,* 4, 49, 1990.
22. **Chakrabarti, P.,** *Biochemistry,* 29, 651, 1990.
23. **Chawla, R.K., Parks, J.S., and Rudman, D.,** *Annu. Rev. Med.,* 34, 519, 1983.
24. **Chakrabarti, P.,** *Protein Eng.,* 4, 57, 1990.
25. **Christianson, D.W.,** *Adv. Prot. Chem.,* 42, 281, 1991.
26. **Christianson, D.W. and Alexander, R.S.,** *Nature,* 346, 225, 1990.
27. **Christianson, D.W. and Alexander, R.S.,** *J. Am. Chem. Soc.,* 111, 6412, 1989.
28. **Christianson, D.W. and Lipscomb, W.N.,** *Acc. Chem. Res.,* 22, 62, 1989.
29. **Christianson, D.W. and Lipscomb, W.N.,** *J. Am. Chem. Soc.,* 110, 5560, 1988.
30. **Christianson, D.W. and Lipscomb, W.N.,** *J. Am. Chem. Soc.,* 109, 5536, 1987.
31. **Christianson, D.W. and Lipscomb, W.N.,** *Proc. Natl. Acad. Sci. U.S.A.,* 83, 7568, 1986.
32. **Coleman, J.E.,** in *Zinc Enzymes,* Bertini, I., Luchinat, C., Maret, W., and Zeppezauer, M., Eds., Birkhäuser, Boston, 1986, 49.
33. **Cunningham, B.C., Bass, S., Fuh, G., and Wells, J.A.,** *Science,* 250, 1709, 1990.
34. **Cunningham, B.C., Henner, D.J., and Wells, J.A.,** *Science,* 247, 1461, 1990.
35. **Cunningham, B.C., Mulkerrin, M.G., and Wells, J.A.,** *Science,* 253, 545, 1991.
36. **Cunningham, B.C., Ultsch, M., De Vos, A.M., Mulkerrin, M.G., Clausen, K.R., and Wells, J.A.,** *Science,* 254, 821, 1991.
37. **Cunningham B.C. and Wells, J.A.,** *Proc. Natl. Acad. Sci. U.S.A.,* 88, 3407, 1991.
38. **Cushman, D.W., Cheung, H.S., Sabo, E.F., and Ondetti, M.A.,** *Biochemistry,* 16, 5484, 1977.
39. **Cushman, D.W. and Ondetti, M.A.,** *Prog. Med. Chem.,* 17, 41, 1980.
40. **Davis, R.P.,** *J. Am. Chem. Soc.,* 80, 5209, 1958.
41. **de Vos, A.M., Ultsch, M., and Kossiakoff, A.,** *Science,* 255, 306, 1992.
42. **Eriksson, A.E., Jones, T.A., and Liljas, A.,** in *Zinc Enzymes,* Bertini, I., Luchinat, C., Maret, W., and Zeppezauer, M., Eds., Birkhäuser, Boston, 1986, 317.
43. **Eriksson, A.E., Jones, T.A., and Liljas, A.,** *Proteins Struct. Funct. Genet.,* 4, 274, 1988.
44. **Eriksson, A.E., Kylsten, P.M., Jones, T.A., and Liljas, A.,** *Proteins Struct. Funct. Genet.,* 4, 283, 1988.
45. **Friedenwald, J.S.,** *Am. J. Ophthalmol.,* 32, 9, 1949.
46. **Galardy, R.E.,** *Biochem. Biophys. Res. Commun.,* 97, 94, 1980.
47. **Galardy, R.E.,** *Biochemistry,* 21, 5777, 1982.
48. **Galardy, R.E. and Grobelny, D.,** *J. Med. Chem.,* 28, 1422, 1985.
49. **Galardy, R.E., Kontoyiannidou-Ostrem, V., and Kortylewicz, Z.P.,** *Biochemistry,* 22, 1990, 1983.
50. **Gandour, R.D.,** *Bioorg. Chem.,* 10, 169, 1981.
51. **Chawla, R.K., Parks, J.S., and Rudman, D.,** *Annu. Rev. Med.,* 34, 519, 1983.
52. **Goffin, V., Norman, M., and Martial, J.A.,** *Mol. Endocrinol.,* 6, 1381, 1992.
53. **Goli, U.B., Grobelny, D., and Galardy, R.E.,** *Biochem. J.,* 254, 847, 1988.
54. **Graham, S.L., Shepard, K.L., Anderson, P.S., Baldwin, J.J., Best, D.B., Christy, M.E., Freedman, M.B., Gautheron, P., Habecker, C.N., Hoffman, J.M., Lyle, P.A., Michelson, S.R., Ponticello, G.S., Robb, C.M., Schwam, H., Smith, A.M., Smith, R.L., Sondey, J.M., Strohmaier, K.M., Sugrue, M.F., and Varga, S.L.,** *J. Med. Chem.,* 32, 2548, 1989.
55. **Graham, S.L., Hoffman, J.M., Gautheron, P., Michelson, S.R., Scholz, T.H., Schwam, H., Shepard, K.L., Smith, A.M., Smith, R.L., Sondey, J.M., and Sugrue, M.F.,** *J. Med. Chem.,* 33, 749, 1990.
56. **Grant, W.M. and Trotter, R.R.,** *Arch. Ophthalmol.,* 51, 735, 1954.
57. **Greenlee, W.J., Allibone, P.L., Perlow, D.S., Patchett, A.A., Ulm, E.H., and Vassil, T.C.,** *J. Med. Chem.,* 28, 434, 1985.
58. **Grobelny, D., Goli, U.B., and Galardy, R.E.,** *Biochem. J.,* 232, 15, 1985.

59. **Grobelny, D., Goli, U.B., and Galardy, R.E.,** *Biochemistry,* 28, 4948, 1989.
60. **Hanson, J.E., Kaplan, A.P., and Bartlett, P.A.,** *Biochemistry,* 28, 6294, 1989.
61. **Holden, H.M., Tronrud, D.E., Monzingo, A.F., Weaver, L.H., and Matthews, B.W.,** *Biochemistry,* 26, 8542, 1987.
62. **Holmes, M.A. and Matthews, B.W.,** *Biochemistry,* 20, 6912, 1981.
63. **Holmquist, B. and Vallee, B.L.,** *Proc. Natl. Acad. Sci. U.S.A.,* 76, 6216, 1979.
64. **Ho, K.Y. et al.,** *J. Clin. Endocrinol. Metab.,* 64, 51, 1987.
65. **Hughs, J.P. and Friesen, H.G.,** *Annu. Rev. Physiol.,* 47, 469, 1985.
66. **Ippolito, J.A., Alexander, R.S., and Christianson, D.W.,** *J. Mol. Biol.,* 215, 457, 1990.
67. **Isaksson, O., Eclen, S., and Jansson, J.O.,** *Annu. Rev. Physiol.,* 47, 483, 1985.
68. **Jacobsen, N.E. and Bartlett, P.A.,** *J. Am. Chem. Soc.,* 103, 654, 1981.
69. **Kanyo, Z.F. and Christianson, D.W.,** *J. Biol. Chem.,* 266, 4264, 1991.
70. **Kaplan, A.P. and Bartlett, P.A.,** *Biochemistry,* 30, 8165, 1991.
71. **Kester, W.R. and Matthews, B.W.,** *J. Biol. Chem.,* 252, 7704, 1977.
72. **Kim, H. and Lipscomb, W.N.,** *Biochemistry,* 29, 5546, 1990.
73. **Kim, H. and Lipscomb, W.N.,** *Biochemistry,* 30, 8171, 1991.
74. **Kinsey, V.E.,** *Arch. Ophthalmol.,* 50, 401, 1953.
75. **Koshland, D.E.,** *Proc. Natl. Acad. Sci. U.S.A.,* 44, 98, 1958.
76. **Krebs, J.F., Fierke, C.A., Alexander, R.S., and Christianson, D.W.,** *Biochemistry,* 30, 9153, 1991.
77. **Kutzelnigg, W.,** *Pure Appl. Chem.,* 49, 981, 1977.
78. **La Bella, F., Dular, R., Vivian, S., and Queen, G.,** *Biochem. Biophys. Res. Commun.,* 52, 786, 1973.
79. **Leung, D.W., Spencer, S.A., Cachianes, G., Hammonds, G., Colins, C., Henzel, W.J., Barnard, R., Waters, M.J., and Wood, W.,** *Nature,* 330, 537, 1987.
80. **Liang, J.-Y. and Lipscomb, W.N.,** *Biochemistry,* 28, 9724, 1989.
81. **Lindskog, S.,** in *Zinc Enzymes,* Spiro, T.G., Ed., John Wiley & Sons, New York, 1983, 78.
82. **Lindskog, S.,** in *Zinc Enzymes,* Bertini, I., Luchinat, C., Maret, W., and Zeppezauer, M., Eds., Birkhäuser, Boston, 1986, 307.
83. **Lipscomb, W.N.,** *Ann. N.Y. Acad. Sci.,* 367, 326, 1981.
84. **Lorenson, M.Y., Robson, D.L., and Jacobs, L.S.,** *J. Biol. Chem.,* 258, 8618, 1983.
85. **Lowman, H.B., Bass, S.H., Simpson, N., and Wells, J.A.,** *Biochemistry,* 30, 10832, 1991.
86. **Mangani, S., Carloni, P., and Orioli, P.,** *J. Mol. Biol.,* 223, 573, 1992.
87. **Mann, T. and Keilin, D.,** *Nature,* 146, 164, 1940.
88. **Maren, T.H.,** *Physiol. Rev.,* 47, 595, 1967.
89. **Maren, T.H.,** in *Biology and Chemistry of the Carbonic Anhydrases,* Tashian, R.E. and Hewett-Emmett, D., Eds., Ann. N.Y. Acad. Sci. 429, 49, 1984.
90. **Maren, T.H.,** *Drug Dev. Res.,* 10, 255, 1987.
91. **Martin, N.F.,** *Aviat. Space Environ. Med.,* 55, 1148, 1984.
92. **Matthews, B.W.,** *Acc. Chem. Res.,* 21, 333, 1988.
93. **Maycock, A.L., DeSousa, D.M., Payne, L.G., ten Broeke, J., Wu, M.T., and Patchett, A.A.,** *Biochem. Biophys. Res. Commun.,* 102, 963, 1981.
94. **McGregor, M.J., Islam, S.A., and Sternberg, M.J.E.,** *J. Mol. Biol.,* 198, 295, 1987.
95. **Menziani, M.C., De Benedetti, P.G., Gago, F., and Richards, W.G.,** *J. Med. Chem.,* 32, 951, 1989.
96. **Menziani, M.C., Reynolds, C.A., and Richards, W.G.,** *J. Chem. Soc. Chem. Commun.,* 853, 1989.
97. **Merz, K.M., Murcko, M.A., and Kollman, P.A.,** *J. Am. Chem. Soc.,* 113, 4484, 1991.
98. **Monzingo, A.F. and Matthews, B.W.,** *Biochemistry,* 21, 3390, 1982.
99. **Monzingo, A.F. and Matthews, B.W.,** *Biochemistry,* 23, 5724, 1984.
100. **Nair, S.K. and Christianson, D.W.,** *J. Am. Chem. Soc.,* 113, 9455, 1991.
101. **Nair, S.K. and Christianson, D.W.,** *Biochem. Biophys. Res. Commun.,* 181, 579, 1991.
102. **Nakagawa, S., Umeyama, H., Kitaura, K., and Morokuma, K.,** *Chem. Pharm. Bull.,* 29, 1, 1981.
103. **Nicoll, N.C., Mayer, G.L., and Russell, S.M.,** *Endocrinol Rev.,* 7, 169, 1986.
104. **Nishino, N. and Powers, J.C.,** *Biochemistry,* 18, 4340, 1979.
105. **Ondetti, M.A., Condon, M.E., Reid, J., Sabo, E.F., Cheung, H.S., and Cushman, D.W.,** *Biochemistry,* 18, 1427, 1979.
106. **Ondetti, M.A. and Cushman, D.W.,** *Annu. Rev. Biochem.,* 51, 283, 1982.
107. **Ondetti, M.A., Rubin, B., and Cushman, D.W.,** *Science,* 196, 441, 1977.
108. **Palmer, A.R., Ellis, P.D., and Wolfenden, R.,** *Biochemistry,* 21, 5056, 1982.
109. **Patchett, A.A. and Cordes, E.H.,** *Adv. Enzymol.,* 57, 1, 1985.
110. **Patchett, A.A., Harris, E., Tristham, E.W., Wyvratt, M.J., Wu, M.T., Taub, D., Peterson, E.R., Ikeler, T.J., ten Broeke, J., Payne, L.G., Ondeyka, D.L., Thorsett, E.D., Greenlee, W.J., Lohr, N.S., Hoffsommer, R.D., Joshua, H., Ruyle, W.V., Rothrock, J.W., Aster, S.D., Maycock, A.L., Robinson, F.M., and Hirschmann, R.,** *Nature,* 288, 280, 1980.
111. **Pauling, L.,** *Nature,* 161, 707, 1948.

112. **Pitts, J.E.,** *Nature,* 346, 113, 1990.

113. **Pocker, Y. and Meany, J.E.,** *Biochemistry,* 11, 2535, 1965.

114. **Pocker, Y. and Sarkanen, S.,** *Adv. Enzymol.,* 47, 149, 1978.

115. **Rees, D.C. and Lipscomb, W.N.,** *J. Mol. Biol.,* 160, 475, 1982.

116. **Root, A.W., Duchett, G., Sweetland, M., and Reiter, E.O.,** *J. Nutr.,* 109, 958, 1979.

117. **Rosenfield, R.E., Swanson, S.M., Meyer, E.F., Carrell, H.L., and Murray-Rust, P.,** *J. Mol. Graphics,* 2, 43, 1984.

118. **Smulevich, G., Mauro, J.M., Fishel, L.A., English, A.M., Kraut, J., and Spiro, T.G.,** *Biochemistry,* 27, 5477, 1988.

119. **Tainer, J.A., Getzoff, E.D., Beem, K.M., Richardson, J.S., and Richardson, D.C.,** *J. Mol. Biol.,* 160, 181, 1982.

120. **Taylor, A.L., Finster, J.L., and Mintz, D.H.,** *J. Clin. Invest.,* 48, 2349, 1969.

121. **Thompson, R.G., Rodriguez, A., Kowarski, A., and Blizzard, R.M.,** *J. Clin. Invest.,* 51, 3193, 1972.

122. **Thorlacius-Ussing, O.,** *Neuroendocrinology,* 45, 233, 1987.

123. **Thorsett, E.D., Harris, E.E., Peterson, E.R., Greenlee, W.J., Patchett, A.A., Ulm, E.H., and Vassil, T.C.,** *Proc. Natl. Acad. Sci. U.S.A.,* 79, 2176, 1982.

124. **Tronrud, D.E., Holden, H.M., and Matthews, B.W.,** *Science,* 235, 571, 1987.

125. **Ullrich, A. and Schlessinger, J.,** *Cell,* 61, 203, 1990.

126. **Vallee, B.L. and Auld, D.S.,** *Biochemistry,* 29, 5647, 1990.

127. **Vedani, A. and Dunitz, J.D.,** *J. Am. Chem. Soc.,* 107, 7653, 1985.

128. **Vedani, A., Huhta, D.W., and Jacober, S.P.,** *J. Am. Chem. Soc.,* 111, 4075, 1989.

129. **Vedani, A. and Huhta, D.W.,** *J. Am. Chem. Soc.,* 112, 4759, 1990.

130. **Vidgren, J., Liljas, A., and Walker, N.P.C.,** *Int. J. Biol. Macromol.,* 12, 342, 1990.

131. **Vonderhaar, B.K. and Ziska, S.E.,** *Annu. Rev. Physiol.,* 51, 641, 1989.

132. **Wallmeier, H. and Kutzelnigg, W.,** *J. Am. Chem. Soc.,* 101, 2804, 1979.

133. **Yarden, Y. and Ullrich, A.,** *Annu. Rev. Biochem.,* 57, 443, 1988.

134. **Allen, F.H., Kennard, O., and Taylor, R.,** *Acc. Chem. Res.,* 16, 146, 1983.

135. **Bernstein, F.C., Koetzle, T.F., Williams, G.J.B., Meyer, E.F., Brice, M.D., Rodgers, J.R., Kennard, O., Shimanouchi, T., and Tasumi, M.,** *J. Mol. Biol.,* 112, 535, 1977.

136. **Silverman, D.N. and Lindskog, S.,** *Acc. Chem. Res.,* 21, 30, 1988.

137. **Alexander, R.S., Kiefer, L.L., Fierke, C.A., and Christianson, D.W.,** *Biochemistry,* 32, 1510, 1993.

138. **Prugh, J.D., Hartman, G.D., Mallorga, P.J., McKeever, B.M., Michelson, S.R., Murcko, M.A., Schwam, H., Smith, R.L., Sondey, J.M., Springer, J.P., Sugrue, M.F.,** *J. Med. Chem.,* 34, 1805, 1991.

139. **De Vos, A.M., Ultsch, M., and Kossiakoff, A.A.,** *Science,* 255, 306, 1992.

140. **Higaki, J.N., Fletterick, R.J., and Craik, C.S.,** *Biochemistry,* 29, 8582, 1990.

141. **Ippolito, J.A. and Christianson, D.W.,** *Biochemistry,* 32, 9901, 1993.

142. **Jain, A., Whitesides, G.M., Alexander, R.S., and Christianson, D.W.,** *J. Med. Chem.,* 1994, in press.

143. **Kiefer, L.L., Krebs, J.F., Paterno, S.A., and Fierke, C.A.,** *Biochemistry,* 32, 9896, 1993a.

144. **Kiefer, L.L., Ippolito, J.A., Fierke, C.A., and Christianson, D.W.,** *J. Am. Chem. Soc.,* 115, 12581, 1993b.

145. **McGrath, M.E., Haymore, B.L., Summers, N.L., Craik, C.S., and Fletterick, R.J.,** *Biochemistry,* 32, 1914, 1993.

146. **Pessi, A., Bianchi, E., Crameri, A., Venturini, S., Tramontano, A., and Sollazzo, M.,** *Nature,* 362, 367, 1993.

147. **Smith, G.M., Alexander, R.S., Christianson, D.W., McKeever, B.M., Ponticello, G.S., Springer, J.P., Randall, W.C., Baldwin, J.J., and Habecker, C.N.,** *Protein Science,* 3, 118, 1994.

Chapter 9

DESIGN OF IMMUNOMODULATORY PEPTIDES BASED ON ACTIVE SITE STRUCTURES

Anil B. Mukherjee and Lucio Miele

CONTENTS

I. INTRODUCTION

For the past several years the potential of biologically active proteins and peptides as therapeutic agents has been realized. Thus, many investigators have focused on delineating the antimicrobial, antienzymatic, and immunomodulatory properties of various natural and synthetic polypeptides. Additionally, in genetic diseases where the pathogenesis is due to a lack or alteration (mutation) of a specific gene product, polypeptides encoded by the wild-type gene have already been successfully used as therapeutic agents. Natural proteins and peptides produced by recombinant DNA technology or by chemical synthesis have been used to exert their characteristic biological effects. However, these substances are currently not suitable for oral administration, and they may have a short half-life *in vivo* because of their susceptibility to proteolytic degradation. In addition, heterologous polypeptides, particularly if they are relatively large and/or poorly soluble in aqueous media, can elicit a humoral and/or cellular immune response in the organism to which they are administered. This, in turn, can result in a rapid degradation or inactivation of the polypeptides *in vivo* and may elicit hypersensitivity reactions following repeated administrations. Because of the great therapeutic potential of specific and potent peptide drugs, there is a mounting interest in developing ways to overcome these problems. New vehicles and routes of administration are being sought to prevent degradation of peptide drugs in the digestive tract. Various chemical modifications can be used to render bioactive peptides more stable and resistant to proteolytic degradation. Another possible approach involves designing relatively small peptides from the active site structures of biologically active proteins. Once reasonably active peptides are identified, nonpeptide organic compounds, "peptide-mimetics", can be designed and synthesized, using the peptides as models, with the ultimate goal of developing therapeutic agents suitable for oral or parenteral administration. Computer modeling plays an important role in this strategy.

One of the fields in which bioactive peptides have found, and will probably continue to find, wide applications is that of the regulation of the immune system and the inflammatory processes. It is now well accepted that the activity of the immune system can be pharmacologically modulated by various natural and synthetic oligopeptides. Many of these peptides are stimulators and some are suppressors of the immune system. In addition, a large number of diverse proteins and peptides play vital roles as physiological mediators of the mammalian immune system. These include cytokines, polypeptide hormones, and receptors mediating cell-cell communication and autocrine regulatory functions, antibodies, other plasma proteins, and peptides and enzymes released during inflammatory reactions. These proteins are all potential targets of studies directed to develop analogs or antagonists with possible immunoregulatory activities *in vivo*.

The complexity of the mammalian immune system is far from being completely understood. However, due to the advent of sophisticated biochemical, cellular, and molecular biological techniques during the past two decades considerable knowledge has amassed in this field. There is compelling evidence to suggest that the immune system is regulated through various levels of intricate interactions among many cellular and humoral components of the so-called "immune network". In Figure 1 a schematic diagram is presented to explain the human immunoregulatory network as it is presently understood. Although this is a highly simplified representation, the diagram looks complicated. As shown in Figure 1, the bone marrow stem cells are the progenitors of two distinct types of cell lineages. These are the T cells or thymus-derived cells and B cells or bone marrow-derived (or bursa-equivalent) cells. Thymic peptides are thought to participate in the differentiation of immature T cells and in the stimulation of T cell functions. The activation of T cells occurs when monocytes/macrophages or other antigen-presenting cells process an antigen and "present" the processed antigen to resting T cells under the form of peptides complexed with membrane proteins belonging to the major histocompatibility complex (MHC) family.[1] The cytokine interleukin (IL)-1, secreted mainly by monocytes/macrophages and potentially by many other cell types, participates in the activation of T cells, and has numerous other immunoregulatory functions.[2,3] The activated T cells in turn secrete another cytokine,[2,3] IL-2. T cells themselves secrete IL-1 and IL-2 and express receptors for both.[2,3] When an antigen binds to their specific membrane receptors, T cells first proliferate and then differentiate to perform various specialized functions. These phenotypically distinct subsets of T cells function as cytotoxic, suppressor-inducer, suppressor, and direct helper cells. The process of proliferation and differentiation of T lymphocytes is stimulated by IL-1 and -2. These specialized T cells play regulatory roles on other cells of the immune system as well as nonlymphoid cells.[1,4,5]

Interferon (IFN)-γ, another cytokine produced by T cells, participates in the activation of monocyte/macrophages.[1,3] These activated macrophages participate in microbicidal/tumoricidal activities[6] and produce proinflammatory soluble mediators such as eicosanoids. IFN-γ also induces activation of cytotoxic T cells and has effects on other cells, including polymorphonuclear leukocytes, erythrocytes, B cells, natural killer cells, endothelial cells, fibroblasts, etc.[3,7,8] Non-MHC-restricted, natural killer-like cells can also be generated by a direct action of IL-2 on LGL (large granular lymphocytes) and on cytotoxic T cells.[1] *In vitro* methods have been developed to generate cytotoxic cells (lymphokine-activated killer cells) which can be used to kill non-T cell tumor cells.[9] T cells also affect the maturation and functions of B cells in several ways: antigen-specific and nonspecific T helper cells can stimulate antigen-induced growth and differentiation, while T suppressor cells can downregulate these processes.[1] The communication between B and T cells may either be mediated by direct cell-cell contact or via cytokines. In a series of steps, in response to cytokines (e.g., ILs 2–7, IFN-γ, etc.) produced by T cells, by the B cells themselves, or by other cells, such as bone marrow stromal cells,[1] the B cells first proliferate and then differentiate into plasma cells which secrete immunoglobulins.

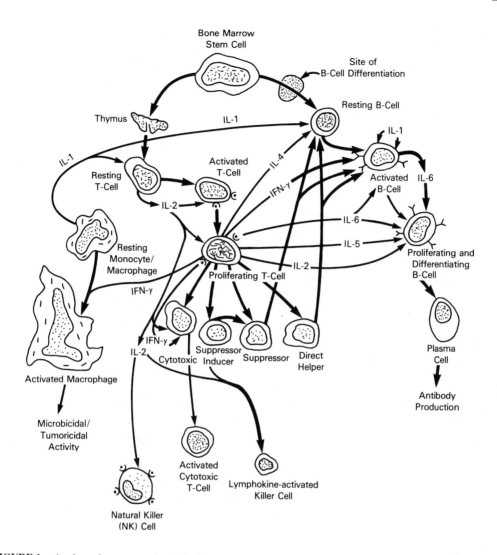

FIGURE 1. A schematic representation of the human immunoregulatory network. See text for explanation. (Redesigned with modifications after References 5 and 12.)

The final objective of the immune system is to eliminate or to inactivate "non-self" antigenic material.[10] This is accomplished through a multistep process. The first step involves binding of the antigen to specific immunoglobulins or T cell antigen receptors. Immunoglobulins can be either free in the extracellular fluids or bound to the surface of effector cells like phagocytes or mast cells. Antigen binding by T cells, cell-bound immunoglobulins, or free immunoglobulins generally leads to the production and/or release of proinflammatory mediators, including cytokines, components of the complement system, components of the contact activation system, histamine, eicosanoids, etc. This ultimately results in the triggering of inflammation. Mononuclear and polymorphonuclear phagocytes recruited or activated by proinflammatory mediators are responsible for the ultimate degradation of an antigen. This process is accompanied by the release of intracellular enzymes, such as proteases and phospholipases, which can cause tissue damage. Most forms of clinically relevant or subclinical acute and chronic inflammation occur as a result of humoral and/or cellular immune phenomena.[10] Although this summary is necessarily brief and incomplete, it illustrates the diverse roles that protein and peptide mediators play in the immune system, and, thus, it clearly indicates the pharmacological potential of peptide immunomodulators.

"An immunomodulator" as defined by Fauci et al.[5] is a "biological or nonbiological substance that directly influences a specific immune function or modifies one or more components of the immunoregulatory network to achieve an indirect effect on a specific immune function". A large number of natural and synthetic peptide agents could be listed among the immunomodulators that have been developed and studied so far. A tentative classification of these peptides is presented in Table 1. Several reviews[11,12] have been published in recent years which describe in detail the best-characterized immunoregulatory peptides presently known and their biological effects. In this chapter we shall briefly discuss some of the general aspects of active site-directed peptide design and its applications to the development of immunoregulatory peptides. We shall then describe in detail how we generated a group of immunomodulatory peptides from the structure of an immunomodulatory protein, uteroglobin. An attempt will also be made to summarize our present understanding of the mechanism(s) of action of these peptides.

II. WHY DESIGN PEPTIDES FROM BIOACTIVE PROTEINS?

Structure and function of proteins are as a general rule intimately related. A complete understanding of the mechanisms through which a protein manifests its biological activities cannot be achieved without at least some knowledge of its three-dimensional structure. Although the past decade has seen tremendous progress in molecular biology, which is still continuing at a very rapid pace, the advances in our understanding of protein folding and structure-function relationship have not been as striking. This relatively slow progress is not due to a lack of interest in the field of protein structure, but to the arduous technical and conceptual problems typical of these studies. One important limiting factor, for instance, is the availability of high-quality protein crystals suitable for high-resolution diffraction studies. In addition, structural studies require a large amount of highly purified native proteins, and some proteins are too unstable or too scarce to allow large-scale purification. Thus, although the primary structures of tens of thousands of proteins have been determined, mostly deduced from their cDNA sequences, the three-dimensional structures of only a few hundreds have been resolved. Nonetheless, there are several reasons to expect an increase in the pace of progress in this field. Recombinant expression is bringing an increasing number of proteins within reach of X-ray diffraction studies. The introduction of heteronuclear multidimensional nuclear magnetic resonance (NMR) spectroscopy[13,14] has allowed the determination of high-resolution protein structures in solution. Recent advances in peptide synthesis and purification methodology,[15,16] as well as in site-directed mutagenesis and in semisynthetic approaches combining chemical synthesis and site-directed mutagenesis,[17–20] allow the rapid production of numerous variant forms of a particular protein. The application of computer modeling to polypeptides of unknown three-dimensional structure is limited only by the relative inaccuracy of current structure prediction algorithms.[21–24] (See the chapters by Kieber-Emmans and McPherson.)

The combination of these powerful tools should allow significant progress in our understanding of "the folding problem" within the next few years. This problem could be loosely defined as the question of how the information encoded in the primary structure of a protein, and ultimately in the nucleotide sequence of its gene, ultimately determines a three-dimensional "native" structure. This is one of the fundamental problems not only of protein chemistry, but also of biology in general. Understanding the precise relationship between protein structure and function will only be possible when the folding problem has been completely solved. Only then will it be feasible to design entirely artificial proteins and peptides with novel biological properties. One of the most powerful approaches to this problem is the synthesis of peptides with defined structural and functional properties. The crystallographic structures of already well-characterized proteins indicate that the seemingly enormous variety of protein structures is based on a surprisingly small number of simple secondary and supersecondary structural motifs.[22,25] Much of this knowledge has been successfully utilized in designing synthetic peptides reproducing these structural motifs.[26–29] It should be emphasized, however, that presently, peptide design is still divided into structure design and function

TABLE 1

Immunomodulatory Peptides and Proteins of Natural and Synthetic Origin

1. Immunomodulators of bacterial and fungal origin
   (a) Muramyl peptides
   (b) Lipopeptides
   (c) Bestatin and derivatives
   (d) Cyclosporin and congeners
2. Natural immunomodulatory hormones and derivatives
   (a) Thymic peptides and synthetic analogs
   (b) Interleukins and derivatives
   (c) Interferons and interferon inducers
   (d) Transfer factor
3. Immunoglobulins and immunoglobulin-derived peptides
   (a) Monoclonal antibodies
   (b) Peptides derived from the complementarity-determining regions of antibodies
   (d) Tuftsin and derivatives
   (e) IgE-derived pentapeptide
4. Peptides derived from other plasma proteins
   (a) Human prealbumin and derivatives
   (b) Human β_2-microglobulin and derivatives
   (c) Miscellaneous plasma proteins, including complement cascade components, fibrinogen fragments, α-fetoprotein, lipoproteins containing the E-apolipoprotein, C-reactive proteins, etc.
5. Phospholipase A_2 inhibitory proteins and derivatives
   (a) Uteroglobin
   (b) Lipocortin-1
   (c) Antiflammins
6. Peptides derived from food proteins
   (a) Peptides derived from ovine colostral immunoglobulins
   (b) Peptides derived from casein
7. Enzyme fragments
   (a) Fragment of bee venom phospholipase A_2
8. Myelopeptides
9. Immunomodulatory placental proteins

Modified after Reference 12.

design.[27] As more information accumulates, however, these two fields are getting increasingly closer.[27] These studies constitute a necessary prelude to the *de novo* design of biologically active polypeptides.

Small and medium-sized peptides pose relatively less severe problems in terms of structure prediction, compared to full-sized proteins. The main difference between small synthetic peptides and full-sized proteins is that intramolecular noncovalent interactions in peptides are generally fewer and less complex than in proteins, so that small peptides generally tend to have relatively random structures in solution.[26,30] However, intermolecular interactions can exert a considerable influence on the structure of small peptides, so that often they assume ordered structures upon interaction with appropriate binding sites.[26,28,30] This has been observed with naturally occurring peptide hormones such as Leu-enkephalin[30-32] as well as with model synthetic peptides.[30,33] For example, synthetic peptides with identical amino acid composition and theoretical potential for helix formation, differing only in their hydrophobic periodicity, can assume different α-helical or β-sheet secondary structures upon interaction with hydrophobic interfaces.[30,33] Thus, studying the effects of intermolecular interactions on the structure of synthetic peptides can give us information on how intramolecular interactions between different regions of a polypeptide stabilize secondary, supersecondary, and tertiary structures. Since biological activity is dependent upon correct intermolecular interactions, bioactive peptides provide a very useful tool for testing hypotheses on structure-function relationships. In addition, unlike larger proteins, relatively small peptides can be easily synthesized, and a number of chemical modifications can be introduced in these molecules to improve their activity or stability, or to introduce conformational constraints, thus allowing the study of large numbers of derivatives.

From a pharmacological perspective, the design of biologically active peptides derived from naturally occurring proteins can yield invaluable information on the structure-function relationship of such proteins. This information can subsequently be used in the rational design of highly specific peptide-mimetic agents. One of the main objections to the direct pharmacological use of bioactive peptides is their unsuitability for oral administration. However, very recent studies suggest that this problem may be circumvented by administration of peptides via the ocular route in appropriate vehicles.[34] This route of administration is claimed to allow very precise dosage and to produce blood levels comparable to those obtained by parenteral administration.[34] While there is no general rule dictating the minimal size of immunogenic molecules, small peptides are generally far less immunogenic than natural proteins. This can be a distinct advantage with respect to some bioactive proteins (e.g., monoclonal antibodies), the effectiveness and applicability of which can be greatly limited by their immunogenicity. Susceptibility to proteolytic degradation can be decreased by introducing "unusual" amino acids and/or conformational constraints in the synthetic molecules. All in all, active site-directed peptide design is a very promising field of investigation, which can: (1) improve our understanding of the physical chemistry of polypeptide folding and macromolecular recognition; (2) lead to the development of potentially useful biologically active molecules; and (3) generate information on which the rational design of organic pharmacological agents can be based.

III. POSSIBLE STRATEGIES FOR ACTIVE SITE-DIRECTED PEPTIDE DESIGN

One of the general properties of biologically active polypeptides is molecular recognition, i.e., the ability to specifically bind other molecule(s). This is true for peptide hormones, cytokines, growth factors, antibodies, enzymes, and enzyme inhibitors. The target molecule can be a receptor, an antigenic determinant, a substrate or an enzyme, but in every case molecular recognition involves a three-dimensional array of noncovalent interactions. The region of a biologically active protein which is directly involved in recognizing and binding the specific target molecule(s) can be loosely defined as an "active site". When detailed information is available on the parent protein from X-ray diffraction, NMR, and/or site-directed mutagenesis, the design of peptides derived from the putative active region is greatly facilitated. In certain instances where the active site of a polypeptide has been identified or its position can be inferred from sequence comparisons with previously characterized proteins, the initial approach involves designing a series of synthetic peptides reproducing the active site and testing them for activity *in vitro* or *in vivo*. If reasonably active peptide(s) are identified by this approach, their structure can subsequently be modified to increase their activity and/or stability. If the structure of a "target" molecule (receptor, enzyme, membrane protein, etc.) is well defined but there is no structural information on natural ligands for the target molecule, one possible approach involves the development of monoclonal antibodies directed against the target molecule.[35] If antibodies with the desired biological activity are obtained, the sequence of their complementarity-determining regions (CDRs) can be obtained through cloning of the light- and heavy-chain cDNAs. The development of biologically active peptides from the CDRs of monoclonal antibodies directed against the reovirus type 3 receptor has been recently described.[35-37] Based on the structure of one of these peptides, a biologically active and chemically stable peptide-mimetic molecule has also been designed and synthesized.[38]

Another potentially promising approach to the development of oligopeptides recognizing other peptides or proteins is based on the "antisense peptides".[39,40] These are peptides encoded by the antisense strand of the gene coding for the target molecule. Specific binding of antisense to sense peptides was first observed with corticotropin (ACTH) and γ-endorphin and their respective antisense peptides.[39,40] Subsequently, this phenomenon was confirmed in several other experimental systems, including the well-characterized pancreatic ribonuclease-S-peptide system,[41] Arg[8]-vasopressin, and neurophysin II.[42] Antibodies directed against antisense peptides have been used to purify the receptors for ACTH,[39] fibronectin,[43] and angiotensin II.[44] It has been suggested that binding between sense and antisense peptides is based on a multisite interaction which is less conformation-dependent than other forms of molecular recognition.[45] This phenomenon has been suggested to be dependent on the hydropathy profiles of sense and antisense peptides, which appear to be complementary.[46] Not surprisingly, sense-antisense peptide interactions show degeneracy both at the sequence and conformational level.[46] This approach can be applied directly when the cDNA sequence coding for the target molecule is known. In molecules where cDNA sequence information is not available, a computer-assisted algorithm has been described which designs putative "antisense" peptides by maximizing hydropathy complementarity between the target sequence and the putative recognition peptides.[47]

Irrespective of the initial strategy, once one or more peptides have been constructed which possess the desired biological activity, the structure of the active molecules can be optimized by means of several chemical tools.[30] A large number of possible modifications can be explored. Conformational constraints can be introduced so that the active conformation(s) are thermodynamically favored. These include: (1) C_α-methylamino acids, such as α-aminoisobutyric acid; (2) N_α-methylamino acids; (3) D-amino acids; (4) α,β-unsaturated amino acids; and (5) covalent cross-links between residues.[30] The latter can be disulfide bonds between Cys residues, or amide bonds connecting the N- with the C-terminus of a peptide, a side chain with one of the termini, or one side chain to another. The effect of such cross-linking is to create macrocycles that decrease the conformational entropy of the peptide. This, in turn, reduces the potential entropy loss upon adoption of the active conformation, thereby making such a conformation thermodynamically more favored. Other constraints include (6) substituted ring systems to mimic turns;[30] and (7) hydrogen-bond covalent analogs, such as *N*-aminomethylamidine or hydrazone-hydrocarbon links.[48]

Particularly when detailed conformational data are available on the active site from which one wishes to derive a peptide, it should be kept in mind that it may not be necessary or even desirable to reproduce exactly the sequence of the active site. Molecular recognition phenomena can generally tolerate considerable variations in the sequence of the ligand (peptide), as long as certain structural elements are conserved. These can be secondary structure elements, such as amphiphilic α-helices, amphiphilic β-strands or turns, the hydropathy and volume of side chains involved in the interaction, the overall charge balance and hydrophobic-hydrophilic balance, and the distribution of charged groups. Pioneering studies of Kaiser and Kezdy[49–52] and other investigations have demonstrated that it is possible to obtain biologically active analogs of naturally occurring peptides which bear very little sequence similarity to their natural counterparts. These synthetic peptides are designed by creating idealized versions of the putative active structure, very often amphiphilic secondary structures, using only a few different types of amino acids.[49–53] This approach has allowed the successful construction of active analogs of apolipoproteins A-I and B, the signal region of *Escherichia coli* alkaline phosphatase, and a number of peptide hormones, including calcitonin, calcitonin gene-related peptide, β-endorphin, growth hormone-releasing hormone, and platelet chemotactic factor IV.[49–53]

A word of caution is in order for the cell biologist and the pharmacologist wishing to study the structure-function relationship of a protein by means of peptide design: synthetic peptides have widely variable physicochemical properties, and can present unexpected problems in terms of solubility and/or stability. This is particularly true in the first stages of active site-directed peptide design studies. Trying to preserve the sequence of the parent protein, and thus the biological activity, as much as possible can lead to the production of peptides which expose residues that in the parent protein were buried in the interior of the molecule. This, in turn, can lead to low solubility and/or to aggregation of the peptide. The tendency of many synthetic peptides to aggregate in aqueous solutions is well known to peptide chemists.[26] Also, amphiphilic peptides can bind to interfaces,[30] including air-water and lipid-water interfaces, and the walls of glass or plastic test tubes. In addition, several residues, and especially Met, Cys, His, and Trp, can be readily oxidized with possible loss of biological activity. Gln can undergo spontaneous deamination to Glu in the presence of water. In general, all synthetic peptides, and particularly those containing unstable amino acid residues, should be stored lyophilized and desiccated below 0°C, if possible in sealed ampules in the presence of an inert gas. Solutions should be prepared immediately before use and quickly diluted to working concentration to prevent aggregation in concentrated solutions. Storage of peptide solutions, even flash-frozen, should be avoided unless preliminary tests show that it does not affect the activity of the particular peptide under investigation. Once an active sequence, however unstable or poorly soluble, has been identified, changes can be gradually introduced to improve its stability and/or solubility. Thus, it is essential to generate several derivatives of the parent sequence and to test whether particularly unstable or highly hydrophobic residues can be replaced without loss of activity. In some cases, "troublesome" amino acids can be replaced with near-isosteric residues, such as norleucine (Nle) for Met, Ser or Ala in place of Cys. Trp can sometimes be replaced by other aromatic residues like Phe or Tyr. These considerations imply that it should not be automatically assumed that a synthetic peptide with a sequence identical to that of the active site of a protein is going to be biologically active or that a derivative with several residues replaced cannot be more active than the parent sequence. Whenever conformational data are available on the active site, the best approach is probably to construct "idealized" secondary structures composed of a few amino acid types, containing only those residues from the parent protein which are necessary for biological activity.

IV. DESIGN OF ANTIFLAMMIN PEPTIDES FROM THE STRUCTURE OF UTEROGLOBIN

A. PHYSIOLOGICAL PROPERTIES OF UTEROGLOBIN

Uteroglobin (UG), or blastokinin, is a 15.8-kDa secretory protein originally identified in the fluid secreted by rabbit endometrium at the time of implantation (for reviews see References 54 and 55). Subsequently, this protein was identified in several other organs and fluids of the rabbit, including the tracheobronchial tree, the gastrointestinal tract, the oviduct, the seminal vesicles, and the prostate.[54,55] Low amounts of UG have also been detected in the uterine and pulmonary venous blood.[56] The synthesis and secretion of UG in the rabbit are under hormonal control.[54,55] Progesterone stimulates UG synthesis in the endometrium, while estrogens have a stimulatory effect in low doses and an inhibitory effect in higher doses. Estradiol stimulates UG synthesis in the oviduct. Androgens can also stimulate the synthesis of this protein in the endometrium, presumably through the progesterone receptor. In the tracheobronchial epithelium, UG synthesis is constitutive, and it is further stimulated by glucocorticoids. DNA sequences which bind progesterone and glucocorticoid receptors have been identified in the promoter region of the UG gene and in its first intron. So far, mucosal epithelial cells in several organs appear to be the main source of secreted UG in the rabbit. It is not yet clear whether nonepithelial cells can produce UG. Proteins belonging to the same evolutionary family as UG have been identified in hares,[57] rats,[58] and humans.[59] The rat and human UG-related proteins seem to be expressed predominantly in the tracheobronchial epithelium. Rabbit UG is by far the best-characterized protein in this group (see below).

UG is a homodimer formed by two identical subunits joined by two disulfide bonds.[60] After reduction of these disulfide bonds, UG binds progesterone and some structurally related steroids.[54,61] This binding, which has been extensively studied, requires the formation of a relatively unstable tetrameric complex of two UG dimers and one molecule of progesterone.[54,62,63] The possible physiological roles of this property of UG are unclear, since the presence of reduced UG *in vivo* has never been unequivocally demonstrated. In fact, the protein forms disulfide bonds even in the reducing environment of the *E. coli* cytoplasm.[64] In addition, the binding of progesterone to UG is relatively weak[54,65] ($K_D \simeq 4.1 \times 10^{-7}$ M at 4°C), and the binding specificity is different from that of the progesterone receptor.[54,65,66] UG does not bind corticosteroids or estrogens.[66] The rat UG-related protein also binds certain toxic polychlorinated biphenyls.[58]

Two other biochemical properties of rabbit UG have been identified and characterized, namely: (1) UG is a substrate of transglutaminases (TG, EC 2.3.2.13) and (2) UG is a potent inhibitor of some phospholipases A_2 (PLA$_2$s, EC 3.1.1.4). TGs, a class of enzymes that includes blood coagulation factor XIIIa, catalyze an acyl transfer reaction between a Gln residue in a protein or peptide and a primary amino group, forming the corresponding substituted γ-carboxamides.[67] The acyl-acceptor amino group can belong to a low-molecular-weight amine, such as the naturally occurring polyamines, or be the ε-amino group of a Lys residue. In the latter case, TGs catalyze the formation of inter- or intramolecular ε(-γ-glutamyl) lysine covalent cross-links. TGs have been shown to: (1) cross-link UG to the surface of cells such as blastomeres or spermatozoa; (2) form high-molecular-weight covalent aggregates of UG; and (3) covalently cross-link polyamines to Gln residue(s) in UG, forming γ-glutamyl polyamine derivatives.[54,55,68–71] TG-catalyzed cross-linking of UG to the surface of certain cells dramatically increases the immunosuppressive effects of UG (see below). The PLA$_2$ inhibitory properties of UG have been studied in detail. In particular, UG has been shown to inhibit porcine pancreatic PLA$_2$[72,73] and three different PLA$_2$ activities from RAW 264.7 mouse macrophage cell line[72] with various substrates, including sonicated phosphatidylcholine vesicles,[72] natural cell membranes from the RAW 264.7 cells,[72] and mixed micelles of deoxycholate and phosphatidylcholine.[73]

UG possesses inhibitory activities on several cellular processes connected with inflammatory and immune phenomena. These immunosuppressive anti-inflammatory effects may be related, at least in part, to its biochemical properties as a TG substrate and as a PLA$_2$ inhibitor. The biological activities of UG have been reviewed.[54,55] Briefly, they can be divided into: (1) effects on the antigenicity of target cells and (2) direct inhibitory effects on the functions of cells involved in inflammation and immune responses, such as mono- and polymorphonuclear phagocytes and platelets. When bound to the surface of blastomeres[69,70] or spermatozoa,[71] UG dramatically reduces the immune response elicited by these cells in allogenic spleen cells, as measured by ^{3}H-thymidine incorporation into lymphocytes. This effect is strongly potentiated when UG is covalently cross-linked to the cell surface by the action of TG.[69–71] It has been suggested that this effect may have the physiological function of protecting spermatozoa and embryos carrying paternal histocompatibility antigens from degradation by maternal immune/inflammatory response.[69–71] It should be noted that TG is physiologically present both in the rabbit prostatic and uterine secretions.[74,75] Uterine TG levels increase during pregnancy.[75] In addition, the TG activity in crude prostatic fluid and crude fluid from pregnant uteri can cross-link UG to spermatozoa and blastomeres, respectively.[69–71]

UG alone directly impairs the function of both polymorphonuclear and mononuclear phagocytes derived from rabbit and human blood. More specifically, UG shows an antichemotactic and antiphagocytic effect on these cells *in vitro*.[76,78] Finally, UG is a powerful inhibitor of thrombin-induced platelet aggregation.[79] This effect was found to be abolished by arachidonic acid. From these observations, it was concluded that UG can inhibit immune and inflammatory responses to allogenic cells by two mechanisms: (1) by protecting them from immune recognition and possibly from phagocytosis of target cells, to the surface of which it is bound or cross-linked by TG; and (2) by directly impairing the function of cells involved in antigen processing, presentation, degradation, and in the amplification of the inflammatory response. These two mechanisms may be related if one demonstrates that UG cross-linked to cells protects antigenic cells from being phagocytosed, thus preventing the initial step of a possible immune response against cell surface antigens. So far, this possibility has not been investigated experimentally.

PLA$_2$s play a very important role in the pathogenesis of inflammatory phenomena.[80–82] Activated phagocytes and platelets release extracellular PLA$_2$s, which in turn can propagate the inflammatory response by releasing from cell membranes bioactive lipid mediators such as arachidonic acid, lysophospholipids, and lyso-PAF (platelet activating factor). Some of these compounds and their metabolites are extremely potent mediators of inflammation and possess chemotactic properties. In addition, soluble PLA$_2$s can be cytotoxic and cause liberation of intracellular proinflammatory substances such as proteolytic enzymes and free radicals.[80,81] Thus, it is possible that at least some of the biological properties of UG may be due to its ability to inhibit PLA$_2$s in the extracellular spaces and to protect cell membranes from soluble PLA$_2$s. The fact that its effect on platelet aggregation is completely abolished by free arachidonic acid lends credibility to this hypothesis. An increase in the activity of intracellular PLA$_2$(s) also accompanies phagocyte activation.[82] It remains unclear whether UG can also inhibit intracellular PLA$_2$s under physiologically relevant circumstances. A possible receptor-mediated internalization of UG by trophoectodermal cells has been described.[83] However, it is still unclear whether or not other cells also have the ability to internalize UG and whether UG internalization can lead to inhibition of intracellular PLA$_2$ activities.

B. STRUCTURE OF UTEROGLOBIN

The structural properties of UG have been investigated in detail, and high-resolution crystallographic data are available.[84–86] UG crystallizes in several different forms, which indicates a certain degree of conformational flexibility. In mature UG the two subunits are joined in antiparallel fashion by two disulfide bonds between Cys 3 and 69', 3' and 69. In the best-characterized form (C222$_1$) each subunit consists of four α-helical segments, with one β turn between Lys 26 and Glu-29. The N-terminus is partially buried inside the molecule while the C-terminus is exposed to the solvent. Overall, the dimer is of globular appearance, with a rather large hydrophobic pocket in central position. Several noncovalent interactions stabilize the dimeric structure, and the dimer is fairly stable after reduction and carboxymethylation of the Cys residues. Interestingly, a calculation of the molecular surface of UG based on the atomic coordinates at 1.34 Å showed that the three-dimensional structure of UG is strikingly similar to that of a dimeric phospholipase A$_2$ (Figure 2).[85] This suggested that UG might act as an enzymatically inactive structural analog of PLA$_2$, and that this might be a possible mechanism for PLA$_2$ inhibition by UG.

C. STRUCTURE-FUNCTION RELATIONSHIP OF UG AS A PLA$_2$ INHIBITOR: DESIGN OF ANTIFLAMMINS

Considering the wealth of structural information available on UG, and the fact that it is a small protein with a relatively simple structural organization, it was concluded that a detailed investigation of its structure-function relationship as a PLA$_2$ inhibitor, including a peptide design study, may yield important information. It was felt that such an investigation may also have considerable pharmacological interest, because of the therapeutic potential of nontoxic inhibitors of extracellular PLA$_2$s as anti-inflammatory agents. The study involved two different approaches: (1) recombinant expression of UG, with site-directed mutagenesis to identify functionally relevant residues; and (2) structural comparison of UG with other PLA$_2$ inhibitory proteins, followed by tentative identification of active site(s) and peptide design.

Both approaches presented potential problems. In fact, to our knowledge no other dimeric protein with disulfide bonds had been expressed in *E. coli* in its native conformation.[64] In addition, the only other known mammalian PLA$_2$ inhibitory protein with biological properties similar to those of UG was lipocortin-1,[87] which had not been crystallized and was poorly characterized from a structural point of view. In addition to the PLA$_2$ inhibitory effect, lipocortin-1 has been shown to share with UG the antichemotactic effect *in vitro*.[77] Subsequently, this protein has been demonstrated to be a substrate of TG.[88]

Our initial approach to generate immunomodulatory peptides involved a comparison of the primary structures of UG and lipocortin-1 by several computer-based algorithms. This approach led to the identification of a general sequence similarity between the UG monomer and the ~70-amino acid nonidentical "repeat" which constitutes the main recognizable structural unit of lipocortin-1[89] (Figure 3a). In particular, UG can be aligned to repeats 2 and 3 of human lipocortin-1 and to repeat 2 of lipocortin-2. Lipocortin-1 repeat 3 (residues 199–275) showed the highest degree of similarity with UG (43%, including identities and conservative substitutions). In addition, a prominent region of local similarity was identified between UG residues 40–46 and lipocortin-1 repeat 3 residues 247–

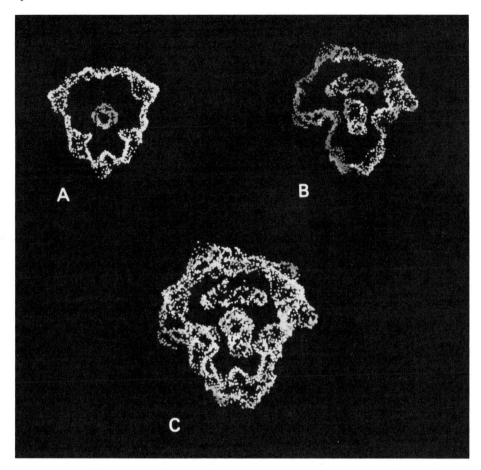

FIGURE 2. The molecular surfaces calculated from the atomic coordinates of UG (A), *Crotalus atrox* dimeric PLA$_2$ (B), and a superimposition of the two (C). Note the presence and symmetry of the central hydrophobic cavity. (Redesigned after Reference 85.)

253. In UG, this region corresponds to the C-terminal half of α-helix 3 (residues 32–47). The exposed face of this helix is accessible to the solvent, and not involved in interchain interactions.[85] The general structural similarity between the UG monomer and lipocortin-1 repeat 3 was also confirmed by hydropathy analysis (Figure 3b). The similarity is evident over the whole sequence, and most remarkable in the amphiphilic region encompassing UG residues 37–52, i.e., the 11 C-terminal residues of α-helix 3, the "bend" between helices 3 and 4 (centered on Pro 49), and the first 3 residues of α-helix 4. Interestingly, the hydropathy profile of porcine pancreatic PLA$_2$ shows a striking resemblance to those of UG and the corresponding regions of lipocortins 1 and 2 (Figure 3b). This is consistent with the proposed similarity in three-dimensional structure between UG and PLA$_2$. Subsequently, by using different protein alignment programs,[73] a sequence similarity between UG and porcine pancreatic PLA$_2$ was also identified. Once again, a region of high local similarity was identified involving UG residues 40–47 and a conserved region (positions 54–60) in type I PLA$_2$s (Figure 4). Thus, several synthetic peptides were synthesized corresponding to the region of highest similarity between UG and lipocortin-1 repeat 3. The nonapeptides corresponding to UG residues 39–47, lipocortin-1 residues 246–254, and a hybrid peptide in which UG Lys 42 was replaced by Asn were active PLA$_2$ inhibitors in nanomolar concentrations.[73,89] A longer peptide corresponding to the entire α-helix 3 of UG was also active, but less potent than the three nonapeptides.[89a] Removal of two residues from the N-terminus and one from the C-terminus of the active peptides abolished their activity. The core tetrapeptide Lys-Val-Leu-Asp, which is conserved

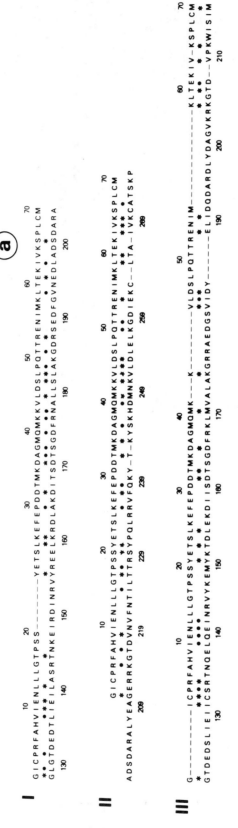

FIGURE 3. (a) Alignment of UG with lipocortin-1 repeat two (I), lipocortin-1 repeat three (II), and lipocortin-2 repeat two (III). Identities are indicated by asterisks and conservative substitutions by dots. Conservative substitutions are defined by pairs of residues falling in the following groups: S, T, A, G, P; N, D, E, Q; R, K, H; M, I, L, V; F, Y, W. (b) Hydropathy profiles of UG, (I), Lipocortin-1 repeat two (II), lipocortin-1 repeat three (III), lipocortin-2 repeat two (IV), and porcine pancreatic PLA₂ residues 24–92 (V). The profiles were obtained using a segment length of seven residues. Positive peaks indicate hydrophobic regions; negative peaks indicate hydrophilic regions. (Redesigned after Reference 89.)

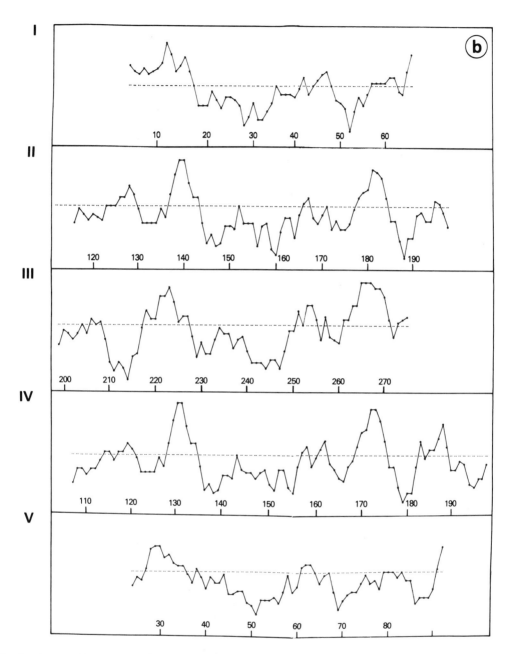

in the three active nonapeptides, was in itself inactive as a PLA_2 inhibitor. When the UG-derived nonapeptide was preincubated with the PLA_2 substrate (mixed micelles of deoxycholate and phosphatidylcholine) rather than with the enzyme, its inhibitory activity was abolished. A control peptide, corresponding to UG residues 1–10, was inactive as a PLA_2 inhibitor. Both parent proteins UG and lipocortin-1 inhibited PLA_2 in this system in a dose-dependent manner comparable to those of the peptides.

The PLA_2 inhibitory peptides were tested for anti-inflammatory activity in the classic model of carrageenan-induced rat footpad edema (see Table 2). In this system, both the UG-derived and the lipocortin-1-derived peptides had very potent and dose-dependent anti-inflammatory effect, which was drastically reduced by the concomitant administration of arachidonic acid. The parent proteins, UG and human recombinant lipocortin-1, also exhibited anti-inflammatory activity in this system, whereas nonspecific protein and peptide controls were ineffective (Table 2). From these results it

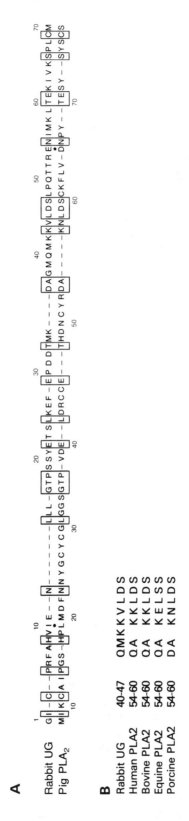

FIGURE 4. (A) Sequence alignment between mature UG and porcine pancreatic PLA₂. Identities are boxed, while conservative substitutions are indicated by dots. (B) Local sequence similarity between UG residues 40–47 and type I PLA₂s from human lung, bovine pancreas, equine pancreas, and porcine pancreas. (Reproduced with permission from Facchiano, A. et al., *Life Sci.*, 48, 453, 1991.)

TABLE 2
Anti-Inflammatory Effect of AF-1 and -2 on
Carrageenan-induced Rat Paw Edema

Treatment	No. of Animals	% Inhibition, mean ± SD
AF-1, 2 mg kg^{-1}	7	82.6 ± 5 ($p < 0.01$)
AF-1, 200 µg kg^{-1}	9	57.5 ± 7 ($p < 0.01$)
AF-1, 20 µg kg^{-1}	9	36.0 ± 8 ($p < 0.01$)
AF-1, 2 µg kg^{-1}	9	31.5 ± 9 ($p < 0.01$)
AF-1, 0.2 µg kg^{-1}	9	24.0 ± 5 ($p < 0.01$)
AF-1, 0.02 µg kg^{-1}	9	n.s. ($p > 0.05$)
AF-2, 2 mg kg^{-1}	8	96.2 ± 2 ($p < 0.01$)
AF-2, 200 µg kg^{-1}	8	30.5 ± 7 ($p < 0.01$)
AF-2, 20 µg kg^{-1}	8	36.3 ± 7 ($p < 0.01$)
AF-2, 2 µg kg^{-1}	8	23.3 ± 3 ($p < 0.01$)
AF-2, 0.2 µg kg^{-1}	8	20.0 ± 4 ($p < 0.01$)
AF-2, 0.02 µg kg^{-1}	8	12.2 ± 3 ($p < 0.05$)
DEX, 10 µg kg^{-1}	5	79.9 ± 6 ($p < 0.01$)
IND, 1 mg kg^{-1}	5	26.1 ± 4 ($p < 0.01$)
AF-1, 2 mg kg^{-1} + AA	4	21.1 ± 5 ($p < 0.01$)
AF-2, 2 mg kg^{-1} + AA	4	20.2 ± 1 ($p < 0.01$)
UG, 100 µg kg^{-1}	4	35.5 ± 7 ($p < 0.01$)
LC-1, 100 µg kg^{-1}	4	24.5 ± 1 ($p < 0.01$)
P7, 2 mg kg^{-1}	4	n.s. ($p > 0.05$)
P7, 10 µg kg^{-1}	4	n.s. ($P > 0.05$)
BSA, 2 mg kg^{-1}	4	n.s. ($p > 0.05$)
BSA, 100 µg kg^{-1}	4	n.s. ($p > 0.05$)
LSZ, 2 mg kg^{-1}	4	n.s. ($p > 0.05$)
LSZ, 100 µg kg^{-1}	4	n.s. ($p > 0.05$)

AF, antiflammin peptide; DEX, dexamethasone; IND, indomethacin; AA,
arachidonic acid; UG, uteroglobin; LC-1, lipocortin-1; P7, peptide 7
(corresponding to UG residues 1–10, used as a negative control); BSA,
bovine serum albumin; LSZ, chicken egg lysozyme. See Reference 89 for
experimental details.

Modified after Reference 89.

was concluded that at least a major mechanism through which these peptides exert their anti-inflammatory activity could be the inhibition of arachidonate release by cellular PLA$_2$(s). The anti-inflammatory peptides are now called antiflammins (AFs) 1 (UG-derived peptide) and 2 (lipocortin-1-derived peptide). Our observations indicated that the exposed residues on the C-terminal half of UG α-helix 3 represent a putative active site for its PLA$_2$ inhibitory and anti-inflammatory activity.

D. BIOCHEMICAL AND BIOLOGICAL PROPERTIES OF AFs

The biochemical mechanism of PLA$_2$ inhibition *in vitro* by these peptides was further character-ized in a separate study.[73] This required a more rigorous standardization of the assay conditions. This was particularly important in the case of PLA$_2$, since this enzyme catalyzes a reaction at a lipid-water interface. In the case of monomeric PLA$_2$s, including the porcine pancreatic enzyme which we used as a model for extracellular PLA$_2$s, the process is composed of several steps:

1. Interfacial recognition or binding of the enzyme to the interface.
2. Interfacial activation, i.e., a dramatic increase in the catalytic activity of the enzyme which takes place upon interface binding, after a lag time which depends on the physical state of the substrate and the temperature.[90–93] This activation has been suggested to be caused by auto-

catalytic acylation of Lys residue(s), followed by dimerization of interface-bound enzyme.[90] In the case of pancreatic PLA$_2$s, autoacylation of Lys 56 triggers dimerization.[94]

3. Actual hydrolysis of the phospholipid substrate.

It is still unclear whether this process is the same for all lipid-water interfaces, but it is well known that variables like the physical form of the interface (e.g., monolayer vesicles, micelles, bilayers, etc.), its charge distribution, the temperature, and the ionic strength have a dramatic influence on the activity of monomeric PLA$_2$s.[90–93] It is likely that the details of the process of interfacial activation are different in different PLA$_2$s.[90–93] In addition, reaction products (lysophospholipid and fatty acid) interfere with the rate of reaction.[92,93]

Because our original observations were obtained with the porcine pancreatic enzyme, we optimized the assay conditions for this enzyme. Also, porcine pancreatic PLA$_2$ can substitute the G-protein-regulated cellular PLA$_2$ in RBL cells.[95] Moreover, this enzyme is a prototype for intra- and extracellular type I PLA$_2$s. Our assay conditions, described in detail in Reference 73, allow reproducible hydrolysis of <1% of the substrate in 30 s. This minimizes the possible interference by reaction products. In addition, with negatively charged deoxycholate-phosphatidylcholine micelles, the binding of PLA$_2$ to the interface is virtually instantaneous, thereby reducing the possibility of artifactual inhibition resulting from the inhibitor preventing PLA$_2$ from binding to the interface. Under these conditions we confirmed the inhibition exerted by AFs on PLA$_2$ and established that the phenomenon is independent of the concentration of phosphatidylcholine (from 2 to 90 μM), i.e., noncompetitive.[73] Additional studies[73] indicated that:

1. AFs and UG decrease the intrinsic fluorescence of PLA$_2$ Trp 3 in solution and cause a red shift of the emission maximum, possibly due to an increase in the polarity of the microenvironment of this Trp residue.
2. The increase in fluorescence which normally accompanies the binding of porcine pancreatic PLA$_2$ to an interface is reduced in the presence of UG or AFs.
3. UG can be cross-linked to PLA$_2$ by glutaraldehyde, and this cross-linking is inhibited in the presence of AFs. AF-1, which is derived from UG, inhibits this cross-linking more effectively than AF-2. A control peptide (UG 1–10) did not interfere with the cross-linking of UG and PLA$_2$.
4. AFs show a considerable sequence similarity with a conserved region of type I PLA$_2$s, which includes Lys 56 (see Figure 4). This residue is involved in the control of PLA$_2$ dimerization (see above).

Taken together, these data indicate that the mechanism by which AFs and UG inhibit porcine pancreatic PLA$_2$ in this system is by interacting with the enzyme. This interaction may prevent autoacylation of Lys 56, PLA$_2$ dimerization, or both.

After the anti-inflammatory activity, the second biological property of AFs to be discovered was their ability to inhibit platelet aggregation. Vostal et al.[96] showed that AFs inhibit thrombin- and ADP-induced platelet aggregation and ADP-induced serotonin secretion in human platelets. In the case of thrombin-induced platelet aggregation, the effect of AFs was found to be weak (IC$_{50}$ $\simeq 10^{-3}$ M) and it is probably due to inhibition of thrombin esterolytic activity by AFs. AFs are potent inhibitors of ADP-induced aggregation (IC$_{50}$ $\simeq 2 \times 10^{-4}$ M for AF-1). When ADP was used as the agonist, arachidonic acid (10^{-4} M) completely abolished the effect of AF-1 on aggregation and dramatically reduced the effect on serotonin secretion, suggesting that these effects are due to inhibition of arachidonate release. Interestingly, the tetrapeptide Lys-Val-Leu-Asp (denominated P4), which does not inhibit PLA$_2$, also inhibited ADP-induced platelet aggregation but not serotonin secretion. This effect is due to a different mechanism than that of AFs; namely, competition with fibrinogen for binding to platelets.[96] This property is not shared by AFs, and the effects of P4 and AF-1 on ADP-induced platelet aggregation are additive.[96] Subsequently, Camussi et al.[97–99] showed that AFs inhibit the synthesis of platelet activating factor (PAF), a potent mediator of inflammation, induced by tumor necrosis factor (TNF) or phagocytosis in rat macrophages and in human neutrophils, as well as by thrombin in endothelial cells. The first step in the synthesis of PAF is the activation of a PLA_2. The peptides inhibit neutrophil PLA_2 activity in cell lysates as well as the release of arachidonate induced by TNF or phagocytosis.[97–99] AFs also inhibited the activation of acetylCoA-lysoPAF acetyltransferase in intact cells.[97–99] The continued presence of AFs in the medium was necessary for their effect on PAF synthesis, and washing the cells with fresh medium reversed the effect. The IC_{50} of AFs for these effects ranged from 50 to 100 nM.

Both AFs were potent inhibitors of neutrophil aggregation and chemotaxis, and of the increase in vascular permeability and neutrophil infiltration induced in rats by intradermal Arthus reaction (induced by injection of C5a or TNF). Since UG is a potent inhibitor of neutrophil chemotaxis and phagocytosis, it was hypothesized[89] that the anti-inflammatory properties of AFs might be due, at least in part, to an impairment of the function of phagocytes. The observations of Camussi et al. are consistent with this hypothesis. The effects of AFs on acute inflammation in rats were further evaluated by Ialenti et al.[100,101] These authors demonstrated that AFs are active in the late phase of carrageenan-induced edema, which is mediated by the release of eicosanoids. The early phase of carrageenan-induced edema and dextran-induced edema, which are mediated by the release of histamine and 5-hydroxytryptamine rather than by arachidonate metabolism, were not inhibited. This profile of activity is similar to that of recombinant lipocortins, and is consistent with the hypothesis that AFs exert their pharmacological effects, at least in part, by preventing arachidonate release and PAF synthesis.

Taken together, these data suggested that AFs or derivatives thereof may be potentially useful pharmacological agents to control inflammation. This possibility was supported by the finding that AFs are pharmacologically active in the topical treatment of endotoxin-induced anterior uveitis (EIU) in rats.[102,103] This experimental disease is a model for human anterior uveitis, and it has been suggested that PLA_2 activation may be an initiating event in the pathogenesis of EIU. Chan et al.[102,103] demonstrated that AFs are potent inhibitors of EIU as assessed by inflammatory cell count and protein determination, as well as by PLA_2 activity measurement in the aqueous humor. In this animal model, as in carrageenan-induced rat paw edema, AFs proved to be as effective as dexamethasone in suppressing inflammation.[102,103] The possible clinical usefulness of topical AF-2 in the treatment of human anterior uveitis is being evaluated through a placebo-controlled clinical trial currently underway at the National Eye Institute.

E. AF DERIVATIVES AND POSSIBLE MECHANISMS OF ACTION OF PEPTIDES OF THE AF FAMILY

In our original study the AFs used were routinely stored below 0°C in glass-sealed vials under anhydrous conditions, and dissolved in ice-cold buffer immediately before use. Concentrated solutions were never stored, and unused portions were discarded. Preliminary experiments[89a] had shown that AFs are inactivated when stored in solution or as powder under nonanhydrous conditions, and that these peptides are completely inactivated by freezing and thawing in solution. Camussi et al.[97–99] studied AF inactivation in detail. These authors found that AFs in general, and AF-1 in particular, are rapidly inactivated in the presence of air due to oxidation of Met residues. The presence of reducing agents such as β-mercaptoethanol or dithiothreitol was necessary to protect AFs from oxidation, particularly in the presence of neutrophils (which produce oxidizing substances such as peroxide and superoxide).[97–99] Furthermore, treatment of AFs *in vitro* with an oxidizing agent completely inactivated AFs. Camussi et al. also observed that freezing these peptides in solution results in complete inactivation. Heating the peptide solution at 45°C for 10 min partially reversed freezing-induced inactivation. The mechanism of this phenomenon is unclear. It is possible that peptide aggregation (a problem well known to peptide chemists)[26] or adsorption to the walls of the tube may explain freezing-induced inactivation of AFs. Thus, it appears that AFs, and particularly AF-1, can be inactivated in solution by at least two mechanisms, one chemical and one physical.

The oxidation-induced inactivation of AF-2 has been circumvented by Tetta et al.[104] These authors replaced the Met residue in AF with Ala, producing AF-2a, or with the near-isosteric residue Nle, producing AF-2n. AF-2n was poorly soluble, most likely due to the more hydrophobic nature of Nle compared to Met or Ala. Thus, Tetta et al. replaced the C-terminal Leu of AF-2n with Ser, producing AF-2ns. The C-terminal Leu is not part of the AF-consensus X-X-Met-X-Lys-Val-Leu-Asp-X, so its replacement was not expected to affect the activity of the resulting peptide. Both AFs 2a and 2ns were as active as AF-2 as inhibitor of PAF synthesis in human polymorphonuclear leukocytes and rat peritoneal macrophages, as well as of arachidonate release from human monocytes.[104] Reducing agents were not necessary to preserve the biological activity of AFs 2a and 2ns. Interestingly, all antiflammins used in this study, including AF-2, were activated by heating at 45°C for 10 min. The mechanism of this activation is still unclear. It is possible that heating destroys large aggregates of peptides. Another possibility is that heating, which favors hydrophobic interaction by increasing the entropic gain due to increased freedom of water molecules, may cause the formation of stable and soluble peptide dimers or multimers. Such complexes may be stabilized by hydrophobic interactions and have hydrophilic residues exposed to the solvent. Like AFs, their derivatives without Met inhibit the synthesis of PAF and the activation, but not the catalytic activity, of acetylCoA-lysoPAF acetyltransferase. The inhibition of PAF synthesis by AF-2 in rat peritoneal macrophages and by AFs 2a and 2ns on PAF synthesis in human PMN was maximal when the cells were preincubated with AFs before the stimulus (phagocytosis of yeast sporae). The same was true for the inhibition of arachidonate release by AF-2 in human PMN. In all these cases, AFs were less inhibitory when added to the cells after the stimulus. These findings, as noted by the authors, are consistent with the hypothesis of Camussi et al. that the AFs interfere with the activation of cellular $PLA_2(s)$ and acetylCoA-lysoPAF acetyltransferase rather than with the catalytic activity of these enzymes. These data are also consistent with the postulated mechanism of porcine pancreatic PLA_2 inhibition by UG and AFs *in vitro.*

Another peptide of the antiflammin family has been recently described by Perretti et al.[105] These authors identified a high sequence similarity between UG residues 39–47 (i.e., AF-1) and residues 204–212 of human lipocortin-5 (LC5), another secreted anti-inflammatory protein of the lipocortin family. A peptide identical to this region of lipocortin-5 was synthesized and tested for biological activity. Both this peptide (LC5 204–212) and AF-2 inhibited contractions of isolated rat stomach strips caused by the application of porcine pancreatic PLA_2, but did not inhibit contractions caused by arachidonic acid or 5-hydroxytryptamine.[105] The basal tone of the organ preparation was not affected by either peptide. The inhibition was reversed after washing the preparation with peptide-free solution. The shortest active peptide in this system was the heptapeptide LC5 206–212. The LC5-derived active peptides, as well as AF-2, inhibited release of prostaglandin E_2 (PGE_2) from human skin fibroblasts stimulated with bradykinin and rat peritoneal macrophages stimulated with opsonized zymosan. Basal arachidonate release was not affected. Bradykinin-induced PGE_2 release from fibroblasts has been shown to be mediated by PLA_2 activation through a specific G-protein. Also the effect of zymosan on macrophages involves activation of PLA_2 (see Reference 105). Carrageenan-induced edema was inhibited by LC5 204–212 with potency and time courses indistinguishable from those for AF-2.

Interestingly, while AF-2 and the LC5-derived peptides inhibited the effect of porcine pancreatic PLA_2 on isolated organ strips, they did not inhibit the same enzyme in an assay using radiolabeled autoclaved *E. coli* cells as the substrate. This is consistent with observations[113] that UG, which is a very potent inhibitor of porcine pancreatic PLA_2 in mixed micellar and phosphatidylcholine vesicle systems, is ineffective against the same enzyme in an assay using autoclaved *E. coli* cells. UG, AFs, and LC5-derived peptides do not possess the consensus sequence common to lipocortins and other Ca^{2+}/phospholipid binding proteins of the endonexin family.[105,106] The mechanism of PLA_2 inhibition by lipocortins in the *E. coli* assay system has been shown to be dependent on Ca^{2+}-mediated phosphatidylserine binding.[107] The suggested explanation for this observation is that lipocortin-1 inhibits PLA_2 by "substrate sequestration", i.e., by covering the interface and rendering it inacces-

TABLE 3
Biologically Active Peptides of the Antiflammin Family

Peptide	Amino Acid Sequence	Ref.
AF-1	Met-Gln-Met-Lys-Lys-Val-Leu-Asp-Ser	89
AF-2	His-Asp-Met-Asn-Lys-Val-Leu-Asp-Leu	89
AF-2a	His-Asp-Ala-Asn-Lys-Val-Leu-Asp-Leu	104
AF-2n	His-Asp-Nle-Asn-Lys-Val-Leu-Asp-Leu	104
AF-2ns	His-Asp-Nle-Asn-Lys-Val-Leu-Asp-Ser	104
LC5 204–212	Ser-His-Leu-Arg-Lys-Val-Phe-Asp-Lys	105
LC5 206–212	Leu-Arg-Lys-Val-Phe-Asp-Lys	105
Consensus	Xxx-Fil-Fob-Fil-Lys-Val-Fob-Asp-Xxx	

Xxx, indeterminate residue; Fil, hydrophilic residue; Fob, hydrophobic residue.

sible to PLA_2.[107] Inhibition of PLA_2 by lipocortin-1 in phosphatidylcholine vesicles (which do not bind the protein) has been ascribed to surfactant effects.[108]

More recently, a monoclonal antibody has been produced which blocks the PLA_2 inhibitory activity of lipocortin-1, but does not affect binding of lipocortin-1 to phosphatidylserine.[109] This indicates that PLA_2 inhibition by lipocortin-1 may not be due simply to substrate sequestration. These conflicting findings may be reconciled by the hypothesis that Ca^{2+} and phospholipid binding induce a conformational change in lipocortin-1 which promotes an interaction with PLA_2 when lipocortin and PLA_2 are lipid-bound. Alternatively, it is possible that different mechanisms, and thus different active sites, contribute to inhibition of PLA_2 by lipocortin-1, and their relative mechanistic importance depends on the assay system used. The presence of distinct pharmacologically active sites on snake venom PLA_2 has already been described.[110] These observations suggest that PLA_2 assays based on autoclaved *E. coli* cells to evaluate the PLA_2 inhibitory properties of proteins and peptides may not be very reliable. Similar considerations have been expressed by Perretti et al.[105] and others.[111]

Despite the apparent simplicity of the *E. coli* assay, it should be noted that the *E. coli* cell wall has a complex supramolecular structure, with two membrane bilayers separated by proteoglycan.[112] The structure is stabilized by divalent cations, such as the Ca^{2+} ions which are required for PLA_2 activity. In addition to phospholipids, the autoclaved cell membranes and cell walls contain denatured proteins, glycolipids, and proteoglycans, as well as products of thermally induced degradation of these molecules. Any of these components can adsorb or bind a putative polypeptide inhibitor, thereby altering its properties. It is unclear whether PLA_2 bound to a bacterial cell wall is freely accessible to inhibitors that are not lipid-soluble or at least lipid-binding. Initial velocity measurements in this assay are extremely difficult at physiological temperatures. Most authors use conditions that result in hydrolysis of 15 to 30% of the available substrate, with significant accumulation of reaction products that can interfere with potential inhibitors. This makes the interpretation of the results very difficult. Nonspecific mechanisms of inhibition due to lipid binding and steric hindrance are indeed possible, and can overshadow other effects that might be evident in a different assay system. These considerations underscore the importance of carefully selected and standardized assay systems to evaluate potential PLA_2 inhibitory peptides, and the absolute need for additional biological evidence (inhibition of arachidonate and prostaglandin release from intact cells, etc.) supporting the possible physiological role of a putative PLA_2 inhibitor.

V. CONCLUSIONS

Our study of the structure-function relationship of UG has led to the generation of a class of very potent anti-inflammatory peptides which could be described as the AF family. The structures of these peptides are shown in Table 3. One of these peptides is currently undergoing a clinical evaluation as a possible therapeutic agent. These studies provide a good example of both the power and the potential pitfalls of active site-directed peptide design. Considerable structural and biological information has been gained from the design of AF peptides. The putative active site for the anti-inflammatory and PLA_2 inhibitory activities of UG has been identified as the exposed residues on the C-terminal half of α-helix 3. From the structures of the AF peptides described so far, it is tempting to speculate that the side chains of Lys 43, Asp 46, and, possibly, Val 47 (positions 5, 8, and 6 in AF peptides) may be indispensable for biological activity. Also the presence of two hydrophobic residues at positions 44 and 45 (7 and 8 in AF peptides) is invariant. Positions 2, 3, and 4 of the AF peptides are consistently occupied, respectively, by a hydrophilic residue with hydrogen-bonding capability, a bulky hydrophobic residue, and a hydrophilic or charged residue.

At the time of our original studies, the three-dimensional structures of lipocortins were not available. Recently, the AF region of lipocortin-5 has been shown to be also part of a completely exposed α-helix (see Reference 105). This strongly supports the hypothesis that, at least in the cases of UG and lipocortin-5, the putative active site is α-helical in secondary structure. It is also worth mentioning that the AF-like region of type I PLA_2s corresponds to the C-terminal part of α-helix C^{92} plus residues 59 and 60. While it is unlikely that short peptides like the AFs may form stable α-helices in solution, it is possible that they might adopt such a structure upon interaction with PLA_2 or other cellular binding sites.

It has been suggested by Pattabiraman[114] that internal salt bridges between Lys 43 and Asp 46 (positions 5 and 8 in AF-1), and/or between Gln 40 and Lys 42 (positions 2 and 4) might stabilize the α-helix formed by AF-1. Similar considerations may also apply for Asp 248 and Asn 250 (positions 2 and 4) in AF-2. The available structural information on the AF peptides should allow the design of more stable analogs, and eventually of peptide-mimetics with AF-like activity. Studies in this direction are currently underway in our laboratory.

Several biochemical and biological properties of the parent proteins, UG, lipocortin-1, and lipocortin-5 have been reproduced with peptides as short as 9, and, in the case of LC5–206–212, 7 amino acids. From the pharmacological data obtained *in vivo* and the *in vitro* studies by several groups, the most likely mechanisms of action of the AFs are (1) the inhibition of arachidonate release and (2) the inhibition of PAF biosynthesis. So far, AFs and derivatives have proved to be ineffective on phenomena which are not mediated by these two biochemical pathways, with the possible exception of their weak effect on thrombin-induced platelet aggregation. These effects do not appear to be specific for a particular cell type or for a particular stimulus. Studies in intact cells by Camussi et al., and our *in vitro* studies with porcine pancreatic PLA_2, support the hypothesis that these peptides prevent the activation, rather than inhibiting the catalytic activity, of PLA_2(s) and acetylCoA-lysoPAF acetyltransferase, respectively. It is still unclear, however, if the same mechanism operates in the inhibition of porcine pancreatic and cellular PLA_2s. The effects of AFs on intact cells are reversible by washout, and the continued presence of the peptides is necessary for the effects studied so far. This might indicate that either (1) AFs do not enter the cells, and they exert their effects on enzymes accessible on the outer surface of cells; or (2) AFs diffuse freely and rapidly in and out of cells, so that washing the cells or the organ preparations removes them easily from the intracellular compartment. The reversibility of the pharmacological effects of AFs and derivatives also suggests that these effects are unlikely to be due to toxic properties of these peptides. So far, no toxic effect of AFs and their derivatives have been reported *in vivo* or *in vitro*. While additional studies are necessary to characterize the complete range of pharmacological effects of these substances, the data available so far suggest that peptides of the AF family can be potentially useful anti-inflammatory agents, and that they can provide useful models for the development of nonpeptide structural analogs with similar biological activity and pharmacological properties.

At least two different mechanisms by which AFs can be inactivated through chemical and physical processes have been described. These observations, as well as the empirical finding that heating AF solutions restores or increases the biological activity of these peptides, illustrate some of the potential problems that can be encountered when characterizing various properties of synthetic oligopeptides. Some of these problems (i.e., the tendency to aggregate in aqueous solution) are much more severe with small peptides than with natural proteins, and might be undetectable by simple analytical procedures such as amino acid analysis or high-performance liquid chromatography (HPLC). Our experience with AF peptides suggests that in the absence of preliminary data confirming the stability of the peptide(s) under investigation it is prudent to adopt every possible measure to prevent peptide inactivation by various mechanisms. In particular, protection from oxidizing environment, storage of lyophilized peptides in sealed glass vials under anhydrous conditions, preparation of peptide solutions just prior to their use, immediate dilution of concentrated solutions to working concentration, and avoiding storage of peptides in solution are simple precautions that can be taken routinely. When biological activity is detected in a peptide, a series of tests can be conducted to establish if the activity is retained under different storage and handling conditions. If one or more causes of instability are detected, it is possible to systematically alter the structure of the active peptide to improve its stability.

These technical problems, as well as our incomplete understanding of the physical chemistry of peptide and protein folding, can make active site-directed peptide design studies rather arduous, and a certain amount of trial and error is still unavoidable. Yet these studies are amply justified by the potential scientific and practical importance of the information they can yield. In fact, active site-directed peptide design represents one of the major routes toward a complete understanding of the relationship between structure and function of natural proteins, and the rational design of biologically active polypeptides.

REFERENCES

1. **Male, D., Champion, B., Cooke, A., and Owen, M.,** Eds., *Advanced Immunology,* 2nd ed., Lippincott, Philadelphia, 1991.
2. **Dinarello, C.A.,** Interleukin-1, *Rev. Infect. Dis.,* 6, 51, 1984.
3. **Meager, A.,** *Cytokines,* Prentice-Hall, Englewood Cliffs, NJ, 1991, chap. 4.
4. **Fauci, A.S., Lane, H.C., and Volkman, D.J.,** Activation and regulation of human immune response: implication in normal and disease states, *Ann. Intern. Med.,* 99, 61, 1983.
5. **Fauci, A.S. et al.,** Immunoregulation in clinical medicine, *Ann. Intern. Med.,* 106, 421, 1987.
6. **Dean, R.T. and Jessup, W.,** Eds., *Mononuclear Phagocytes: Physiology and Pathology,* Elsevier, Amsterdam, 1985.
7. **Lethi, B.-T. and Fauci, A.S.,** Recombinant interleukin-2 and gamma interferon act synergistically on distinct steps of in vitro terminal human B-cell maturation, *J. Clin. Invest.,* 77, 1173, 1986.
8. **Kehri, J.H., Muraguchi, A., Butler, J.L., Falkoff, R.J., and Fauci, A.S.,** Human B-cell activation, proliferation and differentiation, *Immunol. Rev.,* 78, 75, 1984.
9. **Rosenberg, S.A.,** Adoptive immunotherapy of cancer: accomplishments and prospects, *Cancer Treat. Rep.,* 68, 233, 1984.
10. **Gallin, J.I., Goldstein, I.M., and Snyderman, R.,** Eds., *Inflammation: Basic Principles and Clinical Correlates,* Raven Press, New York, 1988, 1.
11. **Werner, G.H.,** Natural and synthetic peptides (other than neuropeptides) endowed with immunomodulating activities, *Immunol. Lett.,* 16, 363, 1987.
12. **Georgiev, V.S.,** Immunomodulatory peptides of natural and synthetic origin, *Med. Res. Rev.,* 11, 81, 1991.
13. **Wüthrich, K.,** Protein structure determination in solution by nuclear magnetic resonance spectroscopy, *Science,* 243, 45, 1989.
14. **Wüthrich, K.,** The development of nuclear magnetic resonance spectroscopy as a technique for protein structure determination, *Acc. Chem. Res.,* 22, 36, 1989.
15. **Barany, G., Kneib-Cordonier, N., and Mullen, D.G.,** Solid-phase peptide synthesis: a silver anniversary, *Int. J. Protein Pept. Res.,* 30, 705, 1987.
16. **Marshall, G.R.,** Ed., *Peptides — Chemistry and Biology, Proceedings of the 10th American Peptide Symposium,* Escom, Leiden, 1988.
17. **Kramer, W. and Fritz, H.J.,** Oligonucleotide-directed construction of mutations via gapped duplex DNA, *Meth. Enzymol.,* 154, 350, 1987.
18. **Kunkel, T.A., Roberts, J.D., and Zakour, R.A.,** rapid and efficient site-specific mutagenesis without phenotypic selection, *Meth. Enzymol.,* 154, 367, 1987.
19. **Higuchi, R.,** Recombinant PCR, in *PCR Protocols, a Guide to Methods and Applications,* Innis, M.A., Gelfand, D.H., Sminsky, J.J., and White, T., Eds., Academic Press, San Diego, 1990, 177.

20. **Wallace, C.J.A., Guillemette, J.G., Hibiya, Y., and Smith, M.,** Enhancing protein engineering capabilities by combining mutagenesis and semisynthesis, *J. Biol. Chem.,* 266, 21355, 1991.
21. **Fasman, G.D.,** Protein conformational prediction, *Trends Biochem. Sci.,* 14, 295, 1989.
22. **Thornton, J.M. and Gardner, S.P.,** Protein motifs and data-base searching, *Trends Biochem. Sci.,* 14, 300, 1989.
23. **Cohen, N.C., Blaney, J.M., Humblet, C., Gund, P., and Barry, D.,** Molecular modeling software and methods for medicinal chemistry, *J. Med. Chem.,* 33, 883, 1990.
24. **Thornton, J.M., Flores, T.P., Jones, D.T., and Swindells, M.B.,** Protein structure: prediction of progress at last, *Nature,* 354, 105, 1991.
25. **Richardson, J.,** The anatomy and taxonomy of protein structure, *Adv. Protein Chem.,* 34, 167, 1981.
26. **Mutter, M.,** Nature's rules and chemist's tools: a way for creating novel proteins, *Trends Biochem. Sci.,* 13, 260, 1988.
27. **Richardson, J.S. and Richardson, D.C.,** The de novo design of protein structures, *Trends Biochem. Sci.,* 14, 304, 1989.
28. **DeGrado, W.F., Wasserman, Z.R., and Lear, J.D.,** Protein design, a minimalist approach, *Science,* 243, 622, 1989.
29. **Altmann, K.-H. and Mutter, M.,** A general strategy for the de novo design of proteins — template assembled synthetic proteins, *Int. J. Biochem.,* 22, 947, 1990.
30. **De Grado, W.F.,** Design of peptides and proteins, *Adv. Protein Chem.,* 39, 51, 1988.
31. **Schiller, P.W.,** Conformational analysis of enkephalin and conformation-activity relationships, in *The Peptides,* Vol. 6, Udenfriend, S. and Meienhofer, R.J., Eds., Academic Press, New York, 1984, 219.
32. **Schiller, P.W. and DiMaio, J.,** Opiate receptor subclasses differ in their conformational requirements, *Nature,* 297, 74, 1982.
33. **DeGrado, W.F. and Lear, J.D.,** Induction of peptide conformation in apolar/water interfaces. I. A study with model peptides of defined hydrophobic periodicity, *J. Am. Chem. Soc.,* 107, 7684, 1985.
34. **Chiou, George C.Y.,** Systemic delivery of polypeptide drugs through ocular route, *Annu. Rev. Pharmacol. Toxicol.,* 31, 457, 1991.
35. **Williams, W.V., Guy, H.R., Rubin, D.H., Robey, F., Myers, J.N., Kieber-Emmons, T.B., Weiner, D.B., and Greene, M.I.,** Sequence of the cell-attachment sites of reovirus type 3 and its antiidiotype-antireceptor antibody: modeling of the three-dimensional structure, *Proc. Natl. Acad. Sci. U.S.A.,* 85, 6488, 1988.
36. **Williams, W.V., Moss, D.A., Kieber-Emmons, T., Cohen, J., Myers, J.N., Weiner, D.B., and Greene, M.,** Development of biologically active peptides based on antibody structure, *Proc. Natl. Acad. Sci. U.S.A.,* 86, 5537, 1989.
37. **Williams, W.V., Kieber-Emmons, T., VonFeld, J., Greene, M., and Weiner, D.,** Design of bioactive peptides based on antibody hypervariable region structure, *J. Biol. Chem.,* 266, 5182, 1991.
38. **Saragovi, H.U., Fitzpatrick, D., Raktabutr, A., Nakanishi, H., Kahn, M., and Greene, M.I.,** Design and synthesis of a mimetic from an antibody complementarity-determining region, *Science,* 253, 792, 1991.
39. **Bost, K.L., Smith, E.M., and Blalock, J.E.,** Similarity between the corticotropin (ACTH) receptor and a peptide encoded by an RNA that is complementary to ACTH mRNA, *Proc. Natl. Acad. Sci. U.S.A.,* 82, 1372, 1985.
40. **Blalock, J.E. and Bost, K.L.,** Binding of peptides that are specified by complementary RNAs, *Biochem. J.,* 234, 679, 1986.
41. **Shai, Y., Flashner, M., and Chaiken, I.M.,** Anti-sense peptide recognition of sense peptides: direct quantitative characterization with the ribonuclease S-peptide system using high-performance affinity chromatography, *Biochemistry,* 26, 669, 1987.
42. **Fassina, G., Zamai, M., and Chaiken, I.M.,** Antisense peptides to the arginine-vasopressin/bovine neurophysin II biosynthetic precursor: characterization of binding properties by analytical high-performance affinity chromatography, *J. Cell. Biochem.,* 35, 9, 1987.
43. **Brentani, R.R., Ribeiro, S.F., Potacnjak, P., Lopes, J.D., and Nakaie, C.R.,** Characterization of the cellular receptor for fibronectin through a hydropathic complementary peptide, *Proc. Natl. Acad. Sci. U.S.A.,* 85, 364, 1988.
44. **Elton, T.S., Dion, L.D., Bost, K.L., Oparil, S., and Blalock, J.E.,** Purification of an angiotensin II binding protein by using antibodies to a peptide encoded by angiotensin II complementary RNA, *Proc. Natl. Acad. Sci. U.S.A.,* 85, 2518, 1988.
45. **Shai, Y., Flashner, M., and Chaiken, I.M.,** Anti-sense peptides — designing recognition molecules, in *Protein Structure, Folding and Design,* Vol. 2, Oxender, D.L., Ed., Alan R. Liss, New York, 1987, 439.
46. **Blalock, J.E. and Smith, E.M.,** Hydropathic anti-complementarity of amino acids based on the genetic code, *Biochem. Biophys. Res. Commun.,* 121, 203, 1984.
47. **Omichinski, J.G., Olson, A.D., Thorgeirsson, S.S., and Fassina, G.,** Computer-assisted design of recognition peptides, in *Techniques in Protein Chemistry,* Hugli, T., Ed., Academic Press, San Diego, 1989, 430.
48. **Arrhenius, T., Lerner, R.A., and Satterthwait, A.C.,** The chemical synthesis of structured peptides, in *Protein Structure, Folding and Design,* Vol. 2, Oxender, D. L., Ed., Alan R. Liss, New York, 1987, 453.
49. **Kaiser, E.T. and Kezdy, F.J.,** Secondary structures of proteins and peptides in amphiphilic environments: a review, *Proc. Natl. Acad. Sci. U.S.A.,* 80, 1137, 1983.
50. **Kaiser, E.T. and Kezdy, F.J.,** Amphiphilic secondary structure: design of peptide hormones, *Science,* 223, 249, 1984.

51. **Kaiser, E.T.,** The design and construction of biologically active peptides, including hormones, in *Protein Engineering, Applications in Science, Medicine and Industry,* Inouye, M. and Sarma, R., Eds., Academic Press, Orlando, 1986, 71.

52. **Kaiser, E.T.,** Studies on peptide conformation in the design of peptide agonists, *Biochem. Pharmacol.,* 36, 783, 1987.

53. **Kaiser, E.T.,** Design of peptides from hormones to enzymes, in *Protein Structure, Folding and Design,* Vol. 2, Oxender, D.L., Ed., Alan R. Liss, New York, 1987, 433.

54. **Miele, L., Cordella-Miele, E., and Mukherjee, A.B.,** Uteroglobin: structure, molecular biology and new perspectives on its function as a phospholipase A_2 inhibitor, *Endocr. Rev.,* 8, 474, 1987.

55. **Mukherjee, A.B., Cordella-Miele, E., Kikukawa, T., and Miele, L.,** Modulation of cellular response to antigens by uteroglobin and transglutaminase, in *Advances in Post-Translational Modifications of Proteins and Aging,* Zappia, V., Galletti, P., Porta, R., and Wold, F., Eds., Plenum Press, New York, 1988, 135.

56. **Kikukawa, T. and Mukherjee, A.B.,** Detection of a uteroglobin-like phospholipase A_2 inhibitory protein in the circulation of rabbits, *Mol. Cell. Endocrinol.,* 62, 177, 1989.

57. **Lopez de Haro, M.S. and Nieto, A.,** Nucleotide and derived amino acid sequences of a cDNA coding for pre-uteroglobin from the lung of the hare (Lepus capensis), *Biochem. J.,* 235, 895, 1986.

58. **Nordlund-Moller, L., Andersson, O., Ahlgren, R., Schilling, J., Gillner, M., Gustafsson, J., and Lund, J.,** Cloning, structure and expression of a rat binding protein for polychlorinated biphenyls: homology to the hormonally regulated progesterone-binding protein uteroglobin, *J. Biol. Chem.,* 265, 12690, 1990.

59. **Singh, G., Katyal, S.L., Brown, W.E., Phillips, S., Kennedy, A.L., Anthony, J., and Squeglia, N.,** Amino acid and cDNA nucleotide sequences of human Clara cell 10 kDa protein, *Biochem. Biophys. Acta,* 950, 329, 1988.

60. **Nieto, A., Ponstingl, H., and Beato, M.,** Purification and quaternary structure of the hormonally induced protein uteroglobin, *Arch. Biochem. Biophys.,* 180, 82, 1977.

61. **Beato, M.,** Physico-chemical characterization of uteroglobin and its interaction with progesterone, in *Development in Mammals,* Johnson, M.H., Ed., Elsevier, Amsterdam, 1977, 173.

62. **Temussi, P.A., Tancredi, T., Puigdomenech, P., Saavedra, A., and Beato, M.,** Interaction of S-carboxymethylated uteroglobin with progesterone, *Biochemistry,* 19, 3287, 1980.

63. **Tancredi, T., Temussi, P.A., and Beato, M.,** Interaction of oxidized and reduced uteroglobin with progesterone, *Eur. J. Biochem.,* 122, 101, 1982.

64. **Miele, L., Cordella-Miele, E., and Mukherjee, A.B.,** High level bacterial expression of uteroglobin, a dimeric eukaryotic protein with two interchain disulfide bridges, in its natural quaternary structure, *J. Biol. Chem.,* 265, 6427, 1990.

65. **Atger, M., Mornon, J.P., Savouret, J.F., Loosfelt, H., Fridlansky, F., and Milgrom, E.,** Uteroglobin: a model for the study of the mechanism of action of steroid hormones, in *Steroid Induced Uterine Proteins,* Beato, M., Ed., Elsevier, Amsterdam, 1980, 341.

66. **Beato, M., Saavedra, A., Puigdomenech, P., Tancredi, T., and Temussi, P.A.,** Progesterone binding to uteroglobin, in *Steroid Induced Uterine Proteins,* Beato, M., Ed., Elsevier, Amsterdam, 1980, 105.

67. **Lorand, L. and Conrad, S.,** Transglutaminases, *Mol. Cell. Biochem.,* 58, 9, 1984.

68. **Manjunath, R., Chung, S.I., and Mukherjee, A.B.,** Crosslinking of uteroglobin by transglutaminase, *Biochem. Biophys. Res. Commun.,* 121, 400, 1984.

69. **Mukherjee, A.B., Ulane, R.E., and Agrawal, A.K.,** Role of uteroglobin and transglutaminase in masking the antigenicity of implanting rabbit embryos, *Am. J. Reprod. Immunol.,* 2, 135, 1982.

70. **Mukherjee, A.B., Cunningham, D., Agrawal, A.K., and Manjunath, R.,** Role of uteroglobin and transglutaminase in self and non self recognition during reproduction in rabbits, *Ann. N.Y. Acad. Sci.,* 392, 401, 1982.

71. **Mukherjee, D.C., Ulane, R.E., Manjunath, R., and Mukherjee, A.B.,** Suppression of epididymal sperm antigenicity in the rabbit by uteroglobin and transglutaminase, *Science,* 219, 989, 1983.

72. **Levin, S.W., Butler, J.D., Schumaker, U.K., Wightman, P.D., and Mukherjee, A.B.,** Uteroglobin inhibits phospholipase A_2 activity, *Life Sci.,* 38, 1813, 1986.

73. **Facchiano, A., Miele, L., Cordella-Miele, E., and Mukherjee, A.B.,** Inhibition of pancreatic phospholipase A_2 activity by uteroglobin and antiflammin peptides: possible mechanism of action, *Life Sci.,* 48, 453, 1991.

74. **Chung, S.I.,** Multiple molecular forms of transglutaminase in guinea pig and humans, in *Isozymes,* Vol. I, Markert, C.L., Ed., Academic Press, London, 1975, 259.

75. **Alving, R.E. and Laki, K.,** Enzymatic stabilization of protein aggregates, *Fed. Proc.,* 26, 828, 1967.

76. **Schiffman, E., Geetha, V., Pencev, D., Warabi, H., Mato, J., Garcia-Castro, I., Chiang, P.K., Manjunath, R., and Mukherjee, A.B.,** Phospholipid metabolism and regulation of leukocyte chemotaxis, in *Asthma: Physiology, Immunopharmacology and Treatment. Third International Symposium,* Kay, A. B. and Frank Austen, K., Eds., Academic Press, London, 1984, 173.

77. **Schiffman, E., Geetha, V., Pencev, D., Warabi, H., Mato, J., Hirata, F., Brownstein, M., Manjunath, R., Mukherjee, A.B., Liotta, L., and Terranova, V.P.,** Adherence regulation and chemotaxis, in *Agent and Action Supplements,* Vol. 12, Keller, H. U. and Till, G. O., Eds., Birkhauser Verlag, Basel, 1983, 106.

78. **Vasanthakumar, G., Manjunath, R., Mukherjee, A.B., Warabi, H., and Schiffman, E.,** Inhibition of phagocyte chemotaxis by uteroglobin, an inhibitor of blastocyst rejection, *Biochem. Pharmacol.,* 37, 389, 1988.

79. **Manjunath, R., Levin, S., Kumaroo, K.K., Butler, J.D., Donlon, J.A., Horne, M., Fujita, R., Schumaker, U.K., and Mukherjee, A.B.,** Inhibition of thrombin-induced platelet aggregation by uteroglobin, *Biochem. Pharmacol.,* 36, 741, 1987.

80. **Vadas, P. and Pruzanski, W.,** Role of secretory phospholipases A$_2$ in the pathobiology of disease, *Lab. Invest.,* 55, 391, 1986.

81. **Pruzanski, W. and Vadas, P.,** Soluble phospholipases A$_2$ in human pathology: clinical-laboratory interface, in *Biochemistry, Molecular Biology and Physiology of Phospholipase A$_2$ and its Regulatory Factors,* Mukherjee, A. B., Ed., Plenum Press, New York, 1990, 239.

82. **Hoffmann, T., Brando, C., Lizzio, E.F., Lee, C., Hanson, M., Ting, K., Kim, Y.J., Abrahamsen, T., Puri, J., and Bonvini, E.,** Functional consequences of phospholipase A$_2$ activation in human monocytes, in *Biochemistry, Molecular Biology and Physiology of Phospholipase A$_2$ and its Regulatory Factors,* Mukherjee, A. B., Ed., Plenum Press, New York, 1990, 125.

83. **Robinson, D.H., Kirk, K.L., and Benos, D.J.,** Macromolecular transport in rabbit blastocysts: evidence for a specific uteroglobin transport system, *Mol. Cell. Endocrinol.,* 63, 222, 1989.

84. **Mornon, J.P., Fridlansky, F., Bally, R., and Milgrom, E.,** X-ray crystallographic analysis of a progesterone-binding protein. The C222$_1$ crystal form of oxidized uteroglobin at 2.2 Å resolution, *J. Mol. Biol.,* 137, 415, 1980.

85. **Morize, I., Surcouf, E., Vaney, M.C., Epelboin, Y., Buehner, M., Fridlansky, F., Milgrom, E., and Mornon, J.P.,** Refinement of the C222$_1$ crystal from of oxidized uteroglobin at 1.34 Å resolution, *J. Mol. Biol.,* 194, 725, 1987.

86. **Buhener, M., Lifchitz, A., Bally, R., and Mornon, J.P.,** Use of molecular replacement in the structure determination of the P2$_1$2$_1$2 and the P2$_1$ (pseudo P2$_1$2$_1$2) crystal forms of oxidized uteroglobin, *J. Mol. Biol.,* 159, 353, 1982.

87. **Wallner, B.P., Mattaliano, R.J., Hession, C., Cate, R.L., Tizard, R., Sinclair, L.K., Foeller, C., Chow, E.P., Browning, J.L., Ramachandran, K.L., and Pepinsky, R.B.,** Cloning and expression of human lipocortin, a phospholipase A$_2$ inhibitor with potential antiinflammatory activity, *Nature,* 320, 77, 1986.

88. **Ando, Y., Imamura, S., Owada, K., and Kannagi, R.,** Calcium-induced intracellular crosslinking of lipocortin I by tissue transglutaminase in A431 cells, *J. Biol. Chem.,* 266, 1101, 1991.

89. **Miele, L., Cordella-Miele, E., Facchiano, A., and Mukherjee, A.B.,** Novel anti-inflammatory peptides from the region of highest similarity between uteroglobin and lipocortin I, *Nature,* 335, 726, 1988.

89a. **Miele, L. et al.,** unpublished data.

90. **Heinriksonm R.L. and Kezdy, F.J.,** A novel bifunctional mechanism of surface recognition by phospholipase A$_2$, in *Biochemistry, Molecular Biology and Physiology of Phospholipase A$_2$ and its Regulatory Factors,* Mukherjee, A.B., Ed., Plenum Press, New York, 1990, 37.

91. **Hazlett, T.L., Deems, R.A., and Dennis, E.A.,** Activation, aggregation, inhibition and the mechanism of phospholipase A$_2$, in *Biochemistry, Molecular Biology and Physiology of Phospholipase A$_2$ and its Regulatory Factors,* Plenum Press, New York, 1990, 49.

92. **Waite, M.,** *The Phospholipases: Handbook of Lipid Research,* Vol. 5, Plenum Press, New York, 1987, chap. 10.

93. **Slotboom, A.J., Verheij, H.M., and deHaas, G.H.,** On the mechanism of phospholipase A$_2$, in *Phospholipids,* Vol. 4, Hawthorne, J.N. and Ansell, G.B., Eds., Elsevier, Amsterdam, 1982, 359.

94. **Tomasselli, A.G., Hui, J., Fisher, J., Zurcher-Neely, H., Reardon, I.M., Oriaku, E., Kezdy, F.J., and Heinrikson, R.L.,** Dimerization and activation of porcine pancreatic phospholipase A$_2$ via substrate level acylation of lysine 56, *J. Biol. Chem.,* 264, 10041, 1989.

95. **Narashiman, D., Holowka, D., and Baird, B.,** A guanine nucleotide-binding protein participates in IgE receptor-mediated activation of endogenous and reconstituted phospholipase A$_2$ in a permeabilized cell system, *J. Biol. Chem.,* 264, 1459, 1990.

96. **Vostal, J.G., Mukherjee, A.B., Miele, L., and Shulman, N.R.,** Novel peptides derived from a region of local homology between uteroglobin and lipocortin I inhibit platelet aggregation and secretion, *Biochem. Biophys. Res. Commun.,* 165, 27, 1989.

97. **Camussi, G., Tetta, C., Bussolino, F., and Baglioni, C.,** Antiinflammatory peptides (antiflammins) inhibit synthesis of platelet-activating factor, neutrophil aggregation and chemotaxis, and intradermal inflammatory reactions, *J. Exp. Med.,* 171, 913, 1990.

98. **Camussi, G., Tetta, C., and Baglioni, C.,** Antiflammins inhibit synthesis of platelet-activating factor and intradermal inflammatory reactions, in *Biochemistry, Molecular Biology and Physiology of Phospholipase A$_2$ and its Regulatory Factors,* Plenum Press, New York, 1990, 161.

99. **Camussi, G., Tetta, C., Turello, E., and Baglioni, C.,** Anti-inflammatory peptides inhibit synthesis of platelet-activating factor, in *Cytokines and Lipocortins in Inflammation and Differentiation,* Melli, M. and Parente, L., Eds., Wiley-Liss, New York, 1990, 69.

100. **Ialenti, A., Doyle, P.M., Hardy, G.N., Simpkin, D.S., and Di Rosa, M.,** Anti-inflammatory effects of vasocortin and nonapeptide fragments of uteroglobin and lipocortin I (antiflammins), *Agents Actions,* 1–2, 48, 1990.

101. **Di Rosa, M. and Ialenti, A.,** Selective inhibition of inflammatory reactions by vasocortin and antiflammin 2, in *Cytokines and Lipocortins in Inflammation and Differentiation,* Melli, M. and Parente, L., Eds., Wiley-Liss, New York, 1990, 81.

102. **Chan, C.C., Ni, M., Miele, L., Cordella-Miele, E., Mukherjee, A.B., and Nussenblatt, R.B.,** Antiflammins: inhibition of endotoxin-induced uveitis in Lewis rats, in *Ocular Immunology Today,* Usui, M., Ohno, S., and Aoki, K., Eds., Elsevier, Amsterdam, 1990, 467.

103. **Chan, C.C., Ni, M., Miele, L., Cordella-Miele, E., Ferrick, M., Mukherjee, A.B., and Nussenblatt, R.B.,** Effects of antiflammins on endotoxin-induced uveitis in rats, *Arch. Ophthalmol.,* 109, 278, 1991.

104. **Tetta, C., Camussi, F., Bussolino, F., Herrick-Davis, K., and Baglioni, C.,** Inhibition of the synthesis of platelet-activating factor by anti-inflammatory peptides (antiflammins) without methionine, *J. Pharmacol. Exp. Ther.,* 257, 616, 1991.

105. **Perretti, M., Becherucci, C., Mugridge, G., Solito, E., Silvestri, S., and Parente, L.,** A novel antiinflammatory peptide from lipocortin-5, *Br. J. Pharmacol.,* 103, 1327, 1991.

106. **Kretsinger, L.H. and Creutz, C.E.,** Consensus in endocytosis, *Nature,* 320, 573, 1986.

107. **Davidson, F.F., Dennis, E.A., Powell, M., and Glenney, J.R.,** Inhibition of phospholipase A_2 by "lipocortins" and calpactins. An effect of binding to substrate phospholipids, *J. Biol. Chem.,* 262, 1698, 1987.

108. **Davidson, F.F., Lister, M.D., and Dennis, E.A.,** Binding and inhibition studies on lipocortins using phosphatidyl-choline vesicles and phospholipase A_2 from snake venom, pancreas, and a macrophage cell line, *J. Biol. Chem.,* 265, 5602, 1990.

109. **Hayashi, H., Owada, M.K., Sonobe, S., Domae, K., Yamanouchi, T., Kakunaga, T., Kitajima, Y., and Yaoita,** Monoclonal antibodies specific to a Ca^{++}-bound form of lipocortin I distinguish its Ca^{++}-dependent phospholipid-binding ability from its ability to inhibit phospholipase A_2, *Biochem. J.,* 269, 709, 1990.

110. **Kini, R.M. and Evans, H.J.,** A model to explain the pharmacological effects of snake venom phospholipase A_2, *Toxicon,* 27, 613, 1989.

111. **Miele, L., Cordella-Miele, E., Facchiano, A., and Mukherjee, A.B.,** Inhibition of phospholipase A_2 by uteroglobin and antiflammin peptides, in *Biochemistry, Molecular Biology and Physiology of Phospholipase A_2 and its Regulatory Factors,* Plenum Press, New York, 1990, 137.

112. **DiRienzo, J.M., Nakamura, K., and Inouye, M.,** The outer membrane protein of Gram-negative bacteria: biosynthesis, assembly and functions, *Annu. Rev. Biochem.,* 47, 481, 1978.

113. **Cordella-Miele, E.,** unpublished data.

114. **Pattabiraman, N.,** personal communication.

115. **Lloret, S. and Moreno, J.J.,** *In vitro* and *in vivo* effects of anti-inflammatory peptides, antiflammins, *Biochem. Pharmacol.,* 44, 1437, 1992.

116. **Umland, T.C., Swaminathan, S., Furey, W., Singh, G., Pletcher, J., and Sax, M.,** Refined structure of rat Clara cell 17 kDa protein at 3.0 Å resolution, *J. Mol. Biol.,* 224, 441, 1992.

117. **Peri, A., Cordella-Miele, E., Miele, L., and Mukherjee, A.B.** Tissue-specific expression of the gene coding for human Clara cell 10 kDa protein, a phospholipase A_2-inhibitory protein, *J. Clin. Invest.,* 92, 2099, 1993.

118. **Mantile, G., Miele, L., Cordella-Miele, E., Singh, G., Katyal, S.L., Mukherjee, A.B.,** Human Clara cell 10 kDa protein is the counterpart of rabbit uteroglobin, *J. Biol. Chem.,* 268, 20343, 1993.

119. **Cordella-Miele, E., Miele, L., and Mukherjee, A.B.,** Identification of a specific region of low molecular weight phospholipase A_2 (residues 21–40) as a potential target for structure-based design of inhibitors of these enzymes, *Proc. Natl. Acad. Sci. USA.,* 90, 10290, 1993.

Index

INDEX

DATE DUE

GAYLORD | | | PRINTED IN U.S.A.